The Economics of Regulation in Agriculture

Compliance with Public and Private Standards

The Economics of Regulation in Agriculture

Compliance with Public and Private Standards

Edited by

Floor Brouwer

LEI, part of Wageningen UR
The Hague, the Netherlands

Glenn Fox

University of Guelph
Guelph, Canada

and

Roel Jongeneel

LEI, part of Wageningen UR
The Hague, the Netherlands

www.cabi.org

CABI is a trading name of CAB International

CABI
Nosworthy Way
Wallingford
Oxfordshire OX10 8DE
UK

Tel: +44 (0)1491 832111
Fax: +44 (0)1491 833508
E-mail: info@cabi.org
Website: www.cabi.org

CABI
38 Chauncey Street
Suite 1002
Boston, MA 02111
USA

Tel: +1 800 552 3083 (toll free)
Tel: +1 (0) 617 395 4051
E-mail: cabi-nao@cabi.org

A catalogue record for this book is available from the British Library, London, UK.

Library of Congress Cataloging-in-Publication Data

The economics of regulation in agriculture : compliance with public and private standards / edited by Floor Brouwer, Glenn Fox and Roel Jongeneel.
p. cm.
Includes bibliographical references and index.
ISBN 978-1-84593-557-3 (alk. paper)
1. Agriculture--Economic aspects. 2. Agriculture--Environmental aspects. 3. Agricultural laws and legislation--Compliance costs. 4. Agriculture and state. 5. Environmental policy. I. Brouwer, Floor. II. Fox, Glenn. III. Jongeneel, R. A.

HD1415.E253 2012
338.1--dc23

2012016960

ISBN: 978 1 84593 557 3

Commissioning editor: Rachel Cutts
Production editor: Tracy Head

Printed and bound in the UK from copy supplied by the authors by CPI Group (UK) Ltd, Croydon, CR0 4YY

Contents

Contributors

David Baldock
Institute for European Environmental Policy (IEEP)
15 Queen Anne's Gate
London SW1H 9BU
United Kingdom

Nicole Beechey
Department of Food, Agricultural and Resource Economics
University of Guelph
Guelph, Ontario
Canada N1G 2W1

Irina Bezlepkina
LEI – Wageningen UR
PO Box 29703
2502 LS The Hague
The Netherlands

Floor Brouwer
LEI – Wageningen UR
PO Box 29703
2502 LS The Hague
The Netherlands

B. James Deaton
Department of Food, Agricultural and Resource Economics
University of Guelph
Guelph, Ontario
Canada N1G 2W1

Matty Demont
Africa Rice Center (WARDA)
BP 96
Saint-Louis
Senegal

Koen Dillen
Centre for Agricultural and Food Economics
K.U. Leuven
W. de Croylaan 42
BE-3001 Leuven
Belgium

Paloma Esteve
Department of Agricultural Economics and Social Sciences
Universidad Politécnica de Madrid
Avda. Complutense s/n, Ciudad Universitaria
28040 Madrid
Spain

Martin Farmer
Institute for European Environmental Policy (IEEP)
15 Queen Anne's Gate
London SW1H 9BU
United Kingdom

Glenn Fox
Department of Food, Agricultural and Resource Economics
University of Guelph
Guelph, Ontario
Canada N1G 2W1

Kaley Hart
Institute for European Environmental Policy (IEEP)
15 Queen Anne's Gate
London SW1H 9BU
United Kingdom

Thomas Herzfeld
Wageningen University
Social Sciences Group
Hollandseweg 1
6706 KN Wageningen
The Netherlands

John Baptist D. Jatoe
Department of Agricultural Economics & Agribusiness
University of Ghana
PO Box LG 25
Legon
Ghana

Roel Jongeneel
LEI – Wageningen UR
PO Box 29703
2502 LS The Hague
The Netherlands

Lynn G. Knight
94 Sand Hill Road
Milton VT 05468
USA

Anton Meister
Department of Economics and Finance
Massey University
Private Bag 11 222
Palmerston North
New Zealand

Sílvia Helena Galvão de Miranda
Departamento de Economia, Administração e Sociologia
University of São Paulo
Av. Pádua Dias n.11
PO Box 132
Piracicaba – SP – 13.400-970
Brasil

Xavier Poux
Applications des Sciences de l'Action
8 Rue Legouve
75010 Paris
France

Predrag Rajsic
Department of Food, Agricultural and Resource Economics
University of Guelph
Guelph, Ontario
Canada N1G 2W1

Elizabeth Ramlal
Agriculture Sector Development Unit
Ministry of Agriculture
Regent Street and Vlissengen Road
Bourda, Georgetown
Guyana

Kees de Roest
Centro Ricerche Produzioni Animali
Corso Garibaldi, 42
42 100 Reggio Emilia
Italy

Vicki Swales
47/2 Queen Charlotte Street
Edinburgh EH6 7EY
United Kingdom

Consuelo Varela-Ortega
Department of Agricultural Economics and Social Sciences
Universidad Politécnica de Madrid
Avda. Complutense s/n, Ciudad Universitaria
28040 Madrid
Spain

Alfons Weersink
Department of Food, Agricultural and Resource Economics
University of Guelph
Guelph, Ontario
Canada N1G 2W1

Jonathan Winsten
Winrock International
2101 Riverfront Drive
Little Rock
Arkansas 72202
USA

Emmanuel K. Yiridoe
Department of Business and Social Sciences
Nova Scotia Agricultural College
Truro
Nova Scotia
Canada

Abbreviations

AAFC	Agriculture and Agri-Food Canada (Canada)
ACR Act	Agricultural Crop Rotation Act (Canada)
AFO	Animal Feeding Operation (USA)
AFS	Assured Food Standards (UK)
ALUS	Alternative Land Use Services (Canada)
APF	Agricultural Policy Framework (Canada)
APHIS	US Animal and Plant Health Inspection Service (USA)
ARMS	Agricultural Research Management Survey
AU	Animal Unit (USA)
AWA	Animal Welfare Act (USA)
AWU	Annual Work Unit
BMP	Best Management Practice
BSE	Bovine Spongiform Encephalopathy
BST	Bovine Somatotropin
CAA	Clean Air Act (USA)
CAFO	Concentrated Animal Feeding Operation (USA)
CAIS scheme	Canadian Agricultural Income Stabilisation scheme (Canada)
CAP	Common Agricultural Policy (EU)
CERCLA	Comprehensive Environmental Response, Compensation, and Liability Act (USA)
CFSQP	Canadian Food Safety and Quality Program (Canada)
CGE model	Computable General Equilibrium Model
CMO	Common Market Organization
CMP	Conservation Management Practices (USA)
CNMP	Comprehensive Nutrient Management Plan (USA)
CRP	Conservation Reserve Program (USA)
CRPA	Centro Ricerche Produzioni Animali (Italy)
CSP	Conservation Security Program (USA)
CWA	Clean Water Act (USA)
DEFRA	Department for Environment, Food and Rural Affairs (UK)
DOC	Department of Conservation (New Zealand)
EAFRD	European Agricultural Fund for Rural Development
EBI	Environmental Benefits Index (USA)
EC	European Commission
ECFP	East Coast Forestry Project (New Zealand)
ECM	Energy Corrected Milk
EEC	European Economic Community

EFP	Environmental Farm Plan (Canada)
EMAS	Community Eco-Management and Audit Scheme (EU)
EPA	US Environmental Protection Agency (USA)
EQIP	Environmental Quality Incentives Program (USA)
ERMA	Environmental Risk Management Authority (New Zealand)
ERS	Economic Research Service (USA)
ESA	Endangered Species Act (USA)
ETS	Emission Trading Scheme
EU	European Union
FADN	Farm Accountancy Data Network (EU)
FAIR Act	Federal Agricultural and Improvement Reform Act (USA)
FAL	Federal Institute for Agriculture (Germany)
FAO	Food and Agriculture Organization
FDA	US Food and Drug Administration
FEMP	Farm Erosion Management Plan (The Netherlands)
FIFRA	Federal Insecticide, Fungicide and Rodenticide Act (USA)
FMD	Food and Mouth Disease
FNVA	Farm Net Value Added
FOTG	Field Office Technical Guide (USA)
FQPA	Food Quality Protection Act (USA)
FSS	Farm Structure Survey
GAEC	Good Agricultural and Environmental Condition
GATT	General Agreement on Tariffs and Trade
GDP	Gross Domestic Product
GFP	Good Farming Practice
GIS	Geographic Information System
GRP	Grassland Reserve Program (USA)
GTAP	Global Trade Analysis Project
HEL	Highly Erodible Cropland (USA)
HNV	High Nature Value (EU)
IFCN	International Farm Comparison Network
IKB	Integrated Chain Governance (the Netherlands)
IPPC	International Plant Protection Convention
I&R	Identification & Registration (of animals)
JECFA	Joint Expert Committee on Food Additives
LEAF	Linking Environment and Food (UK)
LFA	Less-favoured Area
LO3A	Land Overlay 3A (New Zealand)
LU	Livestock Unit
LWG	Local Work Group
MBA	Migratory Bird Act (Canada)
MDS	Minimum Distance Separation
MEA	Multilateral Environmental Agreement
MFE	Ministry for the Environment (New Zealand)
MTR	Mid-term Review (EU)
N	Nitrogen
NAAQS	National Ambient Air Quality Standards (USA)

NAIS	National Animal Identification System (USA)
NDA	Nitrogen Discharge Allowance (New Zealand)
NHF	Nature Heritage Fund (New Zealand)
NIMBY	Not In My Back Yard
NMP	Nitrogen Management Plan (New Zealand and Canada)
NPDES	National Pollution Discharge Elimination System (USA)
NRCS	Natural Resources Conservation Service (USA)
NSP	Non-point Source Pollution
NU	Nutrient Unit (Canada)
NVZ	Nitrate Vulnerable Zone (EU)
NWR	Nga Whenua Rahui (New Zealand)
NZD	New Zealand Dollar
NZU	New Zealand Unit (New Zealand)
OAQPS	Office of Air Quality Planning and Standards
OECD	Organisation for Economic Co-operation and Development
OIE	World Organization for Animal Health
OP	Operation Program (EU)
P	Phosphorus
PDO	Protected Designation of Origin
PEI	Prince Edward Island (Canada)
PGI	Protected Geographical Indication
PM	Particulate Matter
PNEFA	National Program to Eradicate FMD (Brazil)
PO	Producers' Organization (EU)
ppm	Parts Per Million
QEII	Queen Elizabeth II National Trust (New Zealand)
rBST	Recombinant Bovine Somatotropin
RMA	Resource Management Act (New Zealand)
SAPS	Single Area Payment Scheme
SARA	Species at Risk Act (Canada)
SFP	Single Farm Payment (EU)
SIP	State Implementation Plan (USA)
SMA	Special Management Area (Canada)
SMR	Statutory Management Requirement (EU)
SPA	Special Protection Area
SPR	Soil Protection Review
SPS	Single Payment Scheme (EU)
SPS Agreement	Sanitary and Phytosanitary Agreement
STC	Specific Trade Concern
SWAT	Soil and Water Assessment Tool (USA)
SWPoA	Sustainable Water Programme of Action (New Zealand)
TBT Agreement	Technical Barriers to Trade Agreement
TRAIN	UNCTAD Trade Analysis and Information System
TRQ	Tariff Rate Quota
US	United States of America
USDA	US Department of Agriculture
USDA-ARS	Agricultural Research Service of the US Department of Agriculture

WAP	Water Abstraction Plan (Spain)
WFD	Water Framework Directive (EU)
WHO	World Health Organization
WRP	Wetland Reserve Program (USA)
WTO	World Trade Organization
WWTP	Waste Water Treatment Plant

Preface

Primary agriculture in the developed economies is increasingly subject to environmental, human, plant and animal health and animal welfare regulations. Research in this volume was motivated by the use of cross compliance as a means of harmonizing enforcement of these types of regulations in the EU. Cross compliance makes the receipt of farm support payments contingent on meeting regulatory standards. In general, compliance with regulation gives rise to increased costs at the farm level. A commonly expressed concern, by EU farmers but also by farmers in other jurisdictions, is that compliance with regulations has adverse effects on competitiveness and ultimately on farm business viability. As agriculture is increasingly integrated into global trade liberalization, differentials in national regulatory compliance costs, potentially, can have significant effects on trade flows. Comparisons of regulatory compliance costs, even comparison of regulatory standards, across jurisdictions, is complex. Different polities operating with differing constitutional allocations of authorities, can have different agencies or even levels of government being responsible for different categories of regulations and standards. It is not uncommon for the farming industry in any particular jurisdiction, in the face of these challenges, to perceive the standards that it is required to meet as higher than those which its competitors must achieve. This volume investigates the cost implications and competitiveness effects resulting from regulations and standards in the fields of environment, health and animal welfare and production-linked rules of 'good agricultural conditions' for a range of agricultural products.

The book is the product of an international collaboration among researchers from the EU, North America and New Zealand. The work presented in this volume has been (co-)funded by the project 'Facilitating the CAP reform: Compliance and competitiveness of European agriculture' (EU 6th Framework Programme for Research, Technological Development and Demonstration, Priority 8.1, contract no. SSPE-CT-2005-006489). In addition to chapters based on research conducted under this project, we have solicited additional chapters to enrich the range of perspectives in the volume. The content of this book does not represent the official position of the European Commission and is entirely under the responsibility of the authors.

Collaboration on this project has increased our understanding as researchers of the range of modalities and forms of environmental regulation that are being applied to agriculture in the developed countries. We also came to appreciate the importance of differences in production systems and in environmental, social and economic pressures impinging on agriculture in the countries considered in the study. We are grateful to the European Commission for their financial support for this research.

Floor Brouwer, Glenn Fox and Roel Jongeneel
July 2012

Chapter 1

Introduction

Floor Brouwer, Glenn Fox and Roel Jongeneel

Standards, compliance costs and competitiveness[1]

Farmers in developed countries are increasingly subject to regulations and standards for environmental protection (e.g. protection of groundwater, quality of surface water, air and soil; conservation of habitats and species), food safety (e.g. hygiene rules, requirements on animal feed) and animal welfare (e.g. housing standards). These regulations and standards can influence production costs, but the extent of this influence can vary across agricultural sectors, geographical areas and farming systems. Brouwer and Ervin (2002) identified and compared policy approaches that have been used to internalise external costs related to the environment, food safety and animal welfare across developed country agricultures. On-farm constraints were compared, taking into account differences in policy models (form of standards and legislation affecting agriculture, and a mix of national, federal, regional and local measures), farming practices (e.g. prevailing production and farm management systems, technologies available to the farmer, and the investment cycle on the farm) and biophysical conditions of the farm and its surrounding area.

Much has changed in the decade since Brouwer and Ervin's (2002) research was published. The scope of standards and regulations has broadened, the performance standards are generally higher and it is becoming more and more important for farmers to consider the full range of options at their disposal to meet these standards and regulations. The range of measures taken by farmers depends on the nature of the constraints and the management and technology options available to address those constraints. Measures can include the construction of new housing systems to comply with animal welfare rules, buying new fertilization systems to comply with nutrient control measures and introducing milking cooling systems to comply with food safety rules. Examples of the costs of compliance in the field of environment are increased labour requirements, particularly for record keeping, increased use of land (through leasing or buying land to satisfy standards for manure application), higher capital investments (e.g. housing systems to control

[1] Competitiveness is rooted in the concepts of price, quality and market share and broader concepts like innovative capacity of firms in the agrifood chain. From a producers' perspective, competitiveness can be described as 'the ability to supply goods and services in the location, form and place sought by buyers, at prices that are as good or better than those of other potential suppliers, while earning at least the opportunity cost of the return on resources employed' (Porter, 1990).

emissions of ammonia) and increase in variable costs (e.g. for the storage, transport and spreading of livestock manure, feeding of animals, permit costs, training and record keeping). The main cost components involved in meeting higher food safety standards include increased labour (e.g. for registration of animals, administration, audits), capital investments (e.g. computers, cleaning equipment) and higher variable costs (e.g. introduce ear tags, record keeping). With regards to animal welfare, the most important categories of costs are housing and management costs and labour and operational costs.

Modelling framework

In addition to our review and comparison of policies and standards across the jurisdictions represented in this study, we also wanted to investigate the magnitude of competitiveness and trade effects quantitatively. Given the range of commodities and countries included in this study, it was important to have a consistent modelling approach to facilitate comparisons across countries and products as well as with other research. We elected to use the GTAP (Global Trade Analysis Project) model. The standard GTAP model is a multi-region, multi-sector, computable general equilibrium (CGE) model, which assumes perfect competition and constant returns to scale. An important property of the model is its ability to cope with bilateral trade and to account for potential heterogeneity between goods from different origins. GTAP is coordinated by the Center for Global Trade Analysis, which is housed in the Department of Agricultural Economics at Purdue University. GTAP's goal is to improve the quality of quantitative analysis of global economic issues within an economy-wide framework. Since its inception in 1993, the GTAP model has become a common 'language' for conducting global economic analysis. The modelling system offers a variety of products, including: data, models, and utilities for multi-region, applied general equilibrium analysis of global economic issues. We use the GTAP model database calibrated on 2001 data. The sectoral or commodity coverage for our modelling includes dairy, beef, cereals and pork. Limitations on data and modelling capacity precluded using the GTAP model for fruits and vegetables. Our general strategy was to use the 2001 calibration to solve the model without the additional costs of meeting regulatory standards, then to recalibrate the relevant portions of the model to reflect unilateral regulation in EU countries, then to recalibrate the model once again to reflect multilateral regulation in the developed countries. The specific regulatory costs reflected our comparison and review of policies for each specific commodity or industry. Adjustments in the calibration of the GTAP model were based on farm level estimates of compliance costs or estimates obtained from secondary sources.

Sectoral aggregation

Aggregation is the second step in developing the relevant modelling framework to assess the impact of standards on the external competitiveness of European agriculture. The lowest aggregation of the GTAP 6 Data Package is 87 regions, 57 sectors and 5 production factors. The sectors are aggregated to three main sectors, i.e. food, manufacturing and services. In our modelling framework, we will always consider four sectors, i.e. (i) the sector under research (beef, pigmeat, dairy or cereals), (ii) the rest of the food sector, (iii) manufacturing and (iv) services.

It is important to note that the beef sector is not disaggregated in GTAP. In Table 1.1 we compare GTAP sectors (GSC) and standard central product classification (CPC)[2]. GTAP's 'cmt' variable, for instance, includes eight subsectors. A splitcom utility enables splitting existing GTAP sectors for all regions of interest. If the appropriate data are lacking, reasonable assumptions can be made. The program will make the necessary splits based on any input provided.

Table 1.1 Concordance tables between GTAP sectors (GSC) and standard central product classification (CPC)

Number	Code	Code	Description
19	cmt	21111	Meat of bovine animals, fresh or chilled
		21112	Meat of bovine animals, frozen
		21115	Meat of sheep, fresh or chilled
		21116	Meat of sheep, frozen
		21117	Meat of goats, fresh, chilled or frozen
		21118	Meat of horses, asses, mules or hinnies, fresh, chilled or frozen
		21119	Edible offal of bovine animals, swine, sheep, goats, horses, asses, mules or hinnies, fresh, chilled or frozen
		2161	Fats of bovine animals, sheep, goats, pigs and poultry, raw or rendered; wool grease

Source: https://www.gtap.agecon.purdue.edu/databases/contribute/concordinto.asp

Factoral aggregation

We used the full disaggregation[3] of five production factors available in GTAP to model production factors, i.e. (i) land, (ii) unskilled labour, (iii) skilled labour, (iv) capital and (v) natural resources.

Regional aggregation

Defining representative regional aggregation models for each sector is important in GTAP modelling. The regional aggregation of the model should represent the major forces at stake in global trade. In order to identify the major trade partners, we followed the following procedure:

---

[2] The central product classification (CPC) is a classification based on the physical characteristics of goods or on the nature of the services rendered. Each type of good or service distinguished in the CPC is defined in such a way that it is normally produced by only one activity as defined in International Standard Industrial Classification (ISIC). The CPC covers products that are an output of economic activities, including transportable goods, non-transportable goods and services.

[3] Some authors have modelled the effects of standards by shocking two production factors in GTAP factors, i.e. (i) unskilled labour and (ii) capital (Cassells and Meister, 2001).

- Generate the full regional disaggregated trade matrix of 87 regions using GTAP.
- Create a EU-15 aggregate. We used the EU-15 as the EU aggregate because the GTAP database is constructed on 2001 data, i.e. before accession of the 10 New Member States.
- Select the trade partners that represent at least 5% of global exports or imports of the commodity or sector being studied. This step ensures that the major exporters and importers of the global sector are included.
- Aggregate the remaining countries in two regions: (i) the rest of OECD (ROECD) countries and (ii) the rest of the world (ROW).
- Correct for intra-trade. In GTAP, aggregating does not eliminate intra-trade. Therefore, the trade matrix has to be corrected for this by setting the intra-trade of the EU-15, the ROECD and the ROW to zero and recalculating the totals of the matrix.

The organization of the book

The book focuses on the key environmental, human health and animal welfare issues affecting agriculture in the EU, US, Canada and New Zealand. The competitiveness and trade implications of domestic policies and regulations are assessed for a range of agricultural commodities, differentiating between farm, sector and global levels of competitiveness. The first part of the book highlights changes taking place in developed country agricultures.

Chapter 2 documents the EU experience with cross compliance as a means to improve the linkage between the receipt of CAP payments and a minimum level of environmental protection. Kaley Hart and her co-authors review the rationale and objectives of cross compliance within the CAP. This form of cross compliance introduced is likely to increase rates of compliance with existing legislation. Its extension beyond the environmental domain to animal diseases and food safety, as well as farm animal health and welfare legislation means that it provides incentives to increase compliance with a range of EU legislation that is relevant to farming.

Chapter 3 provides the reader with an understanding of US agri-environmental policy including a review of the approaches that have been used and a description of the overall regulatory context. Jonathan Winsten and his co-author present case studies that characterize compliance costs of key regulations and standards affecting the dairy and swine sectors in selected states. Comparisons between US and EU agri-environmental policy strategies are presented.

Chapter 4 examines agricultural environmental policy in Canada. Canadian agriculture, relative to most of the developed world, is a low-input intensity production system. Predrag Rajsic and his co-authors highlight the implications of this for the demand for environmental regulation in Canada compared to the EU. The nature of agricultural production in Canada may explain the voluntary nature of many important aspects of agricultural environmental policy in Canada. Canadian voluntary programmes, such as the Environmental Farm Planning Initiative, have been successful in involving farmers in actively solving environmental problems on their farms. Nevertheless, Canada does have a comprehensive system of federal, provincial and municipal environmental laws functionally equivalent to the EU Directives. Costs of environmental compliance are lower than in most of the EU but should be expected to increase in the future, as the environmental pressure from agriculture and urban-rural conflict increases.

Agricultural environmental issues in New Zealand are discussed in Chapter 5. Anton Meister and Nicole Beechey clarify the decentralized and outcome oriented approach that has been adopted by a country that is proud of its clean and green image. However, more intensive pastoral land use practices continue to put pressures on soil and water quality, and biodiversity. Guidelines for sustainable land management are implemented and sustainable practices need to be adopted by land managers to control effluent discharge. Following a time period of voluntary action and non-regulatory approaches, regulation has gained importance for the management of water quality, soil erosion, the preservation of biodiversity and greenhouse gas emissions.

Farmers are increasingly involved in supply-chain driven certification schemes, targeted towards the quality and safety of products, as well as the production methods adopted regarding environment and animal welfare. In Chapter 6, Roel Jongeneel and Thomas Herzfeld compare voluntary certification schemes and legal standards in the EU. There are similarities between cross compliance and certification schemes, but the benefits of one approach compared to the other depend on the ability to set appropriate standards. Also, membership of a certification scheme may enhance the likelihood of the farmers complying with environmental standards.

Part II of the volume offers a commodity approach linking public concerns with trade and competitiveness.

In Chapter 7, Irina Bezlepkina and her co-authors analyse the impacts of the Nitrates Directive and the Identification & Registration Directive on the EU dairy sector (Germany, France, Italy, the Netherlands and the UK) and its key competitors (the US and New Zealand). The negative impact of the analysed measures on EU imports and exports is less than 1%. When the standards for nitrate pollution control taken by the US and New Zealand are taken into account along with a full compliance assumption in all countries, this would only marginally improve EU exports. The trade impacts obtained under the assumption of no changes in key competitor countries can be interpreted as upper bounds of the likely trade impacts.

Chapter 8 examines two categories of regulatory compliance costs for the beef industry in the EU and the US. Kees de Roest and his co-authors argue that environmental regulations related to protecting water quality can increase farm costs in the beef sector. Among the very different beef production systems existing in the EU and in the US, intensive large scale finishing farms are most affected by this policy. Food safety policies, including regulations on the use of growth hormones, have effects on all types of beef farms. Compliance costs associated with these policies have a larger effect on farm costs than environmental legislation. Both policies have a significant impact on the international competitiveness of beef on the world market. In particular, a ban on the use of growth hormones in the US would reduce US world beef exports by 36%.

Chapter 9 assesses the impact of compliance costs for farms producing cereals. Roel Jongeneel and his co-authors provide insight into the cereal market in the EU and main production regions around the world. The requirements related to Good Agricultural and Environmental Conditions (GAECs) in the EU have a bigger impact on cereal production than do compliance with other environmental standards. The authors conclude that production costs of cereals increase by some 1% as a result of compliance with regulations, resulting in changes in international trade of cereals of less than 2%.

Chapter 10 explores the trade effects of environmental and animal welfare standards in the hog sectors of the EU, US and Canada. Kees de Roest and his co-authors assess the impacts of improved compliance with the Nitrates Directive by hog producers in the EU. Total compliance costs could increase production costs by up to 5%. Given current compliance rates with this legislation, however, the additional costs of full compliance across the EU might be less than 1% of total production costs. Implementation of the Clean Water Act (CWA) in the US could increase production costs by around 1%, and implementation of nutrient management regulations in Canada could increase production costs in the range between 2 and 4%. Such cost increases can have considerable implications for international trade. Cost increases of animal welfare rules in the hog sector of the EU have limited implications for international trade.

Compliance with environmental standards and the adoption of GAECs are important in the fruit and vegetables sector. Consuelo Varela-Ortega and Paloma Esteve compare costs of compliance with the Nitrates Directive and pesticide regulations across farms in parts of Spain. The authors argue that compliance with standards might even reduce costs, mainly in highly productive regions where changes in management practices (e.g. introduction of pest management practices) improve efficiency. The authors conclude that environmental standards present important challenges to the fruit and vegetables sector in Europe. However, they claim that certification schemes could potentially strengthen the competitive position of this sector, largely through better market access, premium prices and possibly cost savings.

Emerging policy perspectives are identified in Part III of the volume. Chapter 12 provides a perspective of developing countries on the role of the WTO in solving SPS agreement and TBT agreement trade conflicts caused by government regulations. The role of private standards impacting trade patterns is also highlighted by Sílvia Helena Galvão de Miranda. Four Brazilian international disputes related to sanitary and technical issues illustrate the complexity of challenges: the risk analysis requirements to import wheat from the US; the Japanese phytosanitary requirements to control fruit-fly in imported mangoes; the European aflatoxins regulation on Brazil nuts; and the regionalization principle in FMD. The ability of developing countries to comply with technical requirements is particularly influenced by investments on technological infrastructure and human capital.

The changing policy context of meeting standards in agriculture is explored by Vicki Swales in Chapter 13. She explores the role of, and need for, policy interventions to promote standards in relation to agriculture and environmental protection in the EU. She concludes that while there are broad similarities in policy approaches in the EU, US, Canada and New Zealand, there are however important differences, possibly due to the complex interactions between agriculture and the environment.

A long-standing concern in environmental policy making is that setting high standards for environmental protection in one jurisdiction might prompt emission intensive industries to relocate to less demanding regulatory environments. This is sometimes referred to as the pollution haven hypothesis. In Chapter 14, Alfons Weersink reviews the results of two studies of this hypothesis for the livestock sectors in North America. The analysis for the US livestock sector suggests that regional production shares for hogs, and to a lesser extent dairy, have increased in regions in relatively more lenient regulatory regimes.

Regions within Canada, the EU, and the US differ in the way they attempt to control pollutants stemming from potato production. Chapter 15 applies the actual agro-environmental policy approaches from each region to the same watershed to simulate differences in profits, costs, and sediment abatement levels. John Jatoe and his co-authors compare the cost-effectiveness of three different policies. Their general finding is that a uniform approach – e.g. requiring the cover crops for all farms – is more cost-effective than approaches targeting additional conservation practices in more erosive regions within the watershed.

Chapter 16 reflects on the environmental compliance costs in developed country agricultures, and its implications for trade and competitiveness. Glenn Fox and Floor Brouwer highlight the importance of enforcement mechanisms and implementation, subjects which often have been undervalued in research on environmental policies. The authors highlight some of the achievements from this volume:

1. Clarification of the relevant reference standards for environmental performance in agriculture;
2. Documentation of the increasing importance of voluntary private standards;
3. Explanation of the importance of functional equivalence in the comparative analysis of environmental regulation; and
4. Examination of the relationship between historical trends in input intensity and demands for environmental regulatory response.

This concluding chapter also compares market and non-market failures as contributing causes of environmental degradation. The authors argue that important legal and institutional dimensions often are overlooked, when examining externalities. They draw attention to Coase's concept of legalized nuisance and its connection to 'Right to farm' policies.

References

Brouwer, F. and D. Ervin (eds) (2002), *Public Concerns, Environmental Standards and Agricultural Trade*, CAB International, Wallingford, UK.

Cassells, S.M. and A.D. Meister (2001), Cost and trade impacts of environmental regulations: effluent control and the New Zealand dairy sector. *Australian Journal of Agricultural and Resource Economics,* 45:257-74.

Porter, M.E. (1990), *The Competitive Advantage of Nations.* Free Press, New York.

Chapter 2

The role of cross compliance in greening EU agricultural policy

Kaley Hart, Martin Farmer and David Baldock

Introduction

Over the past thirty years farming in Europe has become subject to an increasing body of EU legislation, addressing agricultural practices and setting new standards in relation to the environment, food safety, and farm animal health and welfare. This process continues, but has given rise to growing concerns about unsatisfactory levels of compliance, and raises questions about the extent to which these standards are delivering, particularly for the environment. Against this background a debate began in the 1990s about cross compliance as a policy tool for improving conformity with standards at farm level. After an initial period in which Member States could adopt certain forms of cross compliance if they wished, a stronger formulation of the policy was introduced as part of the far reaching 2003 Mid Term Review of the Common Agricultural Policy (CAP). In doing so EU Agricultural Ministers sought not only to improve compliance with basic standards enshrined in EU legislation, but also to avoid land abandonment and to ensure that agricultural land was maintained in good agricultural and environmental condition. Member States were provided with the authority to reduce or withhold direct payments made to farmers under the CAP if the farmer failed to comply with any of these requirements. The same principle was to apply to a further set of standards outside the sphere of EU legislation, designed to conserve soils, habitats and landscape features on agricultural land on which the direct payment, now a central feature of the CAP, was claimed.

This chapter documents the EU experience with cross compliance as a means of addressing environmental issues associated with agriculture. It reviews the rationale and objectives of cross compliance within the CAP, tracing its development from earlier forms of environmental conditionality in the 1990s. To set it in context, the role of this relatively innovative policy tool within the CAP is considered in relation to that of other forms of intervention, notably binding legislation and voluntary measures. The latter part of the chapter provides an assessment of the initial performance of cross compliance during the period in which it has become operational, considering in turn some of the key issues that have been raised in relation to its potential as a tool for integrating environmental concerns within agricultural policy. These include the variable application of cross compliance in the EU-27, the level of costs imposed on farmers, issues relating to compliance and enforcement as well as monitoring and evaluation.

Ways of regulating farming in Europe

Cross compliance is commonly conceptualized as occupying a policy space separate to that of regulation and the sphere of voluntary incentive measures. This section explores the different roles of regulation and incentives in seeking the achievement of environmental and related outcomes in order to situate cross compliance within the architecture of European agricultural policy.

The theoretical basis

In this discussion, the 'reference level' comes into play. It is a concept that was developed by the OECD in the 1990s to define the dividing line between the level of environmental provision that farmers are expected to deliver at their own expense, and an enhanced level of environmental quality, the delivery of which farmers may expect to be paid, for example through agri-environment schemes (OECD, 1998; Scheele, 1999). It builds upon the Polluter Pays Principle (OECD, 1974), and effectively articulates an individual's property rights, stipulating the extent of obligations that apply with respect to the management of the land.

In classical economic theory there is no *prima facie* reason for granting or denying property rights to private farmers as opposed to society as a whole. The question of where to set the reference level is largely a question of political and legal choices and considerations of equity and fairness in different social and historical settings, rather than determined by the desired environmental outcome, scientific evidence or the biophysical characteristics of land. This explains why reference levels often are set at different levels in different countries (Cooper *et al.*, 2009).

In relation to agriculture, the reference level is either enshrined in binding legislation or expressed in standards of good agricultural practice. Cross compliance, therefore, is an interesting case. In so far as it is based on existing legal standards, it forms part of the reference level, alongside any other EU, national or regional legislation that is relevant to farmers but not included within the cross compliance framework. If, however, it introduces standards that apply only to farmers claiming direct payments then it is a form of conditionality on those payments and not embedded in the underlying reference level.

In practice

The majority of standards which apply to farming practices are set out in binding legislation, established by the institutions of the EU and which has legal force in each of the EU's 27 Member States. Regulations are directly applicable in Member State law, while Directives, a common form of EU legislation, need to be implemented at the Member State level within a defined time period. This legislation concerns the environment, public and animal health and animal welfare, and has been gradually introduced since 1979, the point at which the EU's first nature protection requirements came into force. Much of this legislation is not focused exclusively on agricultural activity and farmland. For example, the need to secure the appropriate management of areas protected under the 1979 Birds Directive and the 1992 Habitats Directive extends to a wide range of terrestrial and marine habitats, including many without any agricultural management.

This body of EU legislation is complemented by standards set through legislation at the national level, introduced by the Member States to address particular issues, such as to limit water abstraction or to provide for the protection of landscape features.

Agricultural policy, mainly embodied in the CAP, operated within an (almost) entirely separate policy sphere to this body of legislation until 2005, when mandatory cross compliance was first introduced. Earlier efforts in the 1990s to apply a level of environmental conditionality to the receipt of CAP support payments had resulted in limited impact. In accordance with Council Regulation 1782/2003, cross compliance allows for most of the main types of payments received by farmers under the CAP to be reduced or withdrawn if the recipient is found not to be compliant with specified legislation affecting farming activity. The relevant obligations in this legislation are rearticulated as 'Statutory Management Requirements' (SMRs) to which cross compliance applies. The principle of reducing or withdrawing payments also applies if recipients of direct payments fail to maintain their agricultural land in 'good agricultural and environmental condition' (GAEC). While EU and national legislative requirements apply to all land and the actors responsible for its management, irrespective of whether it is farmed or in receipt of public money, the GAEC standards, in contrast, only apply to agricultural land in receipt of the direct payment, and thus are more limited in their coverage.

The CAP also includes a system of incentive payments for farmers undertaking environmental management activity, introduced progressively since 1985. Farmers may be compensated for the income foregone and costs incurred in undertaking commitments that are more demanding than those which form part of the reference level, under the agri-environment provisions of the Rural Development Regulation (Council Regulation 1698/2005). For example, as the owner of a plot of land that provides a habitat for a rare species, but it is not protected by legislation or standards of good practice, the farmer has the right to plough it for conversion to a more profitable land use, and therefore to destroy the habitat. However, payments can be offered, under the agri-environment measure, to compensate for the income forgone and additional costs incurred, to persuade farmers to undertake the management required to maintain or restore the habitat in order to meet targets for biodiversity conservation.

Outside the sphere of public policy, privately administered quality assurance schemes for food and other products also set a range of standards that farmers must comply with if they wish to sell into certain markets – such as some large supermarket chains. These schemes tend to require conformity with legal standards, and the more demanding schemes will require compliance with more ambitious standards, typically for the environment or farm animal welfare. There is, therefore, some overlap with cross compliance standards. Although participation in such schemes is voluntary, meeting these standards may allow the producer to access desirable or virtually essential markets, take advantage of any associated product branding, and can help to generate a premium price for the product concerned. Quality assurance schemes are explored further in Chapter 6.

The development of cross compliance in the EU

The introduction of cross compliance as part of the mid-term review CAP reform followed from earlier experiences, in the 1990s, of making the receipt of certain CAP payments conditional on meeting specified environmental standards. This development mirrored the growing linkage between the agricultural and environmental policy spheres which had stemmed from the increasing environmental critique of the impact of farming activities on water, soil and biodiversity. In parallel, the EU had also started to articulate its priorities for the environment more clearly.

The development of environmental conditionality within the CAP

The term 'cross compliance' has its origins in the USA, where the notion of cross compliance was first considered as an agricultural policy option in the early 1980s to address the problems of soil erosion and water conservation (Baldock and Mitchell, 1995) and to attempt to integrate the environmental and agricultural policy domains through a focus on compliance with mandatory standards.

The 1985 Food Security Act linked the eligibility for certain agricultural support payments to compliance with environmental standards. Farmers claiming support under a commodity support programme had to meet the rules of that programme and certain obligations embodied in other federal government programs. This link across programs gave rise to the term 'cross compliance' (Benbrook, 1994). Some key lessons emerged from the early experience in the US. These included difficulties in monitoring compliance, the fact that the scale of coverage was limited to those in receipt of support payments, and that the motivation to comply was linked to the degree to which farmers depended on government support (Baldock and Mitchell, 1995).

In the EU, the importance of protecting the environment became an increasingly significant policy issue in the 1990s. More environmental legislation was adopted, much of it, such as the 1991 Nitrates Directive, related to water. There was an increased emphasis placed on articulating the EU's priorities for the environment, alongside the development of an increasing number of targets. The 1992 Maastricht Treaty underpinned an EC commitment to integrate environmental considerations into other areas of EU policy including agriculture. The 1999 Amsterdam Treaty made sustainable development an explicit objective of the EU and the subsequent 2001 EU Sustainable Development Strategy stressed the need for all European policies to have sustainable development as their core concern. These changes occurred amid a growing critique of the environmental impact of agriculture (Baldock, 1989; Taylor and Dixon, 1990; RSPB, 1996). At the same time, it was recognised that agricultural policy, at this time largely targeted at supporting production, could be reformed so as to help secure the appropriate management of the rural environment (Lowe and Baldock, 2000).

The use of environmental conditionality within the CAP was a rather modest element in the 1992 reform but this sowed the seeds for the transition to contemporary cross compliance by the time of the mid-term review of the CAP, as summarized in Table 2.1.

The MacSharry CAP reform package, agreed in May 1992, was partly motivated by the need to address surplus production of some commodities. The guaranteed output prices of the CAP regime at the time had created an incentive for farmers to maintain a high level of production, and as a result of the reform, these were gradually lowered. Farmers were compensated by the introduction of direct payments, made in the form of area based payments for arable producers and increases in headage payment rates for the livestock sector. Set-aside was also introduced as a supply control measure, requiring most cereal farmers to withdraw a portion of their land from production.

Over time, the CAP and the nature of its payments had contributed to an increase in intensive farming practices in parts of Europe which had led to significant levels of environmental degradation (Taylor and Dixon, 1990). For example, headage payments to the livestock sector encouraged overgrazing in Ireland and the UK for example, which in turn harmed soil and water quality, and reduced the value of grasslands for biodiversity. In the arable sector, CAP support payments had led farmers to seek to increase yields,

Table 2.1 Environmental conditionality applied to CAP direct payments 1992-2005

Date	Type of Support	Sectors	Conditionality
1992 MacSharry CAP Reform Implemented from 1994	Livestock headage payments	Beef cattle and sheep	'Appropriate Environmental Measures' • Member States given the option to attach 'environmental measures' to receipt of payments. Maximum stocking densities • Recipients of beef payments required to comply with maximum stocking densities.
	Set-aside	Arable	• Member States to apply appropriate measures to ensure set-aside is managed with a view to protecting the environment.
1999 Agenda 2000 CAP Reform Implemented from 2000	Direct payments based mainly on crop area or livestock numbers	All those in receipt of direct payments	'Environmental Protection Requirements' Member States to introduce appropriate environmental measures, which could include: • General mandatory environmental requirements. • Specific environmental requirements constituting a condition for direct payments. Member States were permitted to introduce penalties (reduction or withdrawal of payments) for not complying with environmental requirements.
		Farmers benefiting from the Less Favoured Area (LFA) Compensatory Allowance or an agri-environment payment	'Usual Good Farming Practice' Farmers benefiting from either payment were obliged to 'apply usual good farming practice compatible with the need to safeguard the environment and maintain the countryside, in particular by sustainable farming'.
2003 Mid Term Review/ Fischler Reform Implemented from 2005	Decoupled direct payments	All those in receipt of direct payments under the Single Payment Scheme and Single Area Payment Scheme	'Cross Compliance' Farmers receiving direct payments required to respect: • Statutory Management Requirements (SMRs). • Standards for Good Agricultural and Environmental Condition (GAEC). Direct payment is reduced or withdrawn if the farmer is found to be non-compliant with SMRs or GAEC.

involving the use of high levels of pesticides and fertilizers, negatively impacting on soil and water quality. In order to increase the productive area, landscape features such as hedgerows, ponds and in-field trees of value for biodiversity, as well as the visual landscape were removed. The response to the growing critique of the environmental impact of agriculture at the time of the MacSharry reform was to introduce environmental conditionality on livestock headage payments in an attempt to limit stocking densities, as well as set-aside to reduce over-production in the arable sector. However, these measures were only partially successful.

First, Member States were permitted to withhold headage payments for beef cattle and sheep if the recipient failed to comply with 'appropriate environmental measures' (Council Regulation 125/93). These measures were to be defined by the Member State. Second, as part of an effort to encourage more extensive beef and veal production, recipients of certain beef headage payments were required to comply with a maximum stocking rate. This was set at 3.5 livestock units per hectare (LU/ha) in 1993, to decline to 2 LU/ha by 1996 for the Suckler Cow Premium and Special Beef Premium as set out by Council Regulation 2066/92. A maximum stocking density of 2 LU/ha was set for the basic beef premia, and those farms that met an average density of 1.4 LU/ha or less could qualify for an additional extensification premium (Lowe and Baldock, 2000). Third, with respect to land entered into the set-aside scheme, Member States were required to apply 'appropriate measures ... to ensure the protection of the environment' (Commission Regulation 2293/92) and to reduce or withdraw benefits received through the scheme for those found not to be compliant.

The measures were limited to the beef cattle, sheep and arable sectors, and not to other sectors, such as milk and sugar that also received substantial support under the CAP commodity regimes, although in a different form. In the targeted sectors these measures were largely ineffective. Many Member States were reluctant to introduce additional requirements, possibly because these would result in further costs for individual farmers. Some Member States failed to establish standards for the appropriate care of set-aside although progress was made in others (Lowe and Baldock, 2000). Only the UK and Ireland took the opportunity to introduce 'appropriate environmental measures' (Dwyer *et al.*, 2000) in relation to the receipt of beef cattle and sheep headage payments (in both cases, to address problems related to overgrazing). The stocking density limits were criticized for, amongst other reasons, being set too high for environmentally beneficial management, and for only applying to the number of animals for which premia were claimed rather than the total number present on the holding which in some cases was considerably greater (Lowe and Baldock, 2000). The maximum stocking rates for the Suckler Cow and Special Beef Premium were also above the average stocking rates on most farms at the time and hence had limited impact in reducing average stocking densities (Alliance Environnement, 2007b). Due to the patchy uptake in the Member States, and the over-elastic standards in the beef sector, these three forms of conditionality had a partial and limited impact in addressing some of the prevailing environmental problems associated with farming practices.

The Agenda 2000 reform of the CAP, agreed in 1999, altered the approach to environmental conditionality under the CAP in two respects: first, with the introduction of 'environmental protection requirements', and second, with the decision that farmers benefiting from payments under the LFA and Agri-Environment measures should meet a set of standards called 'usual good farming practice'.

At the same time, requirements were introduced for all Member States to apply 'appropriate' environmental measures to the receipt of all direct payments under the

European Agricultural Guidance and Guarantee Fund (in accordance with Regulation 1259/99) in order to ensure a certain degree of environmental protection. Known as 'environmental protection requirements', they extended the scope of conditionality beyond the arable cropping and beef and sheep sectors for the first time. Member States were provided with some discretion in determining what sort of requirements to introduce and were permitted to 'take the environmental measures they consider to be appropriate in view of the situation of the agricultural land used or the production concerned and which reflect the potential environmental effects' (Article 3 of Council Regulation 1259/99). These measures could take the form of 'support in return for agri-environment commitments', 'general mandatory environmental requirements', or 'specific environmental requirements constituting a condition for direct payments'. Member States were obliged to determine penalties 'that are appropriate and proportionate to the seriousness of the ecological consequences of not observing' the environmental measures, and could include a reduction or withdrawal of direct payments.

Many Member States believed that existing environmental legislation and voluntary agri-environmental programs were sufficient, and hence did not see the necessity to develop environmental protection requirements and an associated sanctioning system. Denmark, France, the Netherlands and the UK were, however, among a small number of Member States that chose to introduce requirements (Dwyer *et al.*, 2000; Guyomard and Le Bris, 2003).

'Usual Good Farming Practice' was defined as 'the standard of farming which a reasonable farmer would follow in the region concerned' (Article 29 of Commission Regulation 445/2002) and the standards were intended to 'safeguard the environment and maintain the countryside, in particular by sustainable farming' (Article 14 of Council Regulation 1257/99), by ensuring compliance with a set of verifiable standards, which as a minimum had to include 'general mandatory environmental requirements'. As conditions for receipt of payments, they were, however, limited to those in receipt of LFA or agri-environment payments. A more thorough inspection and sanctioning regime was introduced, with Member States required to monitor the compliance of five per cent of farmers receiving either payment each year. Failure to comply could lead to a reduction or withdrawal of the payment. There was no systematic European framework for the definition of what became known as good farming practice (GFP) standards. As a result, Member States chose a diverse range of standards, variously covering nutrient management, pesticide use, irrigation, soil management, animal husbandry and biodiversity and landscape (IEEP, 2006a). Most of these standards were selected from items of national or EU legislation with which farmers already needed to comply (IEEP, 2006a). Despite the diversity of standards at the European level, and the limited number of farmers to whom they applied, GFP increased the pressure on farmers to comply with baseline environmental standards and raised their awareness of environmental legislation.

The incomplete application of environmental protection requirements across Member States and the inconsistency of the implementation of GFP standards raised concerns that farmers were being treated differently in different Member States. This had implications both for the degree of environmental protection afforded in different Member States as well as for the costs faced by farmers in different countries and thus the prospect of distortions in competition. These concerns were to be addressed by the mid-term review with the introduction of cross compliance as a more standardised approach to environmental conditionality, mandatory for all Member States.

Cross compliance as introduced in 2005

Two successive CAP reforms in the 1990s had demonstrated a political commitment to improve the linkage between the receipt of CAP payments and a minimum level of environmental protection. However, on both occasions the majority of Member States were reluctant to introduce environmental conditions on direct payments. There was wider uptake of GFP standards, but little published analysis of their effectiveness exists. During this period, concerns about the environmental performance of farming continued to be highlighted alongside questions about the rationale for direct payments. If the EU were to succeed in marrying respect for baseline standards with receipt of public money under the CAP, it was increasingly apparent that a more structured approach to environmental conditionality would be needed. At the same time, the EU, along with other countries that supported their agricultural sectors with a high level of public expenditure, were coming under increasing pressure from trading partners to liberalize their agricultural markets as part of the World Trade Organization (WTO) Doha Development Agenda.

These various concerns were recognized by the Agriculture Commissioner Franz Fischler, who identified the need for swift ameliorative action and instigated an urgent review of the CAP in order to deal with these and other issues. A proposal for reform was developed that envisaged a more central role for cross compliance as an integral part of a shift to a decoupled system of direct payments. In order to increase the market orientation of farmers, the new system of direct payments largely broke the link between the payment received by a farmer and the level of production of any commodity.

The resulting reforms, implemented in 2005, led to the term 'cross compliance' being used for the first time in a piece of European agricultural legislation (Chapter 1 of Council Regulation 1782/2003). It became compulsory for all Member States for the first time and applied to all recipients of the requisite direct payments. It was no longer confined to the environment and included several standards in the areas of animal welfare as well as public, animal and plant health. In contrast to the preceding environmental protection requirements and the GFP standards, cross compliance was to apply to the receipt of all Pillar 1 direct payments and to eight relevant Pillar 2 measures. As a result, cross compliance superseded GFP in becoming the set of standards on which receipt of payments under many rural development measures was based. These measures included:

- natural handicap payments (as the LFA measure was renamed in Council Regulation 1698/2005);
- Natura 2000 payments and payments linked to the Water Framework Directive (Council Directive 2000/60/EC);
- agri-environment payments;
- animal welfare payments;
- first afforestation of agricultural land;
- Natura 2000 payments for forests; and
- forest environment payments.

The rationale for the introduction of cross compliance remains somewhat ambiguous, with more than one objective in play. In the initial proposals, the emphasis was on 'supporting the implementation of environmental ... legislation' and creating an obligation to 'maintain land in good agricultural condition' (CEC, 2002). It was extended to 'good agricultural and environmental condition' following pressure from environmental interests who stressed the opportunities that cross compliance provided to more fundamentally integrate sustainable land management obligations with the receipt of direct payments. As the negotiations proceeded, it became clear that cross compliance was intended to serve two main purposes. The first was to promote more sustainable agriculture, a goal that can be traced back to the policy heritage established by the 1992 'environmental measures' and the 1999 'environmental protection requirements'. Its second purpose was to act as a flanking measure to address undesirable side-effects of the introduction of decoupled direct payments, most notably the cessation of the active management of farmland and the risk of land abandonment. The decoupling of support payments to farmers from production led to fears that, without the link between the direct payment and the production of agricultural commodities, some farmers would find it more profitable to disengage from any form of agricultural management, yet would still be able to receive payments (IEEP, 2006b). Cross compliance served to thwart this threat by ensuring that those in receipt of the direct payment would need to continue to undertake a minimum level of agricultural management. It was seen as unacceptable for an agricultural policy to support the transition to a land use that would not be suitable for agricultural production, and hence it became an objective of cross compliance 'to avoid the abandonment of agricultural land and ensure that it is maintained in good agricultural and environmental condition'.

Together with the new system of largely decoupled direct payments, known as the Single Payment Scheme (SPS), cross compliance was 'to promote more market oriented and sustainable agriculture' (all quotes in this paragraph are from the preamble to Council Regulation 1782/2003). To work towards this Member States were obliged 'to incorporate in the common market organizations basic standards for the environment, food safety, animal health and welfare and good agricultural and environmental condition'. These 'basic standards' were defined according to a list of Statutory Management Requirements (SMRs) and were composed of selected Articles from European legislation in the areas of the environment, food safety, animal health and animal welfare, and set out in Annex III of the Regulation and shown in Table 2.2. Although the rationale for including existing legislation in cross compliance is not particularly well articulated in the regulatory texts, it is founded on the principle that those who breach their legal obligations should not be entitled to receive the amount of public money for which they would otherwise be eligible.

The legislative framework allowed for the SMRs to be introduced at national level over a number of years, with the pace of implementation slower for those Member States that acceded to the EU in 2004 and 2007. The 17 Member States that operate the SPS (i.e. the EU-15, Malta and Slovenia) were required to introduce a first swathe of SMRs concerned with the environment and the identification and registration of animals in 2005, followed by further SMRs regulating public, animal and plant health and the notification of livestock diseases in 2006. The final SMRs cover animal welfare and were introduced in these Member States in 2007. The remaining eight Member States which joined the EU in 2004 and that operate the Single Area Payment Scheme (SAPS) began introducing the SMRs over a three year period from 2009. Bulgaria and Romania will begin to implement the SMRs from 2012. The SMRs are entirely based on specific articles from pre-existing European legislation as transposed by the Member State. The SMRs are not new standards and are best characterized as providing an additional incentive for complying with current legal requirements. Despite this, there were some initial, largely unrealized fears, that farmers would forgo the direct payment in order to avoid cross compliance inspections.

Table 2.2 EU legislation and relevant articles defined as SMR under cross compliance

Applicable from 1.1.2005	
Environment	
Council Directive 79/409/EEC of 2 April 1979 on the conservation of wild birds	Articles 3, 4(1), (2), (4), 5, 7 and 8
Council Directive 80/68/EEC of 17 December 1979 on the protection of groundwater against pollution caused by certain dangerous substances	Articles 4 and 5
Council Directive 86/278/EEC of 12 June 1986 on the protection of the environment, and in particular of the soil, when sewage sludge is used in agriculture	Article 3
Council Directive 91/676/EEC of 12 December 1991 concerning the protection of waters against pollution caused by nitrates from agricultural sources	Articles 4 and 5
Council Directive 92/43/EEC of 21 May 1992 on the conservation of natural habitats and of wild flora and fauna	Articles 6, 13, 15, and 22(b)
Public and animal health	
Identification and registration of animals	
Council Directive 92/102/EEC of 27 November 1992 on identification and registration of animals	Articles 3, 4 and 5
Commission Regulation (EC) No 2629/97 of 29 December 1997 laying down detailed rules for the implementation of Council Regulation (EC) No 820/97 as regards eartags, holding registers and passports in the framework of the system for the identification and registration of bovine animals	Articles 6 and 8
Regulation (EC) No 1760/2000 of the European Parliament and of the Council of 17 July 2000 establishing a system for the identification and registration of bovine animals and regarding the labelling of beef and beef products and repealing Council Regulation (EC) No 820/97	Articles 4 and 7
Council Regulation (EC) No 21/2004 of 17 December 2003 establishing a system for the identification and registration of ovine and caprine animals and amending Regulation (EC) No1782/2003 and Directives 92/102/EEC and 64/432/EEC	Articles 3, 4 and 5
Applicable from 1.1.2006	
Public, animal and plant health	
Council Directive 91/414/EEC of 15 July 1991 concerning the placing of plant protection products on the market	Article 3
Council Directive 96/22/EC of 29 April 1996 concerning the prohibition on the use in stockfarming of certain substances having a hormonal or thyrostatic action and of beta-agonists, and repealing Directives 81/602/EEC, 88/146/EEC and 88/299/EEC	Articles 3, 4, 5 and 7
Regulation (EC) No 178/2002 of the European Parliament and of the Council of 28 January 2002 laying down the general principles and requirements of food law, establishing the European Food Safety Authority and laying down procedures in matters of food safety	Articles 14, 15, 17(1),18, 19 and 20
Regulation (EC) No 999/2001 of the European Parliament and of the Council of 22 May 2001 laying down rules for the prevention, control and eradication of certain transmissible spongiform encephalopathies	Articles 7, 11, 12, 13 and 15
Notification of diseases	
Council Directive 85/511/EEC of 18 November 1985 introducing Community measures for the control of foot-and-mouth disease	Article 3

Table 2.2 (ctnd)

Council Directive 92/119/EEC of 17 December 1992 introducing general Community measures for the control of certain animal diseases and specific measures relating to swine vesicular disease	Article 3
Council Directive 2000/75/EC of 20 November 2000 laying down specific provisions for the control and eradication of bluetongue	Article 3
Applicable from 1.1.2007	
Animal welfare	
Council Directive 91/629/EEC of 19 November 1991 laying down minimum standards for the protection of calves	Articles 3 and 4
Council Directive 91/630/EEC of 19 November 1991 laying down minimum standards for the protection of pigs	Articles 3 and 4(1)
Council Directive 98/58/EC of 20 July 1998 concerning the protection of animals kept for farming purposes	Article 4

Source: Council Regulation (EC) No 1782/2003

The SMRs are complemented by a framework of standards for GAEC, which are specified in a different way in the Regulation, with considerable scope for national or even regional standards within the EU framework. Annex IV of Council Regulation 1782/2003 provides this framework and lists a total of eleven very broadly phrased standards. Six of these relate to the management of soils, and specifically address soil erosion, soil organic matter and soil structure. Five other standards are set out with the aim of securing a basic level of maintenance of agricultural land (see Table 2.3). Article 5 of Council Regulation 1782/2003 allows Member States to define their own GAEC standards in line with this framework. The standards should be adapted to national circumstances and 'take into account the specific characteristics of the areas concerned, including soil and climatic conditions, existing farming systems, land use, crop rotation, farming practices and farm structures'. The standards should also take into account the management options for which farmers are able to obtain payment under voluntary incentives under agri-environment schemes. Those measures attracting payment should be more demanding than the 'reference level of good agricultural practice'. The introduction of GAEC put the onus on Member States to consider issues of particular relevance to them and whether to initiate new measures to meet Community standards. In the absence of any relevant pre-existing national legislation, Member States would in principle need to introduce new standards in order to meet the requirements of the GAEC framework or address any other environmental problems that were not already satisfactorily dealt with by European or national legislation.

In contrast to the earlier forms of environmental conditionality, which allowed the Member States to decide on appropriate penalties themselves, Council Regulation 1782/2003 introduced a consistent European approach to the inspection, control and application of penalties for breaches of the cross compliance SMRs and GAEC standards. Article 44 of Commission Regulation 796/2004 states that a minimum of one per cent of all farmers receiving direct payments are to be inspected each year, with Article 45 stating that 20% to 25% of farmers should be selected randomly and the remainder according a series of risk based criteria. The penalty applied for non-compliance would be graded according to four categories: severity (in relation to the impacts of failing to comply, expressed in relation to the objectives of the standards); extent (whether the impact of the breach is limited to the farm or has a wider impact); permanence (the period of time that the impact arising from the non-compliance will last, and whether this may be rectified); and repetition (cases where the same breach on a particular farm is identified more than once in a three year period). Articles 66 and 67 of Commission Regulation 796/2004 set out the penalties for non-compliance. Where the breach has occurred as a result of the negligence of the farmer, a penalty of a 1% to 5% reduction to the direct payment may be made, rising to a 15% reduction in the event of repeated breaches. If the breach is judged to have occurred due to intentional actions by the farmer, the penalty may range from a 15% to a 100% reduction. From 2008, Member States have had the option not to apply reductions to farmers if these amount to €100 or less. These penalties are additional to any that may arise as a result of non-compliance with the underlying national legislation.

Table 2.3 The EU framework of Good Agricultural and Environmental Condition

Issue	Standard
Soil erosion: Protect soil through appropriate measures	Minimum soil cover Minimum land management reflecting site-specific conditions Retain terraces
Soil organic matter: Maintain soil organic matter levels through appropriate practices	Standards for crop rotations where applicable Arable stubble management
Soil structure: Maintain soil structure through appropriate measures	Appropriate machinery use
Minimum level of maintenance: Ensure a minimum level of maintenance and avoid the deterioration of habitats	Minimum livestock stocking rates or/and appropriate regimes Protection of permanent pasture Retention of landscape features, including, where appropriate, the prohibition of the grubbing up of olive trees Avoiding the encroachment of unwanted vegetation on agricultural land Maintenance of olive groves in good vegetative condition

Source: Annex IV of Council Regulation 1782/2003, as amended by Council Regulation 864/2004

The system of cross compliance implemented in 2005 was both more comprehensive in the scope of issues it sought to address and more rigorous than previous policies in this domain in requiring action from all Member States. This common approach helped to reduce earlier concerns that farmers across the Union would be treated differently in terms of the standards they would need to meet and with respect to the penalty they would face if identified as non-compliant. This helped to temper arguments put forward by some Member States that their competitive position in the EU would be undermined as a result of the introduction of cross compliance, although considerable differences in GAEC standards arose relatively rapidly.

Cross compliance in practice

The application of cross compliance since 2005 has undoubtedly expanded the area of land subject to basic environmental management requirements. However, it has been subject to criticism from both sides. On the one hand farming organizations have expressed dissatisfaction with the burdens that they feel that it has introduced (Alliance Environnement, 2007a). On the other hand environmental organizations have argued that the potential of the cross compliance mechanism to deliver environmental benefits has not been maximized, with a number of reports setting out its shortcomings in this respect (ECA, 2008; BirdLife International, 2009). In particular, issues have been raised in relation to the variation in GAEC standards implemented in different Member States, levels of compliance and enforcement and the lack of monitoring and evaluation. These issues and the questions of cost burdens for farmers are explored in turn.

The variation in GAEC standards between Member States

As permitted by the legislation, there are considerable variations between Member States both in terms of the total number of GAEC standards being applied and in the level of ambition in using these standards to address environmental problems (Farmer *et al.*, 2007). For example, most GAEC standards are based on the requirements of pre-existing national legislation or have otherwise been adapted from previous GFP standards (Alliance Environnement, 2007a). The soil erosion GAEC standards introduced by the Netherlands and the Slovak Republic are, for example, based on pre-existing national legislation, as are many of the minimum maintenance requirements in Belgium (Wallonie), Finland, Malta and the UK.

Responsibility for determining which GAEC standards should be translated into obligations on farmers rests with the Member State. According to evidence collected by Alliance Environnement (2007a), only twelve Member States, including Austria and the UK, implemented a comprehensive set of GAEC standards, introducing relevant standards for each of the four GAEC 'issues' of soil erosion, soil organic matter, soil structure and minimum level of maintenance. Of the remaining countries, ten Member States introduced GAEC standards for three of the issues, whilst three Member States introduced GAEC standards for two of the issues. The GAEC standards vary from those that are relatively simple, in that they approximate to normal practice on the majority of farms, to those that are more demanding and require the introduction of some management changes. It is difficult to determine the equivalency in the number and scope of standards between Member States, and thus to determine whether the burden of standards is greater in one country than another but it is fairly clear that there are significant variations.

This is not surprising given the freedom provided to the Member States to implement the GAEC framework, the lack of any requirement for Commission approval of the standards introduced, differing institutional capacities to define and enforce compliance with an array of standards across various policy domains (soil, water, biodiversity, animal registration and animal welfare), and varying levels of enthusiasm for cross compliance amongst the Member States, in part explained by differing political priorities and the severity of particular environmental problems.

Variation is not a concern per se, but the issue is whether Member States have too much latitude to adopt a minimalist approach, with adverse consequences both for the environment and for the level playing field. The division of GAEC standards into those that are compulsory for Member States to implement and those which are optional, as part of the CAP Health Check agreement in December 2008, should help address this to some extent. A greater degree of consistency in terms of the type of standards that are implemented in the EU-27 can be expected although the detailed requirements of these standards, and hence the degree to which they are designed to address the specific environmental needs of an area, is still in the hands of the Member States.

Costs to farmers

It is rarely possible to introduce a new policy intervention of this kind without incurring some form of cost. An element of this will be administrative cost but there are also costs to farmers of complying with new conditions. At the time cross compliance was introduced there were concerns among a number of farm lobby groups that cross compliance would result in additional costs over and above the cost implications of existing legislation.

However, at the European level, very few additional costs that arise at the farm level can be attributed to cross compliance. This is because the universe of standards that forms the reference level for agricultural management in the EU did not substantially change after the introduction of cross compliance. Almost all the standards packaged together under the banner of cross compliance were already a feature of the existing legislative environment. The SMRs were entirely based on selected articles of pre-existing pieces of EU legislation and, in many cases, Member States chose to base the GAEC standards on pre-existing items of national legislation. Because of this continuity in standards, cross compliance, for the most part, did not result in any significant additional costs to farmers that they did not already need to meet. The exception was the cost related to complying with these new GAEC standards that did not have a legislative precedent.

For SMRs, the costs of meeting the requisite standards may be significant on some farms, for example as a result of a need to make management changes or investments to achieve compliance. However, these costs arise from the obligations of extant EU legislation, which are applicable to all farms, irrespective of whether they receive direct payments or not. The only costs attributable to cross compliance per se arise from any specific administrative or new inspection burdens introduced. Where cross compliance increased farmers' motivation to comply with such legislation set out in SMRs, this may have given rise to increased costs, but in so doing, they have also avoided penalties under national law.

New GAEC standards may require farmers to adjust their management practices, and thus to incur additional costs, especially in those less common cases where they are relatively ambitious. Austria, Finland, France and England, for example, all introduced a standard that, whilst formulated differently in each case, required the introduction of a narrow buffer strip along certain field margins in order to protect a watercourse or boundary feature from agricultural pollution (Farmer *et al.*, 2008). The compliance costs to farmers of meeting such standards are difficult to estimate, not least because the extent to which farmers need to adjust their management practices in order to comply is difficult to appraise and yields along field margins tend to be below average. Box 2.1 provides examples of how the costs for two GAEC standards, implemented for the first time in England through cross compliance, were assessed.

Box 2.1 Attributing costs to GAEC standards - two examples from England

Costs attributable to the soil protection review (SPR) GAEC standard

- Each farmer in receipt of the Single Payment is obliged to complete a review of the soil problems on their farm, in accordance with a common template, and subsequently implement measures in order to address the problems identified.
- Costs vary depending on the seriousness of the soil problems on the farm. The highest costs are likely to occur on arable, general cropping, outdoor pig farms, dairy farms (with maize) and intensively managed stock farms where these activities coincide with vulnerable soils, slopes and watercourses.
- In many cases compliance costs may be low and are likely to comprise good husbandry which might be undertaken irrespective of cross compliance.
- The Department for Environment, Food and Rural Affairs (Defra) estimated that cost of developing a SPR for a medium-sized farm of 130ha would equate to £2/ha (€2.80/ha) with a 10% variation for smaller and larger farms (reflecting the higher unit costs of overhead management on smaller farms and lower costs on larger farms).

Costs attributable to 2m margins for the protection of hedgerows and watercourses GAEC standard

- To comply with the standard, a strip of land must be left uncultivated adjacent to hedgerows and watercourses. The strip should measure 2m from the centre of a hedge and a minimum of one metre from the top of the ditch bank. Fertilizers, herbicides or pesticides cannot be applied to it. The standard does not apply to fields that are smaller than 2 ha.
- There could be a financial impact on farm businesses from the loss of a crop on the 2m uncultivated strips, although most field margins tend to be less productive than the field as a whole.
- In some areas the total hedge width may be equal to 4m or more and therefore, there would be no need to have an additional uncultivated strip. As a result, areas where hedgerows are traditionally quite narrow with minimal shading will experience the larger yield reductions.
- The impact will also fall disproportionately on those with relatively small fields and/or those farms with a high density of watercourses and hedgerows.
- Based on income foregone, Defra estimated the cost of this measure at £7 (€10) per 100m of 2m margin on arable and dairy farms. This equates to around 0.5% of total farm costs or 2-3% of the Single Payment. For example, a 150 ha cereal farm with 50% of its margins against hedgerows or watercourses could have a cost for the loss of a full 2m of production of around £500 (€700) per annum (Defra, 2004).

Source: Farmer et al. (2006)

Compliance and enforcement

During 2005, the first year of cross compliance controls, 71% of all instances of non-compliance in the EU were for breaches of those SMRs concerned with the identification and registration of cattle, and usually arose due to the absence of eartags. A further 10% of all breaches – the second greatest number of breaches – arose in relation to the SMRs for the Nitrates Directive (European Commission, 2007). Just over two-thirds of all penalty reductions (68%) were made at the 1% level, with only 6% made above the 5% level, indicating that the majority of these breaches are likely to have been relatively minor. From this starting point, it might be anticipated that the number of SMR breaches will decrease over time, as cross compliance raises awareness amongst farmers of the most relevant aspects of EU legislation affecting agricultural management, and as farmers take note that penalties are being applied for identified compliance failures. Alternatively, it may be that the penalty system is not acting as an adequate deterrent. For example, in many cases, the cost of complying with the standards is higher than the potential reduction in payment for non-compliance, thereby reducing the incentive to comply (ECA, 2008). In addition, the European Court of Auditors concluded that 'Member States had not implemented effective control and sanction systems and that as a consequence the control system provides insufficient assurance on farmer compliance' (ECA, 2008). They also demonstrated that the environmental SMRs and GAEC standards are less well checked during inspections than other standards, often due to them being less straightforward to inspect, and therefore the data on breaches may under-represent the real situation. Despite this, a large proportion of Member States have backed calls for changes to cross compliance to introduce more flexible rules in relation to enforcement in line with the simplification agenda (Agra Facts, 2009).

There have also been criticisms about the nature of the sanctions applied, particularly the fact that these are related to the size of direct payments received by the farmer rather than proportional to the severity of the infringement (ECA, 2008; BirdLife, 2009).

Monitoring and evaluation

It is more challenging to assess the performance and impact of cross compliance than many other policies within the CAP. At the political level it has led to a greater focus on implementing EU legislation at farm level by both governments and farmers but it is difficult to appraise the impact of this. Many farmers have become more aware of their environmental obligations than previously and are concerned to avoid penalties which are subject to discussion in the press, for example (Alliance Environnement, 2007a).

The lack of formal assessment, reporting or monitoring requirements in relation to cross compliance is a significant weakness of the system. While GAEC standards can be tailored to local circumstances, priorities and needs, the Commission has provided only limited guidance for Member States on the SMR element of cross compliance. No guidance documents have been provided on the development or implementation of GAEC standards. The lack of guidance, the absence of any formal requirement for such standards to be based on an assessment of needs, together with the lack of checks by the Commission on the quality or appropriateness of the standards introduced by Member States, has led to a mixed pattern of implementation with some Member States taking limited measures.

Although Member States are required to report on the number of cross compliance checks undertaken and the results of these checks, they are not obliged to report on which standards they have introduced, and their success in addressing either the objectives set out in Council Regulation 1782/2003 or more local environmental needs and priorities, where these have been identified. The lack of performance indicators, for example, to measure the extent to which land abandonment has been prevented or land has been maintained in good agricultural and environmental condition, is particularly notable, an issue that has been picked up by the European Court of Auditors (ECA, 2008). Without such information, evaluations of how cross compliance is performing to achieve its objectives is problematic, which raises questions about its effectiveness, particularly in relation to the environment.

Conclusions

The form of cross compliance introduced into the EU in 2005 demonstrates a form of environmental conditionality that is likely to be more successful than its predecessors in improving the management of agricultural land from an environmental perspective (albeit with some variation between Member States and sectors). This will be achieved partly as a result of increasing rates of compliance with existing legislation. Its extension beyond the environmental domain to animal diseases and food safety, as well as farm animal health and welfare legislation means that it incentivizes compliance with a breadth of EU legislation that is relevant to farming. As a sanctioning mechanism, it operates without resulting in many additional costs to farmers that they were not already obliged to meet. It provides an incentive for both governments and farmers to pay greater attention to the implementation of standards, some of which have not been widely adhered to.

However, the significant benefits for the environment that the introduction of GAEC standards has the potential to deliver are yet to be realised in most Member States. Many have demonstrated a very low level of ambition in designing GAEC standards, choosing to reframe national legislation or introduce standards that simply approximate normal farming practice on the majority of farms. This, combined with the fact that the SMRs consist of legislation that farmers already were obliged to comply with, would suggest that criticisms of cross compliance as imposing significant additional costs on farm businesses are not borne out in practice. The lack of formal assessments of the quality and appropriateness of the standards that are put in place and the absence of indicators to aid monitoring of how cross compliance is delivering against its objectives, mean that it is difficult to assess how well cross compliance is performing in practice although there are clearly substantial variations between Member States. Significant efforts continue to be needed to improve the design and delivery of cross compliance if its potential for integrating environmental concerns within agricultural policy is to be maximized.

References

Agra Facts (2009) Farm Council: Policy debate on Simplification & LFAs, & 4 more MFB 'no's' on AOB, No. 31-09, 24/04/09.

Alliance Environnement (2007a) *Evaluation of the Application of Cross Compliance as Foreseen Under Regulation 1782/2003: Part I*, Descriptive Report, Final Report prepared for DG Agriculture, Alliance Environnement, Brussels.

Alliance Environnement (2007b) Evaluation of the environmental impacts of CAP (Common Agricultural Policy) Measures related to the beef and veal sector and the milk sector, Final deliverable Report to DG Agriculture under contract notice 30-CE-0101431/00-04.

Baldock, D. (1989) *Agriculture and habitat loss in Europe*, Institute for European Environmental Policy and WWF, London.

Baldock, D. and K. Mitchell (1995) Cross Compliance within the Common Agricultural Policy. A Review of Options for Landscape and Nature Conservation. IEEP, London.

Benbrook, C.M. (1994) Environmental Stewardship through Cross Compliance: A Review of the US Experience in Agriculture. Report for Agriculture Canada. Benbrook Associates, Washington DC.

BirdLife International (2009) Through the Green Smokescreen: How is CAP cross compliance delivering for biodiversity? Birdlife International, Brussels.

CEC (2002) *Mid-Term Review of the Common Agricultural Policy*, Communication from the Commission to the Council and the European Parliament, COM(2002) 394 final, Brussels, 10.7.2002.

Cooper, T., K. Hart and D. Baldock (2009) *The Provision of Public Goods Through Agriculture in the EU*, Report Prepared for DG Agriculture and Rural Development, Contract No 30-CE-0233091/00-28, Institute for European Environmental Policy, London.

Defra (2004) Final Regulatory Impact Assessment of options for the implementation of cross compliance – Good Agricultural and Environmental Condition. Department for Environment, Farming and Rural Affairs, London.

Dwyer, J., D. Baldock and S. Einschütz (2000) *Cross Compliance under the Common Agricultural Policy.* A Report to the Department of the Environment, Transport and the Regions. IEEP, London.

ECA (2008) *Is Cross Compliance an Effective Policy?* European Court of Auditors Special Report No 8, 2008, Luxembourg.

European Commission (2007) Report from the Commission to the Council on the Application of the System of Cross Compliance (COM 2007 (147) 29.03.07).

Farmer, M., J. Bartley and V. Swales (2006) *Deliverable 5: Mandatory standards in 7 EU countries and 3 non-EU countries (UK Report),* Paper prepared for SSPE-CT-2005-006489 Facilitating the CAP reform: Compliance and competitiveness of European agriculture.

Farmer, M., V. Swales, L. Kristensen, H. Nitsch and X. Poux (2007) *Cross Compliance: Practice, Lessons and Recommendations,* Deliverable D24 of the CC Network Project, SSPE-CT-2005-022727.

Farmer, M., T. Cooper, D. Baldock, G. Tucker, R. Eaton, K. Hart, J. Bartley, M. Rayment, K. Arblaster, G. Beaufoy, P. Pointereau, F. Coulon, M. Herodes, L. Kristensen, E. Andersen, R. Landgrebe, S. Naumann, A. Povellato, A. Trisorio, R. Jongman and B. Bunce (2008) Final Report – *Reflecting Environmental Land Use Needs into EU Policy: Preserving and Enhancing the Environmental Benefits of Unfarmed Features on EU Farmland.* Contract No. ENV.B.1/ETU/2007/0033, report for DG Environment. IEEP, London.

Guyomard, H. and K. Le Bris (2003) The Fischler's Proposals for the Common Agricultural Policy: Paving the Way for the Future? Working Paper 03-05. INRA – ESR, Rennes.

IEEP (2006a) An Evaluation of the LFA Measure in the 25 Member States of the EU - Report for DG Agriculture. IEEP, London.

IEEP (2006b) *Cross Compliance: An Example of Better Regulation?* A Research Paper of the Cross Compliance Network. IEEP, London.

Lowe, P. and D. Baldock (2000) *Integration of Environmental Objectives into Agricultural Policy Making,* in: Brouwer, F. and P. Lowe (eds) CAP Regimes and the European Countryside pp. 31-52. CAB International, Wallingford, UK.

OECD (1974) *The Polluter Pays Principle*, Council Recommendation C(74):223 adopted on 14 November 1974.

OECD (1998) Agriculture and the environment: Issues and policies, OECD Publications, Paris.

RSPB (1996) Birds of Conservation Concern, RSPB, Sandy.

Scheele, M. (1999) *Policies to manage local public goods in an EU context*, in: Virchow, D. and J. von Braun (eds.) Villages in the Future – Crops, Jobs, and Livelihood, Springer, Hannover.

Taylor, J.P. and J. Dixon (1990) Agriculture and the Environment: Towards integration, RSPB, Sandy.

Legislative Texts

Council Regulation (EEC) No 2066/92 of 30 June 1992 amending Regulation (EEC) No 805/68 on the common organisation of the market in beef and veal and repealing Regulation (EEC) No 468/87 laying down general rules applying to the special premium for beef producers and Regulation (EEC) No 1357/80 introducing a system of premiums for maintaining suckler cows Official Journal L 215 , 30/07/1992 P. 0049 – 0056.

Commission Regulation (EEC) No 2293/92 of 31 July 1992 laying down detailed rules for the application of Council Regulation (EEC) No 1765/92 with regard to the set-aside scheme referred to in Article 7 Official Journal L 221, 06/08/1992, P. 0019 – 0021.

Council Regulation (EEC) No 125/93 of 18 January 1993 amending Regulation (EEC) No 805/68 on the common organisation of the market in beef and veal Official Journal L 018, 27.01.1993 P.0001-0003.

Council Regulation (EC) No 1257/1999 of 17 May 1999 on support for rural development from the European Agricultural Guidance and Guarantee Fund (EAGGF) and amending and repealing certain Regulations OJ L 160/80 26.6.99.

Council Regulation (EC) No 1259/1999 of 17 May 1999 establishing common rules for direct support schemes under the common agricultural policy OJ L 160/113 26.6.1999.

Commission Regulation (EC) No 445/2002 of 26 February 2002 laying down detailed rules for the application of Council Regulation (EC) No 1257/1999 on support for rural development from the European Agricultural Guidance and Guarantee Fund (EAGGF) OJ L 74/1 15.3.2002.
Council Regulation 1782/2003 of 29 September 2003 establishing common rules for direct support schemes for farmers, OJ L 270, 21.10.2003.
Commission Regulation (EC) No 796/2004 of 21 April 2004 laying down detailed rules for the implementation of cross-compliance, modulation and the integrated administration and control system provided for in Council Regulation (EC) No 1782/2003 establishing common rules for direct support schemes under the common agricultural policy and establishing certain support schemes for farmers, OJ L 141, 30.4.2004.
Council Regulation 1698/2005 of 20 September 2005 on support for rural development by the European Agricultural Fund for Rural Development, OJ L 277, 21.10.2005.

Chapter 3

State and federal responses to increasing input intensity in the US

Jonathan Winsten and Lynn G. Knight

Introduction

The fundamental economic tenet of comparative advantage, as first described by Ricardo (1891) states that countries should specialize in the production of goods or services for which they are relatively more efficient. The theory follows that production and trade, according to comparative advantage, will increase total world production and wealth. As sound as this theory is, the subsequent issues surrounding international trade require further attention and analysis, as evidenced by the importance of international conventions, such as the General Agreement on Tariffs and Trade (GATT) and the World Trade Organization (WTO).

The heterogeneity across nations in per capita wealth, agricultural productivity, natural resources, and attitudes toward government intervention combine, with numerous other factors, to determine the standards and regulations, or lack thereof, that each nation institutes as the policy context in which its agriculture and food systems operate. The areas of greatest concern within this agricultural policy context tend to include environmental quality, protection of natural resources, human health, and animal welfare. However, the absolute and relative weights and priorities associated with each concern will also vary among nations and, even, regions, states, and provinces.

Agricultural production is inextricably linked to environmental quality, natural resource management, human health, and animal welfare (in the case of livestock agriculture). However, the issues of environmental quality and human health seem to dominate the debate regarding trade in agricultural and food products. Trade in these products includes, as externalities to farm production decisions, risks to a nation's environmental quality and human health. These risks represent disservices (i.e. economic bads) that, while implicit in the trade of goods, are nonetheless transferred across national boundaries (Runge, 1990; Runge and Nolan, 1990).

How nations, states, or other jurisdictions decide to address these risks (or use the threat of such risks to justify protectionist trade policies), is likely to have nontrivial impact on trade patterns, the competitiveness of their agricultural products on the world market, and resulting producer and consumer surpluses, as well as overall welfare. Enforced regulations and standards related to a nation's food production system, whether to protect environmental quality, human health, animal welfare, or other concerns, are likely to

 The Economics of Regulation in Agriculture: Compliance with Public and Private Standards (eds F.M. Brouwer, G. Fox and R.A. Jongeneel)

increase average production costs and potentially reduce competitiveness. This assumption, however, has been refuted by Porter (1990) and others who believe that stricter regulations will increase innovation and reduce costs of production in the longer-term.

In the US, agriculture, specifically farming and ranching, has been largely exempt from regulations and standards. As such, there has not been significant costs imposed on farmers and ranchers for compliance and the resulting impact on competitiveness of US food products has generally been small (Carpentier and Ervin, 2002). There are indications that this relatively lax regulatory approach towards agriculture is changing. Litigation over the past decade has resulted in more comprehensive enforcement of the Clean Water Act and Clean Air Act. As the consequences of this federal litigation trickle down, individual states are starting to impose more rigid compliance requirements on farms and ranches, as is discussed in more detail below.

The intent of this chapter is to explain the approach and main policy tools that are used to address the impacts of agricultural production on society, with a primary focus on environmental impacts, and to quantify the impact of the current regulatory burden on representative dairy and swine operations. The chapter provides an overview of US agricultural production, as well as details on the primary policies and programmes that comprise the policy context in which US farmers operate.

US agriculture: a primer

The intensity of agricultural production in the US has increased greatly since 1950. Some indications of this trend include a many-fold increase in chemical fertilizer and pesticide applications, as can be seen in Figures 3.1 and 3.2. Increases in stocking rates (animal units per area of land) for livestock production have also been dramatic for certain locations and sectors. The darker colours shown in Figure 3.3 indicate the areas where the greatest increase in the number of larger livestock operations occurred from 1982 through 1997. The greatest increases are shown to be in North Carolina (swine), Minnesota (swine), Wyoming (beef), the Western states (dairy and beef), and various other areas scattered throughout the country (Kellogg *et al.*, 2000). Farms with greater than 1,000 animal units that feed animals primarily in confinement are designated by Clean Water Act rules as Concentrated Animal Feeding Operations (CAFOs) and are the focus of water quality regulations at the federal and state levels.

Although the US has federal regulations that apply in all states, each of the 50 US states can and will have its own interpretation and enforcement of these regulations. Additionally, many states have their own regulations that are not part of federal mandates. Because the implementation, enforcement, and farm-level impact of regulations pertaining to agriculture vary from state to state, a description of the leading states for several major commodities is presented below. The regulatory context is described in the following section.

Corn production is dominated by the North-central states, known as the Cornbelt. Table 3.1 shows that over 63% of US corn production in 2007 originated from just 5 states, led by Iowa with over 18%. A very similar production pattern exists for soybeans. With regard to wheat, North Dakota, Kansas, Montana, South Dakota, and Texas led the US in production in 2007, accounting for more than 50% of the total production.

Figure 3.1 Chemical fertilizer use in US agriculture (1960-2006)

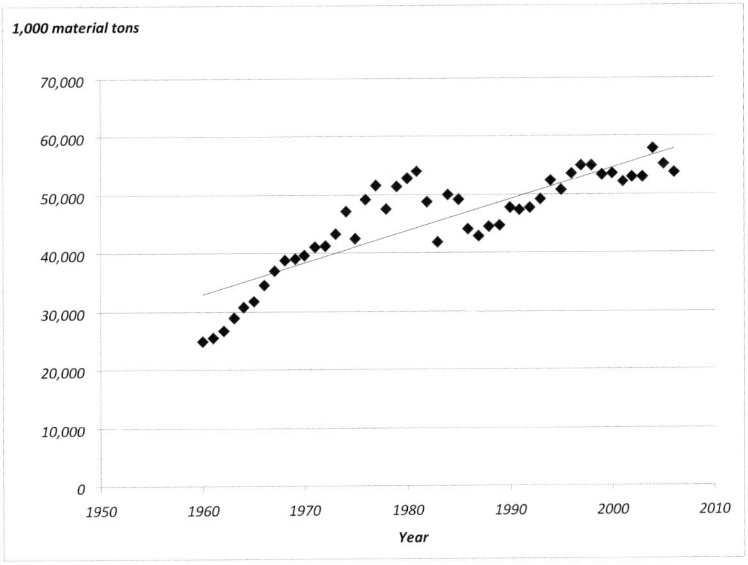

Source: (USDA Economic Research Service, 2007)

Figure 3.2 Index of pesticide use in US agriculture (1960-2006)

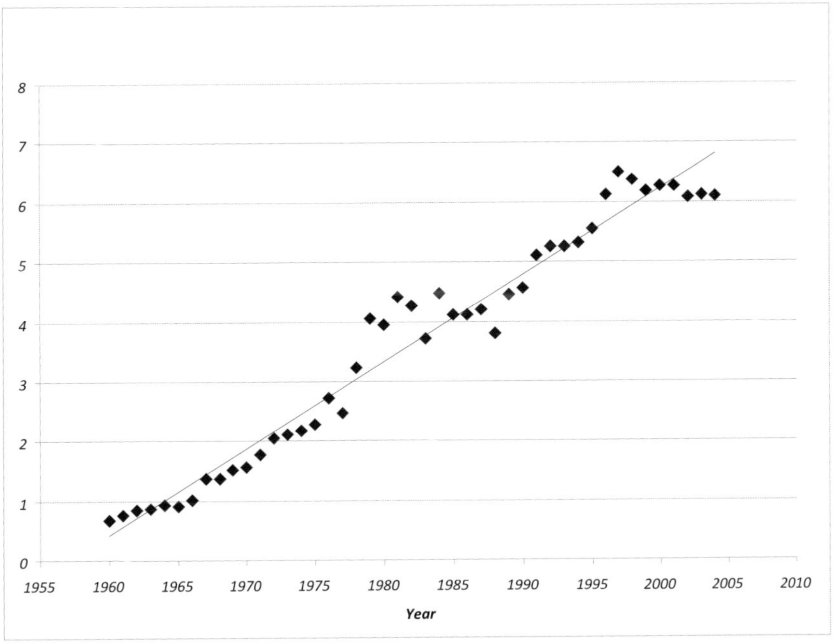

Source: (USDA Economic Research Service, 2004)

Figure 3.3 Change in number of livestock operations

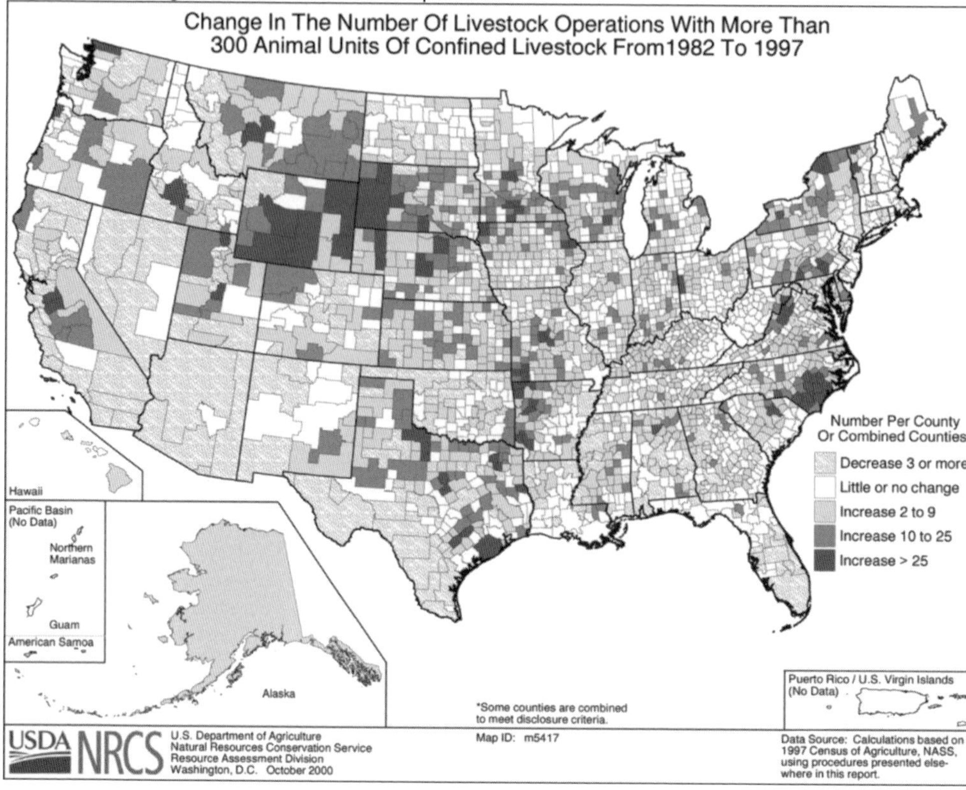

Table 3.1 Top five states in the production of major US commodities. Percentage of US production shown in parentheses

Commodity	Share of US production
Corn	Iowa (18%); Illinois (17%); Nebraska (11%); Minnesota (9%); Indiana (8%)
Soybeans	Iowa (17%); Illinois (14%); Minnesota (10%); Indiana (8%); Ohio (8%)
Wheat	North Dakota (16%); Kansas (14%); Montana (7%); South Dakota (7%); Texas (7%)
Dairy	California (24%); Wisconsin (14%); New Mexico (7%); Idaho (7%); Pennsylvania (6%)
Broilers	Georgia (16%); Arkansas (14%); Alabama (12%); Combined California, Missouri, Tennessee and West Virginia (10%); Mississippi (10%)
Swine	North Carolina (18%); Iowa (16%); Minnesota (10%); Illinois (7%); Oklahoma (6%)
Steer	Texas (17%); Nebraska (14%); Kansas (12%); Iowa (7%); Oklahoma (6%)

Source: USDA National Agricultural Statistics Service (2007)

Dairy production is more dispersed across the US than is production of many of the major crops. In recent years, western states such as California, Idaho, and New Mexico have greatly increased their milk production. California has overtaken Wisconsin, the historical hub of US dairying, to be the largest producer (Table 3.1). Average herd sizes in the western states are vastly greater than those in the traditional Dairybelt (the northern tier, from Maine to Minnesota). The larger herd sizes are the object of recently enforced Clean

Water Act and Clean Air Act regulations, making some western states, particularly California a logical choice for a case study analysis of regulatory costs in the dairy sector.

As can be seen in Table 3.1, poultry production is dominated by the south-eastern and Mississippi Delta states, led by Georgia, Arkansas, and Alabama. Important structural changes over the past 15 years include widespread vertical integration in the poultry industry. The majority of large producers are contract growers for poultry processing and marketing firms. Most often, contract growers do not own the birds; they are supplied with chicks, feed, and management oversight. This arrangement has been the focus of much debate regarding the appropriate level of liability that the integrator (i.e. processor) should assume for environmental degradation resulting from poultry production.

In recent years, North Carolina has overtaken Iowa as the leading swine producing state. Vertical integration is also common in the US swine industry, which, along with poultry, has had the fastest rate of growth of CAFOs. Air quality and odour control have become important issues facing large-scale hog production in many regions of the country. Together, North Carolina and Iowa produced 34% of the US swine production in 2007.

The top steer producing states are Texas, Nebraska, and Kansas. The beef cattle industry in the US has two distinct components. Cow-calf operations produce young stock primarily on pasture and rangeland. This occurs in the western and inter-mountain states, as well as in the southern states. Feedlot operations take yearlings and steers and fatten them for slaughter. It is feedlot operations that are facing increasing regulation as CAFOs. The western plains states, such as Nebraska, Colorado, Wyoming, Oklahoma, and the Texas Panhandle are where the majority of beef CAFOs are located.

US agricultural exports and the world market

Because an objective of this book is to quantify the impact of regulations on the relative competitiveness of trading partners in the world market, data have been assembled on total world trade in the major US agricultural commodities, as well as the percentage of the world market that is supplied by EU and US farmers. Table 3.2 provides an indication of the relative importance of US and EU exports into the world market as of 2004. This analysis aided in the identification of the most relevant commodities for the case study analyses presented in this chapter.

The numbers provided in Table 3.2 show that the US dominates world trade in maize and soybeans. Because maize and soybeans are both very minor exports from the EU, the impact on relative competitiveness between the US and EU is not of paramount importance for these commodities. Of the major crops, Table 3.2 shows that the EU and US each supply about one-quarter of the world market for wheat. As important rivals in this market, wheat could be an appropriate crop for analysing the impact of regulations on competitiveness. However, as discussed below, the majority of the regulatory burden on the US agriculture falls upon the livestock sector.

Of the major livestock products from the EU (dairy, pork, poultry, and beef), dairy comprises the greatest value of world trade by far, with over US$6 billion in 2004. Beef and pork production from the EU each represented slightly more than US$3 billion of world trade in 2004, and poultry represented slightly more than US$1 billion. The US was a much smaller supplier of dairy and beef, representing only 2.5% and 0.61% of the world market, respectively, in 2004. US beef exports were down dramatically in 2004 due to fears over mad cow disease; prior to 2004, the US represented 12-14% of the world market trade in beef. The reduced US exports in 2004 also have the effect of inflating the importance of EU beef exports in Table 3.2. The case study analyses presented in this chapter will focus on the dairy and swine industries.

Table 3.2 Percentages of 2004 world market sales from EU and USA

Exports from:	Commodity	Total value (US $1,000)	World market share (%)
EU (25) excluding intra-trade	Maize	168,327	1.43
USA	Maize	6,137,514	52.12
World	Maize	11,775,519	
EU (25) excluding intra-trade	Soybeans	4,170	0.03
USA	Soybeans	6,692,044	42.97
World	Soybeans	15,575,072	
EU (25) excluding intra-trade	Wheat	4,586,046	23.78
USA	Wheat	5,180,994	26.86
World	Wheat	19,285,389	
EU (25) excluding intra-trade	Milk equivalent	6,558,563	16.71
USA	Milk equivalent	1,000,709	2.55
World	Milk equivalent	39,248,448	
EU (25) excluding intra-trade	Pig meat	3,197,761	14.42
USA	Pig meat	2,092,490	9.44
World	Pig meat	22,174,238	
EU (25) excluding intra-trade	Poultry meat	1,010,287	7.33
USA	Poultry meat	2,210,778	16.04
World	Poultry meat	13,786,754	
EU (25) excluding intra-trade	Beef and veal	3,067,174	83.58
USA	Beef and veal	22,358	0.61
World	Beef and veal	3,669,863	

Source: (USDA Economic Research Service, 2008)

The regulatory context surrounding US agriculture

A handful of federal agencies are in charge of the implementation, oversight, and enforcement of regulations affecting agricultural production in the US. These include the EPA, the USDA, the FDA, and the Department of Interior. However, because many federal regulations are administered at the state level, the aforementioned federal agencies often delegate authority to respective state agencies for implementation and enforcement. The states then have some leeway in how they administer the regulations.

In general, agriculture in the US seems to have fewer mandatory standards imposed upon producers through regulations as compared with the directives that are relevant to agriculture in the EU (Carpentier and Ervin, 2002). It has been stated that US federal policies (mostly environmental) affecting agriculture have generally been created to prevent or alleviate specific problems; they have not been designed to produce positive amenities, such as environmental improvement, maintaining an aesthetic working landscape, or bolstering rural communities (Bernstein *et al.*, 2004). The creation of such positive amenities has been the goal of several policies within the EU and many other nations.

Environmental legislation and regulations

There are several important pieces of federal legislation that require mandatory compliance from certain US farmers. These include the CWA, the CAA, the CERCLA, the FIFRA, the FQPA, and the ESA. It is important to note that the applicability and level of enforcement of these regulations varies widely. Of the aforementioned regulations, the CWA has the largest potential impact on US farmers, but until recent years CWA regulations were largely unenforced for the agricultural sector. The CAA has a large impact, but only on farmers located within certain areas that are designated to have existing air quality problems.

These laws notwithstanding, the primary approach for addressing environmental issues from agriculture in the US has been to seek voluntary participation and compliance that is supported with funding to offset the direct, indirect, and opportunity costs of participation. Such programmes generally fall into two major categories; land retirement and working land conservation programmes. The largest of the land retirement programmes, by far, is the CRP. The Wetlands Reserve Program (WRP) and Grassland Reserve Program (GRP) are smaller, but similar programmes that also pay farmers for voluntarily restricting the use of specific parcels of land. These programmes are described below.

Some regulations have been the subject of recent litigation based on claims that they have not been enforced strictly enough. Legislation regarding the CWA has resulted in the CAFO rule and has impacted the way that USDA delivers some of its voluntary programmes (e.g. Environmental Quality Incentives Program [EQIP]) to help ensure that farmers required to comply with the CWA are doing so.

Federal Water Pollution Control Act Amendments of 1972 (Clean Water Act)

Background and general description

The CWA is the primary law protecting surface water quality in the US. It employs a variety of regulatory and non-regulatory tools to reduce pollutant discharges into the nation's waterways. Initially, the primary focus was on point sources of pollution from industry and wastewater treatment plants (WWTPs), with little attention paid to nonpoint sources. The focus in recent years has been expanded to include nonpoint source pollution (NSP). Agriculture has been identified as the primary source of NSP affecting US surface waters (US Environmental Protection Agency, 2002).

Impact on agriculture

The CWA includes regulation of farming activities. However, it was not until 15 December 2002 that the federal government issued specific rules governing CAFOs. CAFOs are typically defined as animal feeding operations with greater than 1,000 animal units (700 dairy cows, 1,000 beef cows, 2,500 hogs or 100,000 chickens). Smaller operations can also be designated as CAFOs, however, if specific environmental conditions apply. Medium CAFOs are those operations with between 300 and 1,000 animal units (AUs) with a stream running through the confinement area, or a man-made conveyance to surface water. Small CAFOs are those operations with less than 300 animal units (AUs) that, through on-site inspection, have been determined to be a significant contributor of pollutants to surface water (US Environmental Protection Agency, 2003).

Under the recent rule, all farms designated as CAFOs are required to obtain a permit under the NPDES. The NPDES permits have been used to control point sources of pollution since the Clean Water Act was established. The CAFO rule is the first application of the NPDES permit to agricultural non-point sources of pollution. The rationale for this rule is that large farms (CAFOs) are potential point sources of pollution because of their large manure storage requirements.

Obtaining an NPDES permit for a CAFO requires the design and implementation of a comprehensive nutrient management plan (CNMP) for the operation. The CNMP must be specific for the operation and detail the proper management of all animal manure produced. It must address the assimilative capacity of the farm's land for the manure and other nutrients applied to the land. The CNMP must be focused on the control of nitrogen (N) or phosphorus (P), based on state or local priorities. The implementation of CNMPs often results in operations seeking additional land on which to spread manure and/or the use of alternative nutrient control strategies for manure.

It is important to note that the CAFO regulations currently affect a very small percentage of livestock operations. The USDA estimated that CAFOs accounted for 2% of all confined livestock operations in 1997 (Gollehon *et al.*, 2001). Although an updated estimate is not available, the current percentage is likely to be higher, but probably less than 5%. The percentage of CAFOs varies across livestock types, with the greatest percentage of CAFOs in the swine and poultry sectors (Kellogg *et al.*, 2000; Speir *et al.*, 2002). Due to their larger size, the percentage of hectares affected by the CAFO regulations is much greater than the percentage of operations, but is still estimated to be a fairly small percentage.

The CWA also provides regulations concerning the application of sewage sludge to agricultural land. During the 1980s and 1990s, numerous studies and reviews were conducted in the US to ascertain the risks of using sewage sludge as an amendment on land used for food crops. Both the EPA and USDA have been proponents of using bio-solids on agricultural land. The regulations that govern the use of this practice set concentration limits for arsenic, cadmium, copper, lead, mercury, molybdenum, nickel, selenium and zinc. Additionally, proper practices for field storage are provided through this rule.

Implementation and approach

The implementation and enforcement of the CWA is delegated to the states. Each state has an authorized NPDES authority, usually the state agency responsible for natural resources and/or environmental quality. Each state's NPDES authority is responsible for issuing the permits and enforcing the related regulations and standards. This includes implementation of the CAFO rule. In most states, the agency of agriculture is also involved in this process, as it affects some portion of livestock operations. The most relevant issues in the CAFO process are (i) the threshold size of an operation for designation as a CAFO and (ii) the stringency of CNMP standards and enforcement.

Interaction between EPA's CAFO rule and state and local regulations

Most states already have some water quality regulations that affect farms. Some states have more stringent criteria, while in other states EPA's regulations will prove to be more stringent than their existing regulations. The USDA Natural Resources Conservation

Service (NRCS) practice standards are designed to meet federal, state and/or local requirements, whichever are most stringent. State and local NRCS offices must be aware of all applicable regulations and adapt their planning activities as required.

State and local regulations can vary greatly, depending upon how the states chose to define CAFOs and related criteria. These definitions are not consistent among states or with the previous EPA definition. For example, the level of acceptable discharge may be different between existing state and local regulations and the current CAFO rule. Finally, prior to the current CAFO rule, CAFOs that already have implemented practices meeting the 24-hour, 25-year storm event criteria were exempt from the NPDES permit requirement. These farms must now acquire a permit, even though they are most likely meeting the permit requirements.

In the states that have less restrictive local and state regulations, relative to the CAFO regulations, NRCS and its conservation partners in these states can expect to see an increase in technical service assistance requests from farm operators that now must apply for, and comply with, CAFO NPDES permits. To more thoroughly assess the types of assistance that will be required, each state and local office will need to determine and understand the relationship between EPA's CAFO regulations and their state and local laws. If the CAFO rule is more restrictive than current state and local regulations, NRCS will need to assure that their Field Office Technical Guide (FOTG) Quality Criteria and conservation practice standards reflect the CAFO requirements.

Many states are anticipating future changes to the CAFO rule and creating regulations for operations that are smaller than those currently defined as CAFOs in the rule. Maryland, central in the Chesapeake Bay watershed, is one of the few states to require CNMPs for all livestock operations, regardless of herd size. Unfortunately, the issue of stocking rate is not currently used as a criterion for designating farms as CAFOs, even though it may be a more relevant guideline than herd size alone.

Farm-level costs of adhering to regulation

The costs of implementing a CNMP will vary widely across farms and regions of the US. As an example, it was calculated that the average hog CAFO in the Mid-Atlantic region would need to increase the amount of available land for manure spreading by almost 500% in order to comply with a nitrogen-based CNMP. Obviously, the amount of additional land required, and the associated costs, vary widely from farm to farm and region to region. However, the overall effect will most likely be an increase in land prices and rents.

Additional costs for compliance with CWA regulations may include investments in manure storage, treatment, and handling, disposal of milkhouse wastewater and minimizing run-off from the production area. Also, the costs of third party technical service providers to assist farming operations with planning and compliance issues are incurred by many farms. However, to varying degrees across states and sectors, the majority of these costs are often borne by federal and state government programmes.

The USDA has estimated the costs for developing and implementing a successful CNMP. Documentation of these costs can be found in USDA's 'Costs Associated with Development and Implementation of Comprehensive Nutrient Management Plans. Part 1 - Nutrient Management, Land Treatment, Manure and Wastewater Handling and Storage and Recordkeeping'. Table 3.3 provides the typical costs of the various components required in a CNMP by animal sector and farm size in the US. These costs are the basis for estimating the CNMP costs of representative dairy and hog described later in this chapter. Obviously, the actual costs are going to vary greatly from farm to farm, based on each farm's current infrastructure and, hence, the required improvements necessary to comply with the standards in each state.

The Clean Air Act

Background and general description

The CAA was created as Federal law in 1970 in order to regulate air emissions. The law authorizes the EPA to establish National Ambient Air Quality Standards (NAAQS) to protect public health and the environment. The CAA directs the states to develop state implementation plans (SIPs) applicable to appropriate industrial sources in the state. Originally, the goal was to attain the NAAQS in each state by 1975. However, progress was very slow. The law was amended in 1977 primarily to set new goals (dates) for achieving attainment of NAAQS since many areas of the country had failed to meet the deadlines. The CAA was amended again in 1990 to address problems not previously focused on, such as acid rain, ground-level ozone, stratospheric ozone depletion and air toxics.

The CAA requires the EPA to set NAAQS (see Title 40 CFR part 50) for pollutants considered harmful to public health and the environment. The CAA established two types of national air quality standards. 'Primary standards' set limits to protect public health, including the health of 'sensitive' populations such as asthmatics, children and the elderly. 'Secondary standards' set limits to protect public welfare, including protection against decreased visibility, damage to animals, crops, vegetation and buildings.

The EPA Office of Air Quality Planning and Standards (OAQPS) has set NAAQS for six principal pollutants, which are called 'criteria' pollutants. They are Carbon Monoxide, Lead, Nitrogen Dioxide, Ozone, Sulphur Dioxide, Particulate Matter PM_{10} (which has been revoked), and $PM_{2.5}$. PM_{10} pollution includes 'inhalable coarse particles', with diameters larger than 2.5 micrometers and smaller than 10 micrometers and $PM_{2.5}$ includes 'fine particles', with diameters that are 2.5 micrometers and smaller.

EPA uses these six 'criteria pollutants' as indicators of air quality and has established for each of them a maximum concentration above which adverse effects on human health may occur. Areas of the country where air pollution levels persistently exceed the national ambient air quality standards may be designated non-attainment areas. Once these non-attainment areas are determined, states must develop Implementation Plans which outline what actions will be undertaken to address the pollutants and the timeframe needed.

Impact on agriculture

Agricultural production can emit a wide variety of material into the air. These include dust, a variety of nitrogen gases, pesticides, particulate matter from diesel engines, and controlled

Table 3.3 Estimated average costs per farm for implementing a Comprehensive Nutrient Management Plan (CNMP)

Dominant livestock type	Number of farms	Animal units per farm[1] ($/yr)	Record-keeping costs per farm ($/yr)	Nutrient management costs per farm ($/yr)	Off-farm transport costs per farm ($/yr)	Land treatment costs per farm ($/yr)	Manure costs per farm	Total CNMP implementation costs per farm: Average ($/yr)	Low[2] ($/yr)	High[2] ($/yr)	Per animal unit ($/yr)
Fattened cattle	10,159	1,298	142	1,655	4,646	2,613	9,112	18,167	1,026	308,005	14
Dairy	79,318	195	160	2,101	1,619	2,660	3,249	9,788	2,362	97,013	50
Swine	32,955	276	224	1,601	2,450	3,615	4,139	12,029	2,060	75,159	44
Broilers	16,251	183	90	248	1,667	1,220	2,351	5,576	1,128	36,187	30
Farm size											
Large	19,746	1,419	168	1,526	9,679	3,925	15,167	30,465	2,199	252,014	21
Medium	39,437	252	150	1,085	2,281	2,897	3,397	9,809	1,210	64,426	39
Small	198,018	80	106	987	345	1,267	1,070	3,773	161	25,298	47
All farms	257,201	210	117	1,043	1,358	1,721	2,509	6,748	195	67,429	32

Notes: [1]*Represents all animal units on the farm, but does not include animal units for specialty livestock types, which were not estimated.* [2]*The low estimate corresponds to the one-percentile value for the farms in each group, and the high estimate corresponds to the 99th-percentile value.*

Source: USDA Natural Resources Conservation Service (2003)

burning of fields. Potential pollutants associated with agriculture include internal combustion engines, dust from dirt access roads, hydrogen sulphide, ammonia, odours, other volatile organic compounds from animal manure, methane from dairy cows and cattle and nitrogen oxides from fertilized fields. Such pollutants can affect human health, reduce visibility, contribute to global warming and/or be a nuisance to neighbours. When the air quality standard for any of six air pollutants is exceeded, states must inform the US EPA on how they plan to respond. Any farm found to be a major source of regulated emissions in a region where air quality standards are exceeded could be required to apply for and comply with an operating permit.

Like other nonpoint source pollution issues, air pollution emissions from agriculture tend to be generated diffusely over a broad land area, rather than from a single pipe or smokestack. As such, it has not been cost-effective to accurately monitor emissions from individual agricultural sources using current technology. The potential cost of monitoring agricultural air pollution is one reason that agriculture has been lagging behind in compliance with environmental regulations (Ribaudo *et al.*, 1999).

Approach and implementation

Similarly to the CWA and other federal legislation, each state is responsible for implementing the CAA. It makes sense for states to take the lead in carrying out the CAA, because pollution control problems often require special understanding of local industries, geography, housing patterns, etc. Each state has a SIP that explains how the NAAQS will be attained. The EPA must approve each SIP; EPA can take over enforcing the CAA in that state if a SIP is unacceptable.

New regulations in many states are seeking to reduce air emissions from agriculture, particularly from animal feeding operations (AFOs). Air quality regulations seem to be most significantly affecting agriculture in California. Ozone and particulate levels in California's San Joaquin Valley have led to new requirements for agricultural producers, as this region has some of the most polluted air in the country. Farmers in this area must develop management plans showing how they will reduce dust, the burning of crop residue (e.g., rice straw, orchard trimmings) is restricted, and large dairies must manage their manure to reduce ammonia emissions (Ribaudo *et al.*, 1999). Two specific air-quality regulations impacting agriculture in the San Joaquin Valley include rules 4550 (for Particulate Matter 10), and 4570 (for Volatile Organic Compounds, including ammonia emissions). Operations meeting the defined criteria must obtain air quality permits and agree to implement conservation management practices which document reductions in air emissions.

Methyl bromide phase-out

Methyl bromide has been used in the US for decades to control insects, nematodes, pathogens, and weeds. It is used for soil fumigation before planting many fruits, vegetables, ornamentals and agricultural nurseries; for post-harvest fumigation of commodities in storage and prior to shipment; and for government-required quarantine treatment to prevent the spread of regulated exotic pests.

The Montreal Protocol is a treaty designed to protect the stratospheric ozone layer, which protects the earth from harmful solar radiation. The Protocol controls global production and trade of ozone depleting substances. The Parties to the Protocol classified methyl bromide as an ozone-depleting substance and agreed to phase out the use of methyl bromide by 2015, except for critical uses.

The net annual loss to consumers and producers of not having methyl bromide for selected pre-plant uses is estimated to be in the range of $400-450 million (Carpenter *et al.*, 2000). This includes $150-200 million for annuals (strawberries, tomatoes and other vegetables), $140 million for perennial crops, and $100 million for ornamentals and nurseries (excluding forest nurseries). These losses represent 8-10% of revenues for the annual crops (15-20% for strawberries, 4% for tomatoes), 3% for the perennial crops and 15% for the ornamental and nursery crops.

Florida fresh tomato producers use significant amounts of methyl bromide due to the more humid and higher rainfall climatic conditions there. The combination of methyl bromide and chloropicrin has been used extensively on Florida tomatoes to suppress crop pests such as nematodes, diseases/fungi and weeds. In 1992 and 1993, approximately 3,500 to 4,000 metric tons of methyl bromide, approximately 14% of the US total annual methyl bromide consumption was used on the Florida tomato crop (US Environmental Protection Agency). It is estimated that in order to comply with the methyl bromide ban by switching to what EPA considers an equally effective alternative treatment, would cost Florida tomato producers an additional $56 per acre. Additionally, telone, a substitute for methyl bromide, is also a highly regulated chemical that has its own restrictions and considerations to contend with, and several counties in Florida have restrictions on its use.

In California, some methyl bromide is used in the southern coast region (where there is higher rainfall and humidity), but it is not regularly used in the Central and San Joaquin Valleys, where a Mediterranean-type climate limits fungal and bacterial disease problems in tomatoes (Rivara, 2007). However, California strawberry producers have historically been a significant user of methyl bromide in their production practices.

Quasi-voluntary programmes: conservation compliance programmes

Typically known as Sodbuster and Swampbuster, USDA compliance provisions require agricultural producers to implement soil conservation systems on highly erodible cropland (HEL), land converted from permanent sod, and refrain from draining wetlands in order to remain eligible for benefits from selected Federal agricultural programmes, including price support loans, income support payments and USDA conservation programme payments and technical assistance. These programmes are considered to be quasi-voluntary because the potential cost of foregoing income from federal programmes, by violating one or more of these provisions, may represent financial risks beyond what the farm can bear. As such, these programmes may be the most rigorous environmental requirements that US farmers face (Bernstein *et al.*, 2004). The programmes are provisions of the 1985 Farm Act (as amended in 1990, 1996, 2002 and 2008).

On HEL that has been previously cropped, a conservation system must be implemented that provides for a substantial reduction in soil erosion from the level existing before the conservation measures were applied. The conservation system must include treating sheet and rill, wind, ephemeral and classic gully erosion. Erosion on HEL cannot exceed twice the soil loss tolerance level (referred to as T) for the predominant highly erodible soil type in the highly erodible field.

On HEL cropland that is broken out of native vegetation (Sodbuster), a conservation system must be implemented that provides for no increase in soil erosion from the level existing before the land use change occurred. The conservation system must include all treatments and measures needed to meet the HEL requirements.

The swampbuster provision is violated if a farmer converts a wetland to cropland. Violations of either sod- or swampbuster are supposed to result in the loss of all USDA programme payments for that year and for the years that follow. In the case of swampbuster, the penalty can apply until the wetland's functions and values are restored. There are allowances for producers to convert wetlands for production of an agricultural commodity if they compensate (through mitigation) for the wetland functions that are lost.

Animal welfare and food safety regulations

The Animal Welfare Act (AWA) was signed into law in 1966. The original intent of the Act was to regulate the care and use of animals in the laboratory. However, it has since become the accepted minimum standard for animal care – the AWA is currently the only Federal law in the US to regulate the treatment of animals in research, exhibition, transport and by dealers (USDA Animal Welfare Information Center, 2000). It is important to note that the AWA regulates the care and treatment of all warm-blooded animals, except those, such as farm animals, used for food, fibre or other agricultural purposes (USDA Animal Welfare Information Center, 2002). Because the AWA specifically does not apply to farm animals, the Act has a minimal impact on agriculture. However, there are state and local efforts to regulate some farm practices.

National Animal Identification System (NAIS)

The National Animal Identification System (NAIS) is intended to identify animals and track them as they come into contact with animals from other premises of origin. The principal aim of the NAIS is to enable animal health officials to identify locations where infected or exposed animals have been and to track animal movement from those locations. In the event of an animal disease outbreak, it is understood that the NAIS will allow officials to locate exposed animals within 48 hours. The plan for implementing the NAIS was already well underway when a case of bovine spongiform encephalopathy (BSE) was detected in the US in December 2003. As a result, USDA quickly announced that it would expedite the implementation of a verifiable national animal identification system (USDA Animal and Plant Health Inspection Service, 2005).

Currently, there is no mandatory requirement for agricultural producers to participate in NAIS; it is completely voluntary. However, debate over mandatory participation in NAIS by all premises that contain livestock is still underway in the US At this point, it is difficult to ascertain the exact impact of a proposed nationwide, USDA-mandated animal identification system.

The impact of regulations on farm profits: the cases of dairy and swine

This section contains empirical examples of the farm-level costs of compliance with the types of regulations discussed above. The analysis of compliance costs for dairy farming looks at two distinct dairy production systems: smaller, traditional dairy farms that are typical in Wisconsin and throughout the Dairybelt states and large, modern confinement dairy farms that are typical in California and the western states. The analysis of compliance costs for swine production focus on Iowa, which is one of the leading swine producing states in the US.

Dairy

Dairy production represents the second largest agricultural sector in the US in terms of cash receipts at the farm level (Miller and Blayney, 2006). Beef cattle is the only larger sector as of 2006. The structure of the US dairy industry has changed markedly since 1980. The most notable changes have affected farm size and number, as well as the regional distribution of milk production across the country. Since 1980, the number of dairy farms in operation in the US has fallen by 75%. However, the number of dairy cows on farms has only fallen by 16.5% (Miller and Blayney, 2006). The logical outcome of these trends is that the average herd size on dairy farms has increased from 32 to 111 cows.

The decrease in total dairy cows on farms (16.5%) has been more than offset by an increase in average annual per cow milk production of almost 60%. This number reached 18,967 lbs per cow for 2004 (Miller and Blayney, 2006). This rapid increase in productivity is generally attributed to improvements both in technology and management.

Significant changes have also occurred in the regional distribution of milk production in the US. The traditional dairybelt states, from Maine westward to Minnesota, have seen decreasing total milk production since 1990. These declines have been more than offset, and in part caused by, rapidly increasing production in the western and mountain states, particularly California, Idaho and New Mexico. In these states, the representative dairy farm has several hundred to several thousand cows kept in modern confinement facilities. During the 1990s, California eclipsed Wisconsin as the leading dairy state in the US and in 2007 produced almost twice as much milk as Wisconsin (see Table 3.1).

This analysis has estimated the costs of compliance with relevant environmental regulations impacting dairy farms typical of Wisconsin and California. These two states were chosen not only because they are the top two dairy states, but also because they represent two distinct milk production systems. Wisconsin dairy farming is typical of the traditional, smaller dairy farm, with an average herd size of 92 cows per farm in 2006. California dairy farming is typical of the modern, expanded confinement feeding systems found in increasing numbers in various locations throughout the western states, with an average herd size of 980 cows per farm in California. Obviously, the relative burden of complying with environmental regulations is expected to be markedly different between these two types of farms.

Compliance costs associated with the Clean Water Act and the Clean Air Act

As described above, the primary manifestation of environmental regulations for livestock agriculture in the US is the requirement to develop and follow a CNMP in order to comply with the EPA's CWA regulatory requirements. These mandatory regulations apply in Wisconsin and California. With regard to the CAA regulations, Wisconsin does not have any non-attainment areas for PM10 or PM2.5, hence is not required to develop an EPA mitigation plan. Therefore farmers in this area do not have to apply additional measures to comply with the CAA. As described below, however, California does have non-attainment areas in dairy producing regions, so the regulations do apply.

The dairy costs presented in Table 3.3 represent estimates of the average cost of all dairy farms in the US and are presented on a per 1,000 lbs of body weight (AU) basis. Because many of the costs associated with compliance require capital investments in farm infrastructure (e.g. manure storage facilities), herd size and additional factors will impact individual farm costs. In order to account for these variations, more specific actual cost information was obtained for the Wisconsin example. Unfortunately, similar state-level cost information, based on actual data, was not available for California. Therefore, the USDA national compliance costs were modified to reflect the much larger dairy farms found in that state. The costs per AU for larger farms were used to construct farm-level costs for the average California herd size of 980 cows. These numbers were adjusted from a per-animal unit basis to a per-cow basis (using 1.4 animal units per dairy cow), to be consistent with the Economic Research Service (ERS) Agricultural Research Management Survey (ARMS) financial data. Then, a CNMP cost per hundred weight of milk was estimated from the CNMP cost per cow estimate. The procedure used to estimate typical California dairy CNMP costs are discussed in more detail below.

Wisconsin case study

The Wisconsin office of the USDA provided data for three years (2003, 2005 and 2006) of EQIP applications for Wisconsin dairies needing to upgrade their facilities in order to comply with CNMP regulations (Table 3.4). These data include the type of practices needed, total cost to implement those practices, and the number of animal units to be treated, as well as other variables. On average, the farms applying for EQIP funds in Wisconsin were larger than the average Wisconsin dairy farm. Because larger farms have the ability to spread investment costs over a larger number of AUs, the average costs of compliance per AU from this sample of dairies would most likely underestimate the true cost of an average dairy in Wisconsin. As described below, costs of compliance for the average Wisconsin dairy farm were calculated from these data by using a subset of the EQIP dairy farm data. This subset was used to create estimates of compliance costs in order to determine the costs of a farm that is more representative of a typical size dairy farm in Wisconsin.

A total of 262 dairy farms applied for EQIP assistance in the three years of data obtained. The distribution of herd sizes across these farms was heavily skewed toward larger herd sizes. This is demonstrated by the fact that the average (mean) number of animal units for the 262 Wisconsin farms applying for EQIP assistance was 402, while the median number was 246. The average Wisconsin dairy farm has 92 dairy cows or 129 animal units. Therefore, to develop a CNMP cost estimate that is representative of Wisconsin dairy farms, a subset of the data from the 262 farms that applied for EQIP funding to comply with the Clean Water Act regulations was used. Of the 262 farms, 111 had 200 or fewer animal units. Of these 111 farms, the average (mean) number of animal units was 122, which is close to the Wisconsin average of 129.

The total costs for CNMP implementation for the smaller farms over the 3 year period is US$10,685,406, which would provide compliance for a total of 13,702 animal units. Therefore, the average total cost can be calculated to be $780 per animal unit or $1,092 per dairy cow (1 dairy cow = 1.4 animal units).

Table 3.4 Wisconsin dairy farms requesting EQIP funds for CNMP implementation (3 years)

Farm size (in AU)	Number of farms	Livestock units (in AU)	Total cost ($)	Cost-sharing requested ($)	Share of cost-sharing requested (%)	Average total cost ($ per AU)[1]	Average annual total cost ($ per AU)[2]
<=200	111	13,702	10,685,406	6,554,998	61	780	163
>200	151	105,769	36,975,779	13,422,866	36	451	43
Total	262	119,471	47,661,184	19,977,864	42	402	48

Note: [1]*If funded for cost sharing, producer cost would be the difference between total cost and cost share requested. Not all applications were funded. Therefore, in case applications are not funded, the producer may either reapply for cost-share funds in later years, or pay total CNMP costs without assistance.* [2]*Based on total cost, 10 years, 8% interest.*

The USDA CNMP analysis assumes that a typical structural life of a full CNMP is 10 years and an 8% interest rate was used in the annual cost calculations. Therefore, to remain consistent, the same assumptions were used in this analysis. Table 3.3 shows the average annual cost estimate per animal unit (10 year, 8% interest) for all US dairy farms is $50 (or $70 per dairy cow). The three years of EQIP data shows that average CNMP cost for all Wisconsin dairies that applied for EQIP assistance was $48 per animal unit. However, the impact of compliance costs on smaller farms is illustrated by comparing the average costs of the 111 smaller farms (200 or fewer AUs) with the average costs of the larger (>200 AUs) dairy farms requesting EQIP funding in Wisconsin. The average annual cost (10 year, 8% interest) for smaller farms is $163 per animal unit, while the cost for larger farms is significantly less at $43 per animal unit. The difference between these two estimates (smaller EQIP farm applications vs. larger EQIP farm applications) highlights the impact that farm size can have on CNMP costs per AU and per unit of output.

These data lead to the conclusion that compliance with CNMP requirements costs the typical Wisconsin dairy farm $0.83 per cwt (equivalent to 45.4 kg) of milk produced. As shown in Table 3.5, this translates into an increase of 4.19% in total production costs and 7.46% of direct costs.

California case study

Since no actual California-specific CNMP cost data are available for this analysis, the national USDA estimates were used. As illustrated in the Wisconsin EQIP data, the CNMP-related costs per animal unit significantly decrease as farm size increases. Table 3.3 does provide average costs per animal unit for large, medium and small farms for all livestock types, as well as average per animal unit costs for all dairies. Data from USDA indicate that the average US dairy herd is 155 cows. Therefore, a typical California dairy with 980 cows would be considered a large dairy on a national scale.

Table 3.3 illustrates that dairies have the greatest per animal unit CNMP implementation costs of all livestock farms. Additionally, Table 3.3 illustrates that for all US livestock farms, the average costs per animal unit decrease as farm size increases. The estimated CNMP implementation cost for all US livestock farms therefore underestimates the costs for dairy farms. To account for the higher CNMP costs associated with dairies, an adjustment factor was calculated. To estimate the decreasing costs on large dairy farms the adjustment factor was obtained by dividing the $21/AU average annual costs for all US large livestock farms by the $39 average annual costs for all US medium livestock farms in Table 3.3. The resulting adjustment factor of 0.538 was then applied to the $50/AU cost for a typical US dairy to estimate the per animal unit costs of $26.90/AU (or $37.69 /dairy cow) for a typical California dairy to comply with CNMP regulations.

Table 3.5 Milk production costs and returns per 45.4 kg sold (2005)[1]

Item	Wisconsin	California	All US farms
Direct (Operating) costs:			
Feed--			
Purchased feed	3.60	6.61	5.03
Home-grown harvested feed	4.32	1.51	3.02
Grazed feed	0.10	0.05	0.09
Total, feed costs	8.02	8.17	8.14
Other--			
Veterinary and medicine	0.94	0.54	0.78
Bedding and litter	0.23	0.04	0.22
Marketing	0.20	0.28	0.26
Custom services	0.37	0.31	0.41
Fuel, lube, and electricity	0.56	0.41	0.55
Repairs	0.62	0.33	0.56
Other operating costs[2]	0.00	0.00	0.00
Interest on operating capital	0.18	0.17	0.18
Total, direct cost	11.12	10.25	11.10
Allocated overhead costs:			
Hired labour	1.40	1.41	1.47
Opportunity cost of unpaid labour	3.30	0.39	2.30
Capital recovery of machinery and equipment[3]	2.97	1.74	2.83
Opportunity cost of land (rental rate)	0.03	0.00	0.03
Taxes and insurance	0.29	0.14	0.21
General farm overhead	0.69	0.23	0.52
Total, allocated overhead	8.68	3.91	7.36
Total costs listed	19.80	14.16	18.46
Total costs (per 45.4 kg milk) due to CNMP requirement[4]	0.83	0.19	0.37
Percentage of Direct Costs	7.47%	1.84%	3.33%
Percentage of Total Costs	4.20%	1.33%	2.00%

Notes: [1] *Developed from the 2005 Agricultural Resource Management Survey of dairy operations.* [2]*Costs for third party organic certification.* [3]*Machinery and equipment, and housing, manure handling, and feed storage structures, and dairy breeding herd.* [4]*Per cow CNMP costs estimated by pro-rating average dairy CNMP/cow costs by large farm CNMP/AU costs ($70 * 0.538 = 37.68).*

Based on the average milk production per cow on California farms (19,973 lbs), the $37.69 CNMP compliance costs result in a cost per cwt of milk produced of $0.19 (Table 3.5). This translates into an increase of 2% of total production costs and 3.3% of direct costs. Due to the much larger scale of the California dairy farms, the burden of the CNMP requirements is less than 25% of that of the average Wisconsin farm.

California dairies must also comply with various air quality regulations, which are determined by Air Quality Districts. California's San Joaquin Valley is a high dairy production region. This area is a designated air quality non-attainment region for PM $_{2.5,10}$ and ozone. Therefore, all farms within these areas must submit plans to their local Air Quality District. Currently, plans must be developed for PM_{10}, documenting dust control practices, and Ozone plans documenting a 20% reduction in nitrogen oxide and volatile organic compounds. By June 2008, these farms will also be required to have completed a $PM_{2.5}$ plan.

Agriculture within the San Joaquin Valley has recently been determined to be compliant with EPA's PM_{10} standards. To be compliant with these standards every farm must have developed and met a dust reduction (conservation management) plan, which addresses all on-field activities (pre and post harvest) and other activities which include dust control on roads, wind erosion control, equipment yards and staging areas and diesel engine replacement or conversion. The local Air Districts do random spot-checks to assure that every farm is in compliance with its dust reduction plan.

EPA mandates that there must be separate plans for each principal pollutant not meeting standards within an area. That includes ozone in the San Joaquin Valley. Therefore, farms within this area must also complete an 8-hour Ozone plan by June of 2007 and are required to comply with Tier 3 EPA/ARB off-road engine standards by the year 2015. All farms with calculated NO_x and VOC emissions from the agricultural operation that is equal to or greater than 12.5 tons/year (about 350 or more contiguous acres or more than 1,900 milk cows) are required to have an air permit (San Joaquin Valley Air Pollution Control District). The San Joaquin Valley Air Pollution Control District provides a calculator on its website for each producer to self-determine whether they would need to comply with these regulations. Also known as Rule 4550, it requires owners/operators of dairy and feedlot facilities with population equal to or more than 500 mature dairy cows (whether milked or dry) to provide a Conservation Management Practices (CMP) Plan to control dust. Additionally, operators who also grow field crops on land greater than 100 acres of contiguous or adjacent farmland are required to submit CMP applications for their field crops.

The CMP plan is designed to document reductions in PM_{10} emissions. In order to comply, operators must:

- select one CMP for implementation from each of the specified categories;
- implement the selected CMPs starting July 1, 2004 and keep a record of the implementation of CMPs to be available upon request by the District representative;
- complete the applications and submit them to the District by December 31, 2004 for approval. The submitted applications constitute a CMP Plan once approved by the District; and
- keep a copy of each submitted application and each CMP Plan to be available upon request by the District representative.

Dust control, especially on farm and field roads, can be a large expense item for a typical San Joaquin Valley dairy producer to comply with EPA's Clean Air Act PM_{10} regulations. Treatment measures range from daily spraying with water, to spraying a sealant every other year. Both treatments can be expensive, running about \$2/foot, or \$10,000 per treated mile of road (Beyer, 2007).

Draft Rule 4570 would provide CAFO with the flexibility to comply with the Volatile Organic Compound control requirements by choosing any mitigation measures from a list of multiple measures listed by the air district or by implementing district-approved alternative mitigation measures that would achieve equivalent emissions reductions as those listed in the rule. The rule impacts dairies within the San Joaquin Valley with 1,000 or more milk cows (not including dry cows). This rule is believed to only impact in a limited number of dairies (233 dairies out of a total of 1,500 in the region), requiring them to implement additional mitigation measures not already used on the facility.

For those dairies that are required to comply with rule 4570, the costs of compliance are estimated to be $18 per cow. Since our example dairy falls below the 1,000-cow threshold, no additional compliance costs are assigned for compliance with this rule.

Swine

The structure of the US pork industry has changed markedly in the last several years. Although inventory dropped in the late 1990s, it has risen steadily since then to almost 63 million head in 2007 (Figure 3.4). What is most remarkable is the steady decrease in the number of operations, which has decreased to roughly one tenth the number of operations 25 years ago (Figure 3.5). The recent trend toward increasingly large hog operations has resulted in just 2.5% of the operations (those with greater than 5,000 hogs) controlling 54% of the hog inventory in the US (Figure 3.6). Although almost 40% of US hog operations have fewer than 99 hogs, these operations only account for 1% of the hog inventory.

Vertical integration is common in the US swine industry, which has experienced the fastest rate of growth of CAFOs, along with the poultry sector. The greatest concentrations of swine exist in the North-central states, led by Iowa and Minnesota, as well as in North Carolina, where recent breaches of large manure lagoons have caused serious environmental concerns. Air quality and odour control have become important issues facing large-scale hog production in many regions of the country. As of 2007, the top five pork producing states produced 57% of the nation's pork supply. Iowa produced 16% and has recently been eclipsed by North Carolina as the number one pork producing state with a total of 17.8 million hogs and pigs. Iowa's hog production generally represents over $4 billion in cash receipts (USDA National Agricultural Statistics Service, 2003).

The Clean Water Act - Regulations in the Iowa swine industry

This analysis has estimated the costs of compliance with regulations concerning water quality for hog farms typical of Iowa. This state was chosen because at the time of the analysis it was the leading hog producing state in the US Iowa's hog farms represent a more modern, confined feeding operation found in increasing numbers in the top hog producing states. The average hog operation size in this region of the country is approximately 4,300 head. This is well above the threshold size to be considered a CAFO.

The primary manifestation of environmental regulations for livestock agriculture in the US is the requirement to develop and follow a CNMP in order to comply with the EPA's CWA regulatory requirements. The USDA estimated national average costs for developing and implementing a CNMP for a swine operation were used for this case study (Table 3.3). Because many of the costs associated with CNMP compliance require capital investments in farm infrastructure (e.g. manure storage facilities), the number of AUs over which such investment costs can be spread will have an important impact on the compliance costs incurred by individual farms.

Figure 3.4 US quarterly hogs and pigs inventory

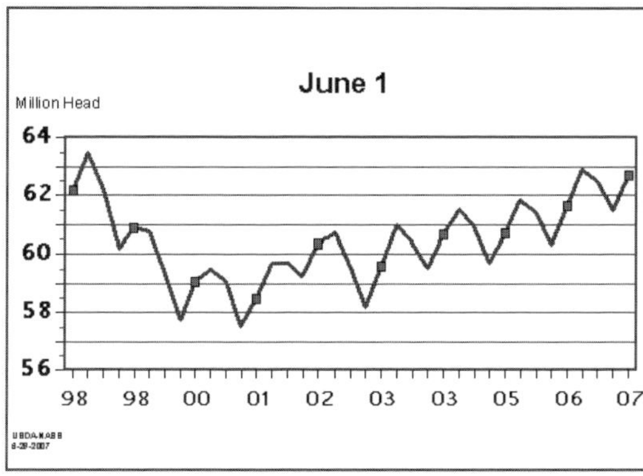

Figure 3.5 Number of hog operations

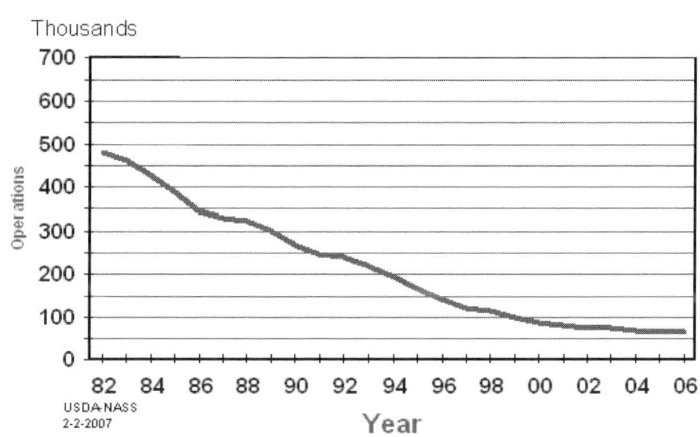

This analysis attempted to secure actual CNMP cost data specific to Iowa, but this was not available. Therefore, the national estimates for the CNMP cost for swine operations (Table 3.3) were adjusted from a per AU basis (US$44) to a per 45.4 kg of gain basis (US$1.57) to be consistent with the available financial data for swine operations. The ARMS data from USDA-ERS was used to provide production costs for a swine operation that is typical of the Heartland region (Table 3.6). The direct and total costs per 45.4 kg of gain are estimated to be $39.03 and $46.67, respectively.

Figure 3.6 US hog operations, number and per cent of inventory, 2006

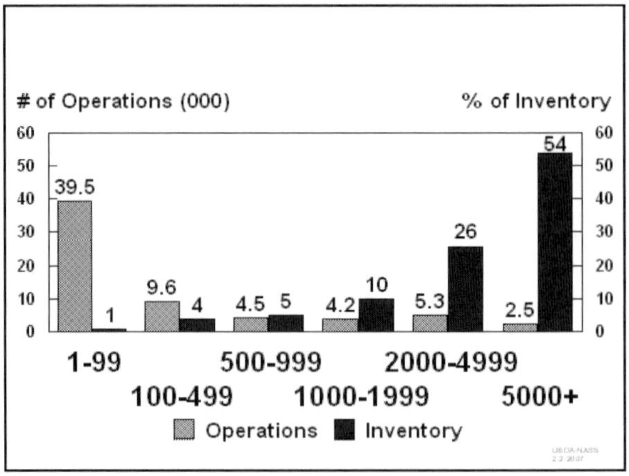

A typical Iowa swine operation produces two cycles of market hogs, or 'turns', per year. This implies that, on average, it takes six months to raise a typical market hog from feeder pig to market. Therefore, at any one time there is only half the number of a farm's total marketed hog capacity on the farm. Feeder pigs generally enter the operation at 50 pounds in weight and leave the operation at 250 pounds. Therefore, over the life of the pig the average weight is 150 pounds, which implies that 1 AU equals 6.67 pigs.

Using the national average cost per AU for CNMP implementation for swine operations of US$44, the average cost per 45.4 kg of gain is estimated to be US$1.65. This is calculated by multiplying the average weight gain per pig of 90.9 kg by 6.67 pigs by 2 turns per year, which results in a weight gain of 1,210 kg (or 2,666 lbs) of gain per AU. This cost of compliance translates into a 3.54% increase in total production costs or a 4.23% increase in direct costs.

Comparing US and EU agri-environmental policy

Differences in US and EU agri-environmental policy have been described in the literature as having a focus on controlling extensification versus intensification, reducing negative consequences versus promoting positive ones and focusing on outcomes versus practices, respectively (Baylis *et al.*, 2005). Although there are valid arguments supporting these assertions, in some instances the assertions can also be shown to be false.

The claim that the focus of US agri-environmental policy is focused on controlling extensification of agriculture is supported by some prominent examples. First, the US Sodbuster policy, created in the 1985 Farm Bill, attempts to control the amount of soil lost from highly erodible lands (HEL) by requiring the use of approved conservation systems. Similarly, the Conservation Reserve Program (CRP) seeks producers to sign voluntary contracts to take HEL out of production. Second, Sodbuster's sibling, Swampbuster, attempts to minimize the loss of wetlands to agricultural production. In both cases, the policy effect is to minimize the intrusion of agricultural production on marginal lands, hence, combating extensification. It is important to realize, however, that Sodbuster and Swampbuster represent only a small portion of US agri-environmental policy and that CRP is a completely voluntary programme that compensates producers for land taken out of production.

Table 3.6 Production costs for a swine operation typical of the Heartland region, 2005

Direct (operating) costs:	**US$ per 45.4kg of gain**
Grain	1.73
Protein sources	1.34
Complete mixes	10.8
Other feed items[1]	0.07
Feeder pigs	21.32
Hired labour	0.52
Veterinary and medicine	0.58
Bedding and litter	0.01
Marketing	0.47
Custom services	0.25
Fuel, lube and electricity	0.82
Repairs	0.48
Interest on operating capital	0.64
Total operating costs	39.03
Allocated overhead costs:	
Opportunity cost of unpaid labour	2.19
Capital recovery of machinery and equipment [2]	4.27
Opportunity cost of land (rental rate)	0.02
Taxes and insurance	0.38
General farm overhead	0.78
Total overhead costs	7.64
Total costs	46.67
Annual CNMP cost [3]	1.65
Percentage of direct costs (%)	4.23
Percentage of total costs (%)	3.54

Note: [1]Milk replacer, milk, milk by-products, antibiotics, and other medicated additives; [2]Machinery and equipment, and housing, manure handling, and feed storage structures, and breeding herd; [3]Annual costs based upon farm capacity = 2 'turns'/year (0.5 number of annual market hogs), $44/animal unit/year, 6.667 feeder pigs (average weight 150 lbs)/AU. These costs are already incorporated into ARMS costs and returns surveys.

Source: USDA Economic Research Service, 2006

The assertion that US policy is focused on reducing negative consequences from agriculture, as opposed to EU policy that is focused on promoting positive consequences seems to have been, until recently, completely valid. Promoting positive consequences from agricultural production is the core meaning of the term multifunctional agriculture. This term has been an important part of the lexicon of EU agri-environmental policy since the end of the twentieth century, but neither this term nor the concept has ever taken residence in the US capitol. The Conservation Stewardship Program (started as the Conservation Security Program in the 2002 Farm Bill) of the 2008 Farm Bill, however, does partially embody the concept of promoting positive outcomes from agriculture. As this programme is implemented nationwide, starting in 2009, US policy starts its migration toward rewarding positive outcomes from agricultural production.

Although the agri-environmental programs being implemented in the US have an ultimate focus on environmental outcomes and make use of tools such as the Environmental Benefits Index (EBI) to prioritize programme spending, the programmes remain almost completely prescriptive in nature. This means that as the programmes are implemented the focus is on getting a specific practice or set of practices implemented on the ground. There is an obvious reason to use a practice-based approach when dealing with nonpoint source pollution from agriculture; it is extremely hard to measure or monitor emissions. However, there are two important drawbacks to this approach. First, there is very little quantification of the actual environmental impact of the policies and programmes being implemented. Second, the set of practices (BMPs) that are approved for cost-sharing represent only a fraction of the possible actions that a producer could take on their farm and fields. This greatly reduces the incentive for farmers to seek out or implement other more appropriate or cost-effective actions for their farms. If US (or EU) agri-environmental policy were to be more focused on outcomes or performance, then appropriately designed incentives could encourage producers to incorporate environmental management into their farm business decision-making (Winsten, 2009).

Lastly, the EU is often viewed as having a more precautionary approach to environmental issues. In many instances this may be true (prominent agricultural examples include hormones used in livestock production and the use of genetically-modified organisms), but not in all instances (Wiener and Rogers, 2002). The US policy response to the threat of mad cow disease provides an example of the precautionary principle in use.

As of 2009, the concept and use of market-based approaches to addressing agri-environmental issues in the US is small, but increasing. The concepts of ecosystem services and environmental credit trading are, although still on the fringes of the policy discussions, gaining ground. By 2009, the USDA has created the Office of Ecosystem Services and Markets. Although still in its infancy in the US this and other related efforts will help to move policies and programs in this direction.

Summary and conclusions

The size and diversity of the US agricultural sector renders a comprehensive *and* concise description of agri-environmental policy extremely difficult. An analysis of US agri-environmental programmes points to the continued use of voluntary adoption of BMPs and idling selected tracts of land, coupled with adequate cost-sharing as the primary approaches for addressing agri-environmental issues across the US In many areas environmental quality problems persist despite over US$2 billion in spending per year on these issues. The increasing frequency of litigation over agri-environmental problems has led to greater enforcement of existing CAFO regulations under the federal Clean Water Act. Although CAFO regulations only apply to a very small percentage of US farming operations, increased enforcement will result in increasing compliance costs.

For most US farming operations there are likely to be some level of costs associated with compliance to environmental regulations. The case studies presented in this chapter indicate that the cost of regulatory compliance in the dairy industry is estimated to be 4.2% of total farm operating costs for traditional dairy farming typical of Wisconsin and 1.3% for large confinement operations typical of California. For the swine industry, a case study presented for the state of Iowa estimates that the cost of regulatory compliance is 3.5% of total farm operating costs.

In the context of this book there are two important questions related to the costs of regulatory compliance. First, what is the impact of regulatory compliance costs on the competitiveness in the world market for US agricultural products? The answer to this question obviously depends on the relative cost burden of regulatory compliance placed on farms in the nations that represent the US's main trading partners. Other chapters in this book provide estimates of compliance costs from many of these countries. Second, what value to society is being gained from the current suite of regulations being imposed on US farms and are the regulatory compliance costs justified by these benefits? This question represents an important area for further research.

References

Baylis, K., G.C. Rausser, and L.K. Simon (2005). Including non-trade concerns: the environment in EU and US agricultural policy, *International Journal of Agricultural Resources, Governance and Ecology*, 4 (3/4): 262 - 76.

Bernstein, J., J. Cooper, et al. (2004). Agriculture and the Environment in the US and EU. In US - EU Agriculture and Food Comparisons. Agriculture and Trade Report: WRS04-04. USDA Economic Research Service. Pages 66-77. http://www.ers.usda.gov/publications/WRS0404/WRS0404.pdf (accessed August 20, 2010).

Beyer, J. (2007). Phone Communication. J. Winsten, NRCS California State Air Quality Specialist.

Carpenter, J., L. Gianessi and L. Lynch (2000). The economic impact of the scheduled US Phaseout of methyl bromide. National Center for Food and Agriculture Policy. Washington, DC. http://www.ncfap.org/documents/mb.pdf (accessed August 20, 2010).

Carpentier, C. L. and D. E. Ervin (2002). USA. In: F. Brouwer and D.E. Ervin (eds), Public Concerns, Environmental Standards and Agricultural Trade, Wallingford, CAB International, 95-139.

Gollehon, N., M. Caswell, M. Ribaudo, R. Kellogg, C. Lander, and D. Letson. (2001). Confined Animal Production and Manure Nutrients. USDA-Economic Research Service. Agriculture Information Bulletin No. (AIB771). Washington, DC.

Kellogg, R., C. Lander, D. Moffitt, and N. Gollehon (2000). Manure nutrients relative to the capacity of cropland and pastureland to assimilate nutrients: spatial and temporal trends for the US. Washington, DC, USDA.

Miller, J. J. and D. P. Blayney (2006). Dairy Backgrounder. Electronic Outlook Report from the USDA Economic Research Service.

Porter, M.E. (1990). The Competitive Advantage of Nations. Free Press. New York, New York.

Ribaudo, M.O., R.D. Horan, and M.E. Smith (1999). Economics of Water Quality Protection from Nonpoint Sources: Theory and Practice. Agricultural Economic Report 782. US Dept of Agriculture, Economic Research Service, Washington, DC. Website: http://www.ers.usda.gov/publications/aer782/ (viewed August 20, 2010).

Ricardo, D. (1891). Principles of Political Economy and Taxation. George Bell and Sons. London, UK.

Rivara, C. (2007). Phone communication. J. Winsten, California Tomato Research Institute.

Runge, C.F. (1990). Environmental regulation and agricultural trade competitiveness. Agriculture and water quality: International perspectives. Edited by John B. Braden and Stephen B. Lovejoy. pp. 183-196. Boulder, Colo.

Runge, C. F. and R.M. Nolan (1990) Trade in disservices: Environmental regulation and agricultural trade. *Food Policy*, 15 (1): 3-7.

Speir, J., M.-A. Bowden, D. E. Ervin, J. McElfish, and R. Perez Espejo (2002). Comparative standards for intensive livestock operations in Canada, Mexico, and the US, Commission for Environmental Cooperation, Montreal, Canada.

US Environmental Protection Agency. 'Telone® C-17 and Tillam Use on Florida Fresh Market Tomatoes', from www.epa.gov/spdpublc/mbr/casestudies/volume1/telc17ti.html.

US Environmental Protection Agency (2002). National Water Quality Inventory: Report to Congress. Office of Water. Washington, D.C.

US Environmental Protection Agency (2003). EPA - Producers' Compliance Guide for CAFOs. Washington, DC.

US Fish and Wildlife Service (2008). ESA Basics: 30 years of protecting endangered species. Endangered Species Program. Washington, DC.

USDA Animal and Plant Health Inspection Service (2005). The Evolution of the National Animal Identification System in the US. V. Services. Washington, D.C.

USDA Animal Welfare Information Center (2000). 'Questions and Answers About the Animal Welfare Act and its Regulations for Biomedical Research Institutions'. http://www.nal.usda.gov/awic/pubs/Legislat/regsqa.shtml (accessed 20 August, 2010).

USDA Animal Welfare Information Center (2002). 'The Animal Welfare Act' www.aphis.usda.gov/lpa/pubs/awact.html.

USDA Economic Research Service (2004). Agricultural Productivity in the US Data Set, USDA Economic Research Service.

USDA Economic Research Service (2006). Farm Business and Household Survey Data: Customized Data Summaries from ARMS. Agricultural Research Management Survey (ARMS), USDA Economic Research Service.

USDA Economic Research Service (2007). US Fertilizer Use and Price Data Set, USDA Economic Research Service.

USDA Economic Research Service (2008). Foreign Agricultural Trade of the US (FATUS) Data Sets, USDA Economic Research Service.

USDA National Agricultural Statistics Service (2007). NASS Data Base, USDA National Agricultural Statistics Service.

USDA National Agricultural Statistics Service (2003). Agricultural land values and cash rents. Washington, DC, USDA.

USDA Natural Resources Conservation Service (2003). Costs Associated with Development and Implementation of Comprehensive Nutrient Management Plans. Part 1 - Nutrient Management, Land Treatment, Manure and Wastewater Handling and Storage, and Recordkeeping: 105, Table 38.

Wiener, J.B. and M.D. Rogers (2002), Comparing precaution in the US and Europe, *Journal of Risk Research*, 5 (4): 317-49.

Winsten, J.R. (2009), Improving the cost-effectiveness of agricultural pollution control: The use of performance-based incentives, *Journal of Soil and Water Conservation*, 64 (3): 88A-93A.

Chapter 4

Canadian agricultural environmental policy: from the right to farm to farming right

Predrag Rajsic, Elizabeth Ramlal and Glenn Fox

Introduction

This chapter describes key features of Canadian agricultural environmental policy, assesses the effectiveness of the Canadian approach and compares this approach to agricultural environmental policies in the EU and the US. The chapter provides an overview of key environmental issues involving primary agriculture in Canada and also describes the policy instruments that have been applied in response to those issues. Two main themes are developed. First, Canadian agriculture, relative to most of the developed world, is a low-input intensity production system. This has important implications with respect to the demand for environmental policy targeted at the sector, especially in comparison with some high input intensity European countries. Secondly, as a federal state, environmental policy that is applied to agriculture in Canada is exercised by the federal government, the provincial governments, municipal and regional governments. In addition, agricultural industry associations and potential common law or civil liability can and have had a significant effect on the environmental performance of Canadian agriculture.

Connections to and comparisons with relevant literature

The environmental performance of Canadian agriculture is coming under increasing scrutiny. Brubaker (2007) and Boyd (2003) have linked the adverse environmental effects of Canadian agriculture, especially intensive livestock agriculture, to the adoption of 'right to farm' legislation at the provincial level. They argue that right to farm legislation has absolved farmers from liability for what would otherwise be nuisances, in the form of contamination of water or air, against their neighbours. This situation is an example of what Coase (1960) called legalized nuisance. In the absence of legislative interference, a property owner against whom a trespass or nuisance had been committed, would be able to seek an injunction or damages. Right to farm legislation forestalls that option. As long as a farmer has acted in a manner consistent with current industry standards, no liability attaches. Both Brubaker and Boyd attribute at least part of the expansion of intensive large scale livestock production facilities in Canada to right to provincial right to farm statutes.

Montpetit (2002), on the other hand, maintains that what he sees as the inferior performance of Canadian agricultural environmental policy, relative to its counterpart in the

US, can be attributed to the contrasting structures of federalism and to differences in the structure of constitutional law between the two countries. Montpetit describes the situation in Canada, which he observes is similar to many OECD countries, as 'agricultural exceptionalism', indicating that agriculture in Canada has been exempted from general environmental policies until relatively recently. Although he predicts that agricultural exceptionalism is coming to an end, he expects that this process will be politically contentious and ultimately less successful in Canada, as a consequence of the less federally centralized structure of agricultural policy relative to the US. Also, the unique treatment of agriculture during the sustained period of multilateral trade liberalization following World War II, which amounted to granting an exemption to agriculture, left developed countries to pursue divergent agricultural policies unconstrained by international trade rules. As a result, the particular structure of agricultural policies in the US differs in important ways from Canadian agricultural policy.

The US experience, shaped significantly by federal farm legislative processes beginning with the Agricultural Adjustment Act of 1933, and reformulated frequently in subsequent farm bills, established the US Federal government as a leading institution in the development of agricultural policy. Establishing a strong federal presence in agricultural policy in the US created both a large federal bureaucracy dedicated to the sector and also created an incentive for national farm organizations to develop and to establish themselves as participants in policy development in Washington. One of the artefacts of the Agricultural Adjustment Act arose because of the differing market shares of the US and Canada for many agricultural commodities. Montpetit does not acknowledge this factor, but it is critical to the emergence of agricultural environmental policy in the US. For many agricultural commodities, the US has what economists call a large country effect. Variations in US demand or supply often are large enough to have an impact on international commodity prices. In many instances, because of the smaller size of its agricultural economy relative to global commodity markets, variations in Canadian demand or supply are not large enough to have a significant impact on international prices. This means that curtailing commodity production in the US can be used to support farm prices and incomes. This is not generally the case for Canadian farm production. Consequently, US farm policy has often relied on land set asides, which took agricultural land out of production in an attempt to reduce supply and increase prices. Land set asides have not been used in Canadian agricultural policy, given the relatively smaller market share of Canada in international agricultural commodity markets.

The use of land set asides created an opportunity to transform agricultural policy into conservation and ultimately environmental policy. Since farmers were receiving payments from Washington to keep agricultural land out of production, it was a short step to make these payments conditional on the use of conservation and environmental protection measures, a process that culminated in the conservation reserve policy of the Reagan administration. The absence of a farm payment and land set aside system in Canada meant that the federal government had less leverage and that there was less incentive for interest groups to organize efforts to control the levers of federal policy, either in the cause of supporting farm incomes or in the cause of environmental policy.

Montpetit concludes that the Canadian federal system, influenced by constitutional allocation of authority for natural resources to the provinces, has produced a lower level of environmental protection when it comes to agriculturally based pollution sources when compared to the US experience. Part of the basis for this conclusion is the widespread reliance on provincial codes of practice and a range of voluntary programmes and initiatives. We will discuss the role of codes of practice later in this chapter. But Fox and Kidon (2002) have previously discussed the role of such codes in Canadian environmental policy. While codes of practice often appear to be weak voluntary guidelines, in practice they often form the basis for local building approval processes and they serve as the industry standard in the determination of negligence in liability issues. As a consequence, the impression of limited enforcement of these codes is more apparent than real. In any case, most provinces in Canada have developed nutrient management policies and regulations that do apply to especially larger livestock operations.

General features of Canadian agriculture

There were 67.5 million hectares (mha) of farmland in Canada in 2001. The Prairie Provinces, consisting of Manitoba, Saskatchewan and Alberta accounted for about 81.5% of that total. The Western Provinces (British Columbia and the Prairies) accounted for about 85.3% of total farmland. Central Canada (Ontario and Quebec) accounted for about 13.2% and the Atlantic Provinces accounted for the remaining 1.5% of total farmland. Cultivated land and pasture land represent about 61% and 30% of the total farm area in Canada respectively. Average farm area (size) varies substantially across Canada. The average farm area in Ontario is 92 hectares (ha) while the average farm size in Saskatchewan is 519 ha. There is a general trend towards increasing scale of operation and consolidation in agricultural production reflecting the large productivity gains that have occurred in agricultural labour and capital inputs. Over the last 60 years, farm size has been steadily increasing in terms of land area and the average farm size in Canada is currently about 273 ha. Farm size is also increasing in terms of herd size. For example, the average number of dairy cows per farm has more than tripled over the last 30 years while the average number of pigs per farm has increased by more than tenfold. This expansion in farm size, however, has not been obtained at the expense of the family farm. It is still the case that 98% of all farms are family owned and operated in Canada.

Canadian agriculture is diverse. As can be seen from Table 4.1, red meats, grains and oilseeds and dairy are the most important commodities. They contribute about 80% of the $36.5 CDN billion (€24.3 billion) in market receipts for the major agricultural commodities. However, the importance of the different commodity groups varies by region. In British Columbia, market receipts are balanced across a range of commodities, but the most important are fruit and vegetables, and floriculture and nursery. In the Prairies, red meats and grains and oilseeds account for over 80% of market receipts. In Ontario and Quebec, red meats and dairy dominate market receipts. In Atlantic Canada, potatoes, eggs, apples and dairy are the most important commodities. Table 4.1 also indicates that grain and red meat are the major Canadian export commodities. In 2008, more than a third of the domestic production of beef, about 60% of pork production, about three quarters of wheat production and virtually all of the flax seed, dry beans and lentils were exported. The export dependency of red meat producers almost doubled between 1991 and 2001 from 30% to 55%. Dairy and poultry producers are largely dependent on the domestic market for their income. Dairy and poultry are managed under a policy of supply management, with farm level production quotas, formula prices and import controls.

Table 4.1 Total cash receipts and provincial distribution of cash receipts in 2008 for 20 major agricultural commodities in Canada

Commodity	National farm cash receipts (million €)	Exports as % of current year national production	Percent of national cash receipts by province/region (%)						
			British Columbia	Alberta	Saskatchewan	Manitoba	Ontario	Quebec	Atlantic provinces
Cattle	3,886	38	4	50	15	8	16	7	1
Dairy products	3,532	-	9	9	3	4	32	37	6
Canola (rapeseed)	3,262	7	0.4	38	41	21	0.4	0.2	-
Wheat, excluding durum	2,310	76	0.1	34	34	19	12	1	0.2
Hogs	2,118	59	1	12	8	22	26	30	1
Hens and chickens	1,322	16	16	9	4	4	32	27	6
Floriculture, nursery and sod	1,235	-	24	7	1	2	48	12	4
Vegetables	1,218	28	22	5	0.1	2	51	18	7
Corn	1,028	8	-	0.2	-	3	59	38	0.3
Soybeans	749	63	-	-	-	6	74	19	0.2
Potatoes	657	13	5	16	4	20	9	13	33
Barley	516	53	0.2	38	46	11	1	3	1
Lentils	421	110	-		100	-	-	-	-
Dry peas	419	64	-	19	78	3	-	-	-
Total eggs	404	9	12	8	3	9	41	17	10
Oats	376	71	1	5	54	33	1	5	0.3
Turkeys	258	14	13	9	4	7	44	20	4
Flaxseed	227	116	-	3	77	20			-
Dry beans	130	113	-	17	-	32	47	4	-
Apples	112	10	26	-	-	-	44	22	8
Blueberries	106	18	41	-	-	-	2	23	34
Total	24,287		4	24	21	12	21	15	3

Notes: *Canadian Dollars translated into € using the 10-year average European Central Bank exchange rate (1.503 CAD/€). Exports as % of current year national production were calculated using reported quantities in metric ton; Atlantic provinces include Nova Scotia, New Brunswick, Newfoundland and Prince Edward Island.*

Source: Statistics Canada (2009)

Environmental issues in Canadian agriculture

The quality of Canadian soil and air has generally improved in the last two decades (McKitrick, 2005; Lefebvre *et al.*, 2005). As shown in Table 4.2, of the four agricultural environmental quality indicators examined by Lefebvre *et al.* (2005), soil quality and greenhouse gas emissions are indicating improvement, while nature conservation and biodiversity and risk of water contamination with nitrogen indicators are showing unfavourable trends. Table 4.2 summarizes major contributors to the changes in each of the four indicators. Nature conservation and biodiversity are deteriorating due to increases in the percentage of farmland being used for cropping. However, there have been some offsetting effects from reduced summer fallow area, increased area under woodlots and wetlands, and conversion of marginal croplands in the Prairies into tame pastures. Increases in fertilizer use, in area under legume crops and in livestock numbers has led to a higher risk of water contamination by nitrogen. On the other hand, reduction in summer fallow, increases in the share of cropland under reduced tillage, no-till and forages contributed to better soil quality and lower greenhouse gas emissions.

Table 4.2 Agri-environmental and economic indicators for agriculture in Canada and Provinces, 1981-2001

Indicator (status)	**Factors contributing to status**
Nature conservation and biodiversity (deterioration)	Improvements: decrease in the species-impoverished summer fallow area; increases in the area of forage crops, woodlots and wetlands; and conversion of about half a million hectares of marginal cropland on the Prairies to tame forages.
	Negative impact: increase in the percentage of farmland used as cropland and the decline in species-rich natural pastures.
Risk of water contamination with nitrogen (increasing)	Negative impact: increase in fertilizer use especially in the Prairies; increase in area of legume crops; and increase in livestock numbers (more manure N applied to land).
Soil quality (improving)	Improvements: reduction in the area under summer fallow; increase in the share of cropland under reduced tillage or no-till; and increase in the areas under forages.
Greenhouse gas emissions (decrease)	Improvements: reduction in summer fallow and increased conservation tillage contributed to increase in soil carbon sequestration.
	Negative impact: increases in nitrous oxide and methane emissions from increased use of nitrogen fertilizer and increases in beef cattle populations.

Source: Lefebvre et al. (2005).

Even though the risk of water contamination by nitrates from agriculture is increasing, Canada is still at relatively low levels of nitrogen application intensity, compared to other OECD agricultures. Table 4.3 summarizes authors' estimates of nitrogen additions from commercial fertilizer and manure and removal by crops in Canada by province. On average, 71 kilograms of nitrogen from manure and commercial fertilizer were applied per hectare in

Table 4.3 Nitrogen additions from commercial fertilizer and manure and removal by crops in Canada by province/region (2001 and 2006)

	Province							
	British Columbia	Alberta	Saskatchewan	Manitoba	Ontario	Quebec	Atlantic provinces	Canada
Additions	114	80	46	85	94	105	123	71
Removal	35	54	42	56	65	58	57	50
Balance	79	27	5	30	29	46	67	21
			Change compared to 2001 (kg/ha)					
Additions	-14	-1	2	1	3	-16	9	0
Removal	-6	15	11	9	16	-1	1	12
Balance	-8	-16	-10	-9	-14	-16	8	-12

Canada in 2006. Most provinces were above the average but Saskatchewan, where more than 1/3 of Canadian agricultural land is concentrated, brings the average down since only 46 kg/ha were applied in this province. The estimated nutrient balance (i.e. nitrogen not used by the crops), on the other hand is about 21 kg/ha/yr overall. While in Saskatchewan it is only 5 kg/ha/yr, other provinces are around or above 40 kg/ha/yr. It should be noted that the estimated figures are likely higher than the true values because nitrogen use data include non-agricultural uses for lawns and greenhouses. It should, however, be noted that there are differences not only across provinces, but also within provinces. Agriculture and Agri-Food Canada (AAFC) (2007) identifies areas with high levels of residual nitrogen (above 40 kg/ha) in all provinces, with largest areas being in southern Manitoba, south-western Ontario and parts of Quebec.

International comparison of nutrient input intensity of Canadian agriculture

Figue 4.1 compares the nutrient intensity of Canadian agriculture to an international sample of developed and developing country agricultures. The figure shows total fertilizer use per hectare of agricultural land per year on the horizontal axis and livestock units per square kilometre of agricultural land on the vertical axis. The total nutrient intensity, as a national average, would be represented by the area of the rectangle bounded by the origin and the point in the interior of the graph identified for each country. The larger the area of this rectangle, the higher the intensity of nutrient application, in the form of livestock manure and commercial fertilizer application, for a given country. By this measure, for the countries included in this comparison, only Australia, Argentina and Latvia have less nutrient intensive agricultures than Canada. Several of the countries whose agricultures are discussed elsewhere in this book, including the US, New Zealand and several EU Member States (e.g. Spain, Poland and Italy) are more nutrient intensive than Canadian agriculture, but still fall below 100 kg/ha/yr of average fertilizer application and 100 livestock units per square kilometre. But, the axes need to be extended to 500 units in each dimension to plot the highly nutrient intensive agricultures of Belgium and Luxemburg, the Netherlands and even the United Kingdom, Germany, Denmark and France. Admittedly, this two dimensional comparison abstracts from differences in yields, which would result in different crop removal rates for nutrients, as well as a number of geophysical factors such as growing season length and average precipitation, both of which influence input intensity of agriculture and also condition the environmental risks of excessive nutrient use, and there are important regional variations in nutrient intensities within countries, but the broad comparisons in this figure do suggest that the demand for agricultural environmental regulation in countries like the Netherlands and Belgium might be much more intense than would be the case in relatively low nutrient intensity agricultures like Australia, Argentina and Canada. These differences in intensity should be taken into account in international comparisons of regulatory systems and also in comparisons of environmental compliance or abatement costs.

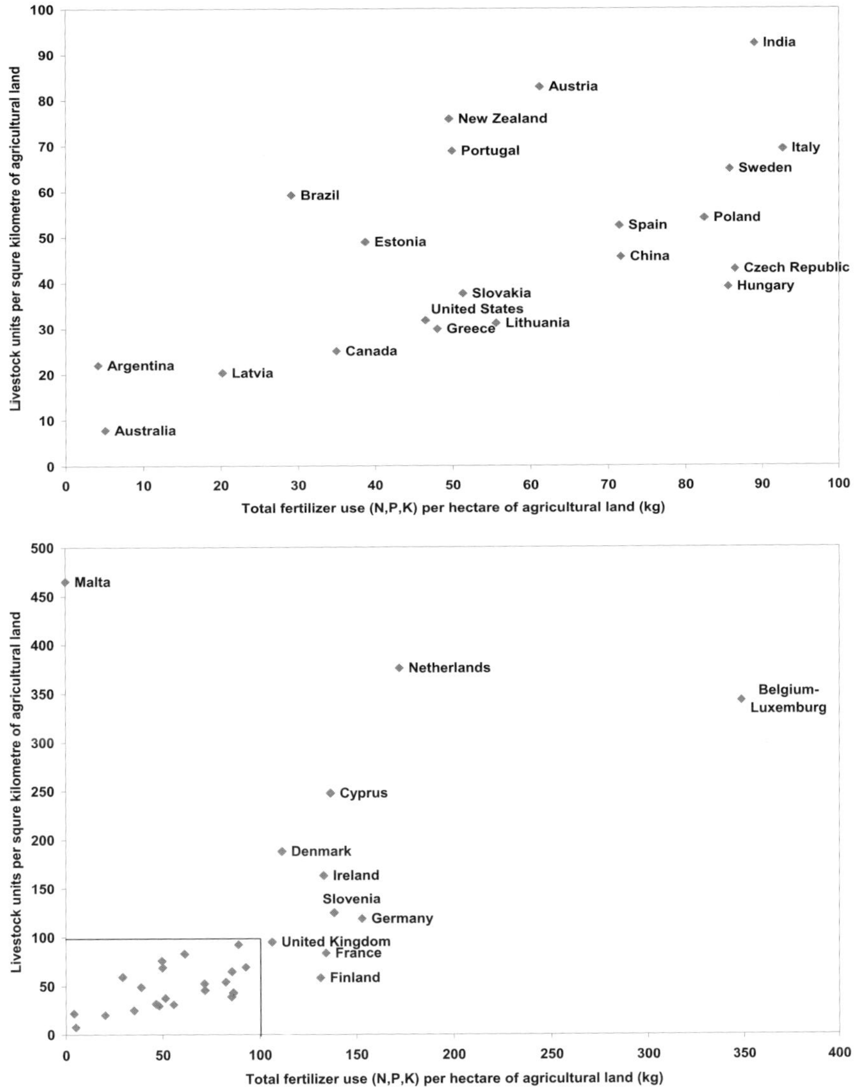

Figure 4.1 Comparison of total fertilizer use per hectare of agricultural land and livestock units per square kilometre across the EU countries, Canada, New Zealand, Australia, US, Argentina, Brazil, China and India

Comparison of indices of environmental pressure across Canada

Large differences in environmental pressure, indicated by human and livestock population density, can be observed within Canada as well. Figure 4.2 compares livestock and human population densities (per km^2 of agricultural land) between 1981 and 2006 across Canadian provinces. Animal population density represents an approximate measure of nutrient emissions, odour, and other adverse effects of livestock farming while human population represents both negative environmental effects of human activity and a measure of conflict between agriculture and non-agricultural population. Quebec and Ontario are the most densely populated provinces, both in terms of human and animal populations. The highest human population density in Quebec was at above 200 persons per square kilometre of agricultural land, while the livestock population was about 160 animal units. For Ontario, human population density was slightly higher while for livestock it was lower. In comparison the remaining three major agricultural provinces – Manitoba, Alberta and Saskatchewan – are far below these numbers. In Alberta and Manitoba, there are about 10 persons and between 20 and 60 livestock units per square kilometre of agricultural land. Saskatchewan is at the bottom of the list, with human and livestock population densities of about 5 and 15 per square kilometre respectively, making it a low land intensity economy by any standards. These differences in intensity, both internationally and within Canada, should be taken into account in international comparisons of regulatory systems and also in comparisons of environmental compliance or abatement costs.

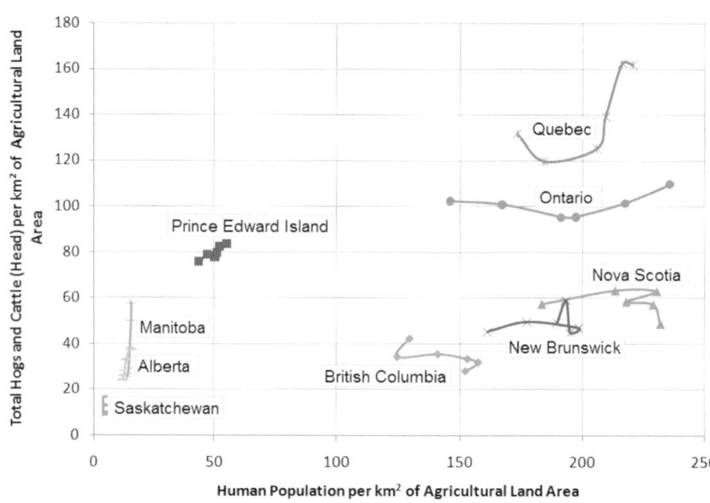

Figure 4.2 Comparison of human population density and livestock (hogs and cattle) numbers across 9 Canadian provinces

Note: Newfoundland and Labrador is excluded due to a relatively high human population per km^2 of agricultural land (over 1,000) resulting from a small agricultural area in this province.

Regulations and standards

Canadian farmers are subject to a growing number of federal and provincial environmental regulations. Canadian farmers also participate in voluntary industry or government-led standards that aim to mitigate the negative impacts of agricultural production practices on the environment. These voluntary standards are embodied in voluntary programmes, often in partnership with governments, which may be certified by a government agency or otherwise. These standards often are expressed as industry codes of practice. While these codes are not generally backed directly with legislative enforcement, they often have considerable influence on farm practices because they have become the standards used to review applications at the municipal level to modify or expand ongoing operations, especially livestock operations, or to construct new facilities. Industry codes of practice are also widely used to determine standards of negligence in the determination of civil or regulatory liability under general common law actions or under general environmental protection legislation. This raises an important analytical point that needs to be emphasized in international comparisons of agricultural environmental compliance costs of the type undertaken in this volume. Environmental regulation and agricultural policy are undertaken by different jurisdictions in different countries. Canada is a federal state, consisting of a national government and also provincial or territorial governments. Under Canadian constitutional law, agriculture is a shared jurisdiction between the federal government and the provincial governments. Non-marine natural resources, on the other hand, are more generally areas of provincial jurisdiction. And, especially in recent years, municipal governments have been playing a more active role in local permitting, land use policy and enforcement issues related to environmental protection. Of course, as a mostly[1] anglo-american common law country, civil liability for trespass and nuisance can also trigger environmental sanctions. One of the implications of the complex division of jurisdiction for environmental protection in federal nation states is that international comparisons of regulations need to examine actions and policies of multiple levels of government and also the effectiveness of civil litigation as a means of protecting environmental values.

Federal legislation

Regulations related to water quality

There is no national counterpart of the EU Nitrates Directive in Canada. However, both federal and provincial regulations and guidelines are aimed at protecting the environment from agricultural pollutants such as nitrates (i.e. *Canada Water Act,* 2009 and provincial nutrient management regulations). The water quality guideline for nitrate-nitrogen concentration is 10 milligrams per litre (Fairchild *et al.*, 2000). This is equivalent to 50 milligrams of nitrates (NO_3-N) per litre.

Regulations related to nature conservation and biodiversity

Several pieces of federal legislation are aimed at biodiversity conservation. The *Migratory Birds Convention Act* (2009), first passed in 1916, and the Migratory Bird Sanctuaries

[1] Quebec is a civil code jurisdiction within the Canadian federation.

(MBSs) established under that Act are aimed at protecting migratory birds and their nests against physical disturbance and hunting but they did not protect habitat or species of wildlife other than migratory birds. The *Canada Wildlife Act* (2009) was passed to authorize the establishment of sites known as National Wildlife Areas (NWAs) in which migratory birds and other wildlife, particularly species at risk and more importantly, their habitat, could receive protection.

The *Migratory Birds Convention Act* (2009) establishes a regime for the protection of migratory birds, whose habitat occurs on agricultural lands. The *Migratory Birds Regulations* stipulate the conditions for the conservation and protection of migratory birds. These regulations control, for example, hunting and possession of migratory game birds; sale, purchase, or shipment of migratory birds, their nests or eggs; scientific collection, aviculture, and taxidermy; activities designed to reduce the damage migratory birds cause to crops or other property and the danger they pose to aircraft; and requirements for hunters to use non-toxic shot for most migratory game bird species.

The *Migratory Birds Sanctuary Regulations* prohibit disturbance of migratory birds, their eggs, and their nests within an MBS (Migratory Bird Sanctuary). There are 92 Migratory Bird Sanctuaries in Canada. The regulations also prohibit disturbance of migratory birds' habitat when MBSs are established on federal Crown land.

The *Species at Risk Act (SARA)* (2009) protects wildlife found on federal lands as well as their critical habitat. The accord for the Protection of Species at Risk, which allows the Government of Canada to work collaboratively with the provinces and territories in the development of a joint species at risk protection approach in Canada, including laws and complementary programmes for the protection of habitats and the recovery of species and the Habitat Stewardship Program, which encourages Canadians to join their efforts under a landscape level approach in order to protect habitats, inter alia.

A number of other laws and agreements are also in force to help in the effort to protect and recover species at risk. These include the *Canada Fisheries Act* (2009), the *Canada National Parks Act* (2009), the *Wild Animal and Plant Protection and Regulation of International and Interprovincial Trade Act* (2003) and the *Convention on International Trade in Endangered Species* (1979). The *SARA* also complements the work done by provincial and territorial governments under the Accord for the Protection of Species at Risk. This recognizes that species protection is a shared responsibility.

Public, animal and plant health

There are several pieces of federal legislation concerning plant and public health that apply to agriculture. The purpose of the *Plant Protection Act* (2005) is to protect plant life and the agricultural and forestry sectors of the Canadian economy by preventing the importation, exportation and spread of pests and by controlling or eradicating pests in Canada. The Act includes provisions for compensation for lost plants, inspection and treatment costs, given that the provisions of the Act have not been breached by the farmer.

The primary objective of the *Pest Control Products Act* (2006) is to prevent unacceptable risks to people and the environment from the use of pest control products by defining and setting standards for registration of pest control products. Unless an authorization to use an unregistered product has been made by the Minister, using an unregistered product represents a punishable offence. The Act also sets export controls on pest control products.

The *Food and Drugs Act* (2008) sets standards aimed at preventing sale of food that has in or on it any poisonous or harmful substances, is unfit for human consumption, consists in whole or in part of any filthy, putrid, disgusting, rotten, decomposed or diseased animal or vegetable substance, is adulterated, or was manufactured, prepared, preserved, packaged or stored under unsanitary conditions.

Provincial legislation

As we explained earlier, under Canadian constitutional law, most environmental issues related to agriculture are managed mainly on the provincial and municipal levels. The extent to which agricultural production is regulated at the provincial level is illustrated in Table 4.4. This Table illustrates the number of provincial legal acts related to environmental protection under 9 thematic areas, specifically general environmental legislation, water use and conservation policy, wildlife habitat protection, land use planning and animal health protection in Saskatchewan, Alberta, Manitoba, Ontario and Quebec. Each asterisk indicates a provincial statute related to the policy issue in question. A detailed examination of each of the pieces of legislation and its associated regulations is beyond the scope of this chapter, but the table does show an important pattern. In each of these thematic areas, the primary agricultural provinces in Canada have at least one and often multiple provincial legislative measures to address different aspects of agricultural environmental policy issues within its jurisdiction.

Table 4.4 Number of provincial legal acts addressing agriculture related issues with potentially most significant effect on farms in Saskatchewan, Alberta, Manitoba, Ontario and Quebec

General policy area	Saskatchewan	Alberta	Manitoba	Ontario	Quebec
General environmental issues	**	*	*	****	*
Water resources and conservation	***	***	****	******	*****
Wildlife and habitat protection	***	***	****	******	*******
Land use and planning and conflict resolution	***	***	****	******	*****
Soil quality	*	*	*		*
Crop production and protection	**	***	***	***	***
Crop residue, manure, and waste product management	**	**	***	***	****
Animal health, welfare and identification	*****	*****	***	****	*
Food quality and safety	***	****	****	*****	*

Note: Each asterisk represents one provincial statute applying to some aspect of the given general policy area

Source: Canadian Legal Information Institute (2009)

Another inference that can be made from the inventories of legislation in Table 4.4 is that the combination of federal and provincial legislation in Canada does, taken together, span similar types of issues as the EU Directives. It could be said that there is a sort of functional equivalence between Canadian agricultural environmental policy and EU policy. Table 4.5 aligns Canadian federal and Ontario provincial legislation with the EU Directives on the basis of functional equivalence. For example, the *Canada Water Act* (2009) generally applies to issues covered by the Protection of Groundwater Directive, the Sewage Sludge Directive and the Nitrates from Agriculture Directive. Additionally, in Ontario, the *Ontario Nutrient Management Act* (2009), the *Drainage Act* (2007), the *Safe Drinking Water Act* (2002), *Guidelines for the Utilization of Biosolids and Other Wastes on Agricultural Land* (1996), the *Environmental Protection Act* (2009), and the *Lakes and Rivers Improvement Act* (2006) cover more specific issues related to the same areas of concern – water quality, land application of agricultural and other waste materials.

The idea of functional equivalence generally applies across provinces as well. In most provinces, regulations regarding the design of livestock operations and storage and application of agricultural source materials have been emerging in recent years. Intensive livestock operations in Saskatchewan are regulated under the *Agricultural Operations Act* (1995) and the *Environmental Management and Protection Act* (2007) and Regulations (i.e *Environmental Spill Control Regulations* concerning agricultural liquid). In Alberta, the *Environmental Protection and Enhancement Act* (2008) (i.e. under *Waste Control Regulation: Agricultural Waste*), *Standards and Administration Regulation* (i.e. under *Manure Storage and Nutrient Management* and *Minimum Distance Separation Standards*) and the *Agricultural Operation Practices Act* (2007) and Regulations (i.e. under *Livestock Manure*) address the issues of design of livestock operations and storage and application of agricultural source materials. Similar acts and regulations can also be found in other provinces with significant field crop and livestock production – Manitoba, Ontario and Quebec. The *Ontario Nutrient Management Act* (2009) and new nutrient regulations in Manitoba will be examined in more detail in the compliance cost section.

A case of nutrient management regulations in Ontario

The *Ontario Nutrient Management Act* (2009) sets out standards for land application of agricultural source materials. Existing, new and expanding operations with more than 300 livestock units ($\geq$ 300 LU) or located within 100 metres of a municipal well must obtain a certified Nutrient Management Plan and a certified Nutrient Management Strategy if they apply nutrients on their farm unit. All generators of prescribed materials must complete certified and approved Nutrient Management Strategy. In addition, all farms are subject to construction and siting standards and setbacks from wells and surface water. Brethour *et al.* (2006) identified the following standards applicable to all farms set by the Ontario Nutrient Management Regulations:

- Liquid manure application within 15 metres of a bank of a river if the maximum sustained slope of the soil is 25% and more is prohibited.
- Application of agricultural source materials within 15 metres of a drilled well (with depth of at least 15 metres and watertight casing) or within 30 metres of any other well or within 100 metres of a municipal well is prohibited.
- Application of nutrients to a field adjacent to surface water unless there is a vegetated buffer zone in the field that lies between the surface water and where the nutrients are applied is prohibited.

- Using a high trajectory irrigation gun capable of spraying liquid more than 10 metres to apply manure or non-agricultural source materials to land except if the material being applied is an aqueous solution or suspension containing more than 99% water by weight is prohibited.
- Having a manure storage within 15 metres of a drilled well (with depth of at least 15 metres and watertight casing), or within 30 metres of any other well or within 100 metres of a municipal well is prohibited.
- Manure storage facilities must be sufficient to contain 240 days of manure.
- On land which is subject to flooding or land where water collects during a rainstorm or thaw and flows directly into surface water, applying manure during the period beginning on December 1 of one year and ending on March 31 of the following year or at any other time when the soil is snow-covered or frozen is prohibited[2].

Table 4.5 Comparison of Canadian Federal and Provincial Legislation with EU Directives

EU Directive	Federal Legislation	Examples of Provincial Legislation
Conservation of wild birds	Migratory Birds Convention Act; Species at Risk Act	Endangered Species Act (Ontario)
Protection of groundwater	Canada Water Act	Nutrient Management Act; Drainage Act; Safe Drinking Water Act; (Ontario)
Sewage sludge		Guidelines for the Utilization of Biosolids and Other Wastes on Agricultural Land - Environmental Protection Act (Ontario)
Nitrates from agriculture		Nutrient Management Act; Lakes and Rivers Improvement Act (Ontario)
Conservation of natural habitats, wild flora and fauna	Canada Wildlife Act; Species at Risk Act	Fish and Wildlife Conservation Act; Wilderness Areas Act (Ontario)
Identification and registration of bovine animals	Canadian Cattle Identification Program	Animal Identification Act and Regulations (Saskatchewan)
Identification of bovine animals, labelling of beef	Health of Animals Regulations; Food and Drugs Regulations	Livestock Identification and Brand Inspection Act and Regulations (Alberta)
Identification and registration of ovine and caprine animals	Health of Animals Regulations; Canadian Sheep Federation Guidelines	
Placing of plant protection products on the market	Plant Protection Act; Pest Control Products Act	Plant Diseases Act; Weed Control Act; Pesticides Act (Ontario)
Use of hormones	Food and Drugs Act (Food and Drugs Regulations)	Food Safety and Quality Act and Regulations (Ontario); The Dairy Act and Regulation (Manitoba);
Requirements of food law		Livestock Medicines Act and Regulation; Milk Act and Regulations; Food Safety and Quality Act and Regulations (Ontario)
Prevention, control and eradication of spongiform encephalopathies	Health of Animals Act (Reportable Disease Regulation)	Animal Health Protection Act and Regulations (Quebec)
Control of foot-and-mouth disease		The Animal Care Act and Regulation (Manitoba)

[2] Applying manure on land that is not subject to flooding is permitted but under very restricted conditions (see Brethour *et al.*, 2006).

Tabel 4.5 (ctnd)

Control of swine vesicular disease		Animal Health Protection Act and Regulations (Quebec)
Control of bluetongue		The Animal Diseases Act and Regulation (Manitoba)
Standards for the protection of calves	Health of Animals Act; Meat Inspection Act; Criminal Code	Livestock, Poultry and Honey Bee Protection Act and Regulation (Ontario); Ontario Society for the Prevention of Cruelty to Animals Act and Regulation
Standards for the protection of pigs		
Protection of animals kept for farming purposes		

Ontario Ministry of Agriculture and Rural Affairs (OMAFRA, 2009a), along with other provincial governments' ministries of agriculture, provides formulae for calculating Minimum Distance Separation (MDS) between new or expanding livestock facilities and other land uses. The formulae are incorporated into provincial legal acts in some provinces (i.e. Alberta, Quebec, Manitoba and New Brunswick) or provincial codes of practice and municipal bylaws (i.e. Ontario, Saskatchewan, Prince Edward Island). The MDS is a function of the size of expansion, livestock type, manure storage method and nature of the neighbourhood. The purpose of MDS requirements is to minimize land use conflicts and nuisance complaints related to odour but not related to dust and noise. Table 4.6 compares the authors' calculated values of MDS for a mid-sized livestock facility near a rural residential area for beef, dairy, hog and laying hen farms in selected provinces. It is evident that there is quite a bit of variation across provinces as well as within some provinces (i.e. Saskatchewan). The details of MDS in Ontario have been covered at some length by Fox and Kidon (2002).

Municipal regulations

Municipal bylaws affect farming operations mainly through zoning and manure management standards. The degree to which nutrient management and siting of livestock operations is divided between the provincial and municipal jurisdiction varies across provinces. Brethour *et al.* (2006) summarize bylaws for selected municipalities with high livestock populations in each province related to siting and construction of livestock facilities and manure management. They note that in the Ontario county of Huron, which is composed of nine municipalities, farmers have to obtain building permits from the municipal authorities for any new facilities. However, the *Ontario Nutrient Management Act* (2009) for new operations greater than 5 livestock units supersedes local bylaws on nutrient management. On the other hand, in the Manitoban Rural Municipality of Hanover, any new livestock operation with more than 600 livestock units has to be approved by the Livestock Technical Review Committee. In addition, land parcels for livestock production have to be at least 40 acres (16.2 hectares) in size. Operations with more than 250 livestock units have to acquire at least 160 acres (64.6 hectares) of land. The barn on these operations must not be less than 400 meters and the earthen manure storage must not be less than 500 meters away from nearby dwellings. Manure cannot be stored within 100 metres of any body of water.

Table 4.6 Minimum distance separation requirements or recommendations for mid-sized beef, dairy, hog and laying hen new livestock facilities located next to a rural residential area in selected Canadian provinces

Livestock type	Alberta	Saskatchewan	Manitoba	Ontario	Quebec	New Brunswick	Prince Edward Island
Beef (100 cows)	219	300 to 800	530	423	208	338	274
Dairy (100 cows)	361	300 to 800	530	406	182	338	231
Hogs (500 pigs or 50 sows)	376	300 to 800	530	421	348	675	365
Laying hens (10,000 birds)	304	300 to 800	530	382	242	675	342

Notes: In Saskatchewan, municipal by-laws are used and thus a range of minimum separation distance values from 300 to 800; liquid non-earthen open liquid or semi-solid manure storage is assumed for all types of animals and all provinces

Sources: Agricultural Operation Practices Act (2007); Rural Municipality of North Qu'Appelle No. 187, Bylaw 03-02 (2003); Rural Municipality of Loreburn No. 254, Bylaw No. 01/98 (1998); Rural Municipality of Rudy No. 284, Bylaw No 4-94 (1994); Rural Municipality of Grayson No. 184, Bylaw 2004-01 (2004); The Saskatoon Planning District Development Plan Bylaw (1997); Planning Act (2005); Ontario Ministry of Agriculture and Rural Affairs (2009b); Livestock Operations Act (1999); Prince Edward Island Department of Agriculture (2009).

The compliance mechanism

The Ontario Ministry of Environment is responsible for compliance under the *Environmental Protection Act* (2009), the *Ontario Water Resources Act* (2009), the *Pesticides Act* (2009) and the *Nutrient Management Act* (2002). The compliance mechanism is based on working together with farmers to develop appropriate farm management strategies. Ontario Ministry of Agriculture, Food and Rural Affairs (2009b) provides an overview of this process. Farm operations could roughly be divided into three groups: (1) operations that are already in compliance with the law, (2) operations with minor violations, and (3) operations with major, repeated violations. Operations that are in compliance with the law may still benefit from additional Best Management Practices. The compliance programme is designed to move the operations with minor violations into compliance. The process is generally initiated through voluntary self-reporting or a neighbour complaint. An Agricultural Environmental Officer (a provincial officer with specialized agricultural training) would work with farmers to achieve compliance with the legislation. For operations with major repeated violations, enforcement action is an option but the aim of the compliance programme is still to provide help through problem solving to move these operations into compliance.

Promoting compliance: Environmental farm plans

In Canada, management practices similar to the Good Agricultural and Environmental Condition practices in the EU are implemented through various cost share programmes. These agricultural practices are known as Beneficial (or sometimes Best) Management Practices (BMPs). Agriculture and AgriFood Canada (AAFC, 2006) defines BMPs as 'farm management practices that: minimize and mitigate impacts and risks to the environment, by maintaining or improving the quality of soil, water, air and biodiversity; ensure the long term health and sustainability of natural resources used for agricultural production; and, support the long-term economic and environmental viability of the agriculture industry'. In addition to management practices not legally required, BMPs also include Nutrient Management Plans, which are required under most provincial nutrient management regulations.

In order to help agricultural producers develop and implement BMPs, the Government of Canada initiated Canada's National Environmental Farm Planning Initiative under the 2003 five-year Agricultural Policy Framework (APF) through provincially delivered Environmental Farm Plan (EFP) programmes. The programme was first introduced in Ontario and soon after in Atlantic Provinces as a pilot project in 1993. The new Growing Forward federal policy framework that took effect March 31, 2009 retained the EFP as its integral part in Ontario, while in other provinces different elements of the previous EFP are offered as separate programs (AAFC, 2010).

EFP is a voluntary programme consisting of several stages. Participating farmers may opt out of the programme at any stage. Farmers first attend an Environmental Farm Plan workshop and complete a workbook designed to assess the current state of the farm and identify areas of concern. Confidentiality of identified environmental on-farm issues is guaranteed. The second workshop is designed for identifying possible actions and learning how to develop a realistic plan. The next stage is developing an action plan for addressing the areas of concern. The action plan is confidentially reviewed by a group of locally appointed farmers (i.e. Peer Review Committee). Once the Peer Review Committee approves the Action Plan, the EFP is completed and a farmer can participate in a Cost-Share Program that helps cover a portion of the costs of implementing eligible projects from the Action Plan (OMAFRA, 2009b; Robinson, 2006).

Under the previous 2003 AFP framework, producers were eligible to apply for cost-share incentives through the Canada Farm Stewardship Program, Greencover Canada, and the Canada Water Supply Expansion Program. These incentives generally apply to initial capital investments necessary for implementing BMPs. There were 36 BMP categories, each containing several practices eligible for funding. The federal government covered up to 60% of the cost of implementing eligible practices. Many practices covered through federal cost share programmes were also eligible for funding under different provincial cost share programmes. As a result up to 90% of the total project cost could be covered by combining federal and provincial funds. The coverage varied depending on farmer eligibility, province and type of Best Management Practices. In Ontario, funding was available through the Nutrient Management Financial Assistance Program, Wetland Farm Stewardship Incentive Program, Oak Ridges Moraine Environmental Enhancement Program and Greenbelt Farm Stewardship Program.

The new Growing Forward framework includes 28 eligible BMP categories. The majority of the practices from the preceding version of EFP was retained. While some practices were completely excluded from the new plan, some were redistributed among the remaining 28 categories. Some new BMP categories, such as 3 categories related to energy conservation and production of renewable energy, were added to the list of eligible practices. All practices are covered under the Canada Farm Stewardship Program and various provincial programmes up to 100% of the total eligible cost. Table 4.7 outlines examples of Best Management Practices that are eligible for cost-sharing under federal and provincial programmes. Provincial Ministries provide a further detailed description of eligible farmer activities, expenses and investments for each of the listed practices.

Table 4.7 Examples of Best Management Practices eligible for cost-sharing under various federal and provincial programmes

BMP Category	Eligible practices
Riparian area management	- Develop livestock watering systems - Establish a buffer (greater than 10 metres) - Construct fences to manage grazing and improve riparian conditions - Restore or establish native rangeland - Manage grazing in uplands surrounding riparian areas - Improve stream crossings
Erosion control structures (riparian)	- Contour terraces - Grassed waterways (earthwork, seedbed preparation, seed, outlet structures) - Bank stabilization works - Drop inlet structures and in-channel control structures - Retention ponds and erosion control dams - Works to improve infiltration of concentrated water flow - Engineering design work
Erosion control structures (non riparian)	- Contour terraces - Gully stabilization/grassed waterways - Bank stabilization works - Retention ponds and erosion control dams - Drop inlet structures and in-channel control - Works to improve infiltration of concentrated water flow - Mechanical wind screens
Land management for soils at risk	- Establish a perennial or annual barrier crop - Grassed waterway construction - Perennial forage establishment on severely erodible or saline soils - Strip cropping - Straw mulching - Grazing management in critical erosion areas not associated with riparian zones
Improved cropping systems	- Equipment modification - Chaff collectors and chaff spreaders installed onto combines - Precision farming applications
Cover crops	- Establishment of a non-economic crop - Equipment modification to facilitate inter-row seeding of relay crop within an existing row crop
Shelterbelt establishment	- First year establishment costs such as site preparation, planting, weed control, irrigation, and temporary fencing - Cost of purchasing tree or shrub seedlings

Source: Ontario Soil and Crop Improvement Association (2008)

Recently, there have been initiatives for EFP framework to include provision of environmental goods and services in addition to legally required level of environmental stewardship. Kristen Thompson, the coordinator for the Alternative Land Use (ALUS)

project, reports (personal communication, August 2009) that the ALUS pilot project, launched in Ontario's Norfolk county in 2003, is aimed at compensating farmers for providing services such as creation and expansion of wetland habitat; riparian buffer establishment; creation of tall grass prairie and oak savannah; reforestation using native Carolinian species; pollinator hedgerows; and grassed windbreaks. A similar project was launched in Manitoba in 2005 and is currently at the evaluation stage. The process of establishing pilot projects is currently underway in Saskatchewan and Alberta as well. In Prince Edward Island (PEI), ALUS has been part of the provincial policy since 2008. The programme is administered through the PEI Department of Environment, Energy and Forestry (Hager, 2009). As a general statement of future ambitions, the ALUS organization aims at full integration of the alternative land use approach into agricultural and environmental programmes that are currently in place on a provincial and national level. According to Bailey (2008), ALUS can 'build on and incorporate existing environmental farm programs such as the EFP and allow farmers to do more for the environment on their land.'

BMP adoption levels

Levels of adoption of environmentally beneficial management practices across Canada vary. A recent survey shows that the level of adoption for most of the practices that were examined is between 40 and 50% (Brethour *et al.*, 2007). For some practices, such as variable rate fertilization, adoption is fairly low, at only about 10%. For others, it is quite high. Conservation tillage and soil testing are examples of practices that were being used by close to 80% of surveyed farmers. Next were nutrient application timing and using the services of a certified crop advisor, at about 65% and somewhat less than 60% respectively. Adoption rates for buffer strips, no-till, cover crops, and Nutrient Management Plan ranged between 40% and 50% (Brethour *et al.*, 2007).

Even though the implementation of Beneficial Management Practices is partly subsidized by the federal and provincial governments, it is not costless to the farmers. Based on data for farmer cost shares under different programmes and the provincial expenditures on the implementation of the 36 BMPs available from the Ontario Soil and Crop Improvement Association (2008) and Conservation Ontario (2006), we estimated the farmers bore about a third (about $44 million) of the cost of implementation of the management practices eligible for funding between 2005 and 2007. More than 11,000 of 57,211 Ontario farmers implemented or were in the process of implementation of BMPs in this period. Together with more than 14,000 farmers that implemented an EFP prior to 2002 (Robinson, 2006), this amounts to almost half of all Ontario farmers.

Lessons learned from the EFP experience

The voluntary adoption of sometimes costly BMPs indicate that farmers see value in such a choice. One explanation is that farmers value environmental stewardship. Smithers and Furman (2003) offer some support for this claim. In a survey of 123 farmers in four counties, they found that farmers' 'intrinsic concerns', such as attitudes toward soil and water quality, human health and desire for more knowledge about the farm environment, were deemed more important by farmers than community image, fear of penalty and financial reward. This indicates that there is a demand for environmental management education within the farming community. The initial stages of EFP program are designed as an educational programme whose main purpose is to help farmers learn how to manage their farm in a more environmentally responsible manner. Thus, EFP could be seen as a response to an existing demand for such a programme. But, in this case, the demand was not coming only from the general public but from the farming community as well.

For those farmers who do not see as much value in environmental stewardship, a possibility of statutory liability if found in non-compliance with the existing regulations through third party complaint (i.e. fees or imprisonment) often makes voluntary identification of a problem the least-cost option for farmers. While completing an EFP does not absolve farmers of legal liability in cases of violations of federal, provincial or municipal environmental regulations it is generally considered an indication of due diligence. On the other hand, farm assessment results under the EFP are confidential, and it cannot be a source for a legal suit. Due to the confidentiality and the due diligence attributes of the EFP, participation in the programme often reduces the overall risk of legal liability for environmental violations.

Another feature of the EFP identified by farmers to be an important contributor to farmer confidence is that the programme is designed ‘by farmers for farmers’. EFP was largely developed and facilitated by farmer organizations. Robinson (2006) identifies this as the key distinguishing feature between the Canada EFP and environmental policies in the EU, which were characterized by a top-down design. Farmer surveys and direct interviews indicated that farmers viewed positively farmer involvement in the identification of on-farm environmental issues.

In summary, there are several key features of the EFP that play a role in shaping farmer incentives. First, the programme is responsive to farmers’ desire for learning. Second, it ensures that farmers are not punished for admitting they have an environmental problem. In addition, there is a reward in the form of the due diligence clause in case of legal liability. Finally, involvement of the farming community in programme design and administration raises the level of farmer trust in the programme.

Available estimates of environmental compliance costs in Canada

An earlier survey of agricultural environmental compliance costs was compiled by Fox and Kidon (2002). Their assessment was that then existing modelling and estimation work indicated that compliance costs for environmental performance in Canadian agriculture were generally in the range of 2% to 4% of overall costs. More recent work has confirmed this earlier finding. Lintner and Weersink (1999) applied a multi-contaminant pollutant control model with endogenous transport coefficients to an Ontario agricultural watershed to determine the effectiveness of policy instruments for achieving multiple environmental objectives (i.e. nitrate and phosphate emission reductions). The 280-hectare study watershed was chosen for the Pilot Watershed Study by Agriculture Canada to examine the effects of alternative farming practices on soil erosion and phosphorus contamination (Agriculture Canada, 1990 as cited by Lintner and Weersink, 1999). While nitrogen water concentrations did not exceed the legal limit, phosphorus concentrations did and the authors estimated that a 20% reduction in phosphate concentrations would be sufficient to meet the phosphorus standards. This could be achieved at a cost of about $11/ha, which is equivalent to 1.5% of total costs.

De Laporte (2007) used a combination of GIS and economic modelling to estimate the opportunity cost of wildlife preservation and erosion reduction in south-western Ontario. The economic model was based on actual costs of wetland drainage, representative crop mix for the observed watershed, yields, prices and costs of production for south-western Ontario. De Laporte's estimates varied depending on the achieved reduction in erosion and wildlife habitat preservation. The maximum estimated reduction in sediment accumulation (67%) and preserving of about 83% of the watershed wetland area from draining would involve foregoing a net farm income equivalent to about 3.7% of total farm cost[3], given that conservation tillage is used. Introducing conservation tillage as an option in addition to wetland preservation can lower the cost of sediment abatement in areas where conservation tillage is a low cost erosion control option compared to wetland preservation. This is consistent with the finding of Yang *et al.* (2005), who estimated the effect of conservation tillage on erosion and greenhouse gas emissions for the Fairchild Creek Watershed and found that, in order to achieve a relatively low reduction (20% to 35%) in sediment loading compared to De Laporte's estimate by using only conservation tillage, similar costs at the watershed level need to be incurred (3.6 to 4% of total cost).

Oginskyy (2005) estimated the cost of compliance with *The Ontario Nutrient Management Act* (2002) for a typical Ontario swine farrow-to-finish farm by using a linear programming technique. The model allows changes in the land base, herd size and composition of ration as means of achieving compliance with phosphorus and nitrogen emission standards. If a farm must comply only with a nitrogen standard, the opportunity cost of compliance ranges from 0 to 6.1% of total farm cost, depending on the selected level of nutrient standard, pig diet (involving the use of lysine and phytase) and farm size. As expected, lower nutrient level standards involve lower compliance costs. Similarly, adding lysine and phytase to the pig diet reduces compliance costs. Cost increases with farm size. While tightening the nitrogen to the current 224 kg/ha involves no cost for smaller farms (400 sows), it does create opportunity costs equivalent to a 3.8% increase in cost of production for larger farms (600 sows). However, adding phytase reduces this cost to about 1.8%. In order to satisfy the phosphorus standard, a maximum opportunity cost of about 8.1% would be incurred and this is higher than the cost for either N or K standard. This indicates that phosphorus is a limiting nutrient (i.e. satisfying the P-standard automatically satisfies the N and K standards but not vice versa). Consequently, running the model after imposing a combined N-P-K restriction produces results similar to the phosphorus restriction. However, adding phytase reduces the P content of manure so that in the case of a combined N-P standard, nitrogen becomes the limiting nutrient, and in the case of a combined N-P-K, potassium is limiting.

Salvano *et al.* (2006) estimated the effect of the proposed phosphorus-based regulations on farrow-to-5 kg and on finishing swine operations in two most swine-intensive townships in Manitoba - La Broquerie and Hanover. When sufficient land was available for manure application on the farm, compliance costs ranged from 0.2% to 0.35%. Compliance costs increased when sufficient land base was not available and manure either had to be transported or treated. The increase in costs in this situation varied depending on the method of transportation and distance as well as the type of operation and the presence of phytase in the ration. The increase in cost ranged from 0.31% to 5.75% for transport with

[3] The percentage of total cost was calculated by dividing the monetary cost values estimated by De Laporte by the average total cost for a corn-wheat-soybean rotation in Ontario in 2007.

tractor-drawn tanker over a 20-km distance and from 0.25% to 4.66% for transport with a truck over a 40-km distance and from 0.15% to 2.72% for a 20-km distance. Aerobic manure treatment when not enough land is available seems to be the most expensive compliance strategy, with increases in cost ranging from 3.48% to 8.42% depending on the size and type of operation. Liquid separation is about 3 times cheaper than aerobic treatment while construction of an additional storage facility due to no winter application involves even lower increases in costs of 0.41% to 1.87%.

Concluding remarks

Constitutionally, Canada is a federal state in which a range of agricultural and environmental issues are areas of shared jurisdiction between the federal and provincial governments. It is important to remember this constitutional arrangement when conducting international comparisons of agricultural environmental policy processes, institutions and standards in Canada with those that have been developed elsewhere. This chapter has argued that a functional equivalency approach is needed in constructing international comparisons of agricultural environmental policies. It is the sum of actions by all levels of government, in concert with standards established in civil litigation and with what is sometimes called industry self regulation that results in environmental outcomes.

Canadian agriculture is a relatively extensive agriculture, especially compared to the input-intensive agricultures that exist in many EU countries. This variation in input intensities has important implications for the environmental effects generated by agricultural production and also for the demand for environmental policies.

Available estimates of environmental compliance costs for Canadian agriculture indicate that compliance costs are relatively low, especially when producers have flexibility in the means that they can use to achieve compliance. Low compliance costs, coupled with relatively low input intensities, may account for the effectiveness of voluntary environmental measures that are used widely in Canada. In addition, the provincial and federal governments have offered cost-sharing arrangements of various types to assist producers in meeting environmental policy targets.

References

Agricultural Operations Act (1995), *Revised Statutes of Saskatchewan*, Regina, SK: Queen's Printer for Saskatchewan, Chapter A-12.

Agricultural Operation Practices Act (2007), *Revised Statutes of Alberta*, Edmonton, AL: Queen's Printer for Alberta, Chapter A-7.

Agriculture and Agri-Food Canada (2002), 'Canadian Fertilizer Consumption, Shipments and Trade, 2000/2001', Report is distributed as a working paper.

Agriculture and Agri-Food Canada Online (2006), 'National Farm Stewardship Program', Available: http://www4.agr.gc.ca/AAFC-AAC/display-afficher.do?id=1181580600540&lang=eng [Accessed: Mar 29, 2009]

Agriculture and Agri-Food Canada Online (2007), 'Nitrogen Use Efficiency', Available: http://www4.agr.gc.ca/AAFC-AAC/display-afficher.do?id=1182179116194&lang=eng [Accessed: Aug 21, 2009]

Agriculture and Agri-Food Canada Online (2010), 'Programs and Services', Available: http://www4.agr.gc.ca/AAFC-AAC/display-afficher.do?id=1204137480722&lang=eng#e [Accessed: Aug 16, 2010]

Bailey, R.O. (2008), 'What's Next for ALUS in 2009: Carbon Trading, Water Conservation and Delivering Ecological Goods and Services.' Press Release, unpublished.

Boyd, D.R. (2003), *Unnatural Law: Rethinking Canadian Environmental Law and Policy*, Vancouver: UBS Press.

Brethour, C., B. Sparling, B. Cortus, T. Moore, and D. Bucknell (2006), 'Environmental and Economic Impact Assessments of Environmental Regulations for the Agriculture Sector: A Case Study of Hog Farming', *George Morris Centre,* Guelph, ON, (Report).

Brethour, C., B. Sparling, B. Cortus, M. Klimas, and T. Moore (2007), 'An Economic Evaluation of Beneficial Management Practices for Crop Nutrients in Canadian Agriculture', *George Morris Centre,* Guelph, ON, (Report).

Brubaker, E. (2007), *Greener Pastures: Decentralizing the Regulation of Agricultural Pollution*, Toronto: University of Toronto Press.

Canada Fisheries Act (2009), *Revised Statutes of Canada,* Ottawa: Queen's Printer for Canada, Chapter F-14.

Canada National Parks Act (2009), *Revised Statutes of Canada,* Ottawa: Queen's Printer for Canada, Chapter 32.

Canada Species at Risk Act (2009), *Revised Statutes of Canada*, Ottawa: Queen's Printer for Canada, Chapter 29.

Canada Water Act (2009), *Revised Statutes of Canada,* Ottawa: Queen's Printer for Canada, Chapter C-5.

Canada Wildlife Act (2009), *Revised Statutes of Canada*, Ottawa: Queen's Printer for Canada.

Canadian Fertilizer Institute (2007), 'Retail Sales Statistics', Available: http://www.cfi.ca/files/PDF/2006_(June30th)_Retail_Stats_Report_(revised_June_12_2007).pdf [Accessed: Aug 20, 2009]

Canadian Legal Information Institute Online (2009), Available: http://www.canlii.org/en/index.php [Accessed: Aug 20, 2009]

Coase, R. (1960), 'The Problem of Social Cost', *Journal of Law and Economics,* 3 (October): 1–44.

Conservation Ontario (2006), 'Environmental Cost-Share Opportunities for Farmers on the Oak Ridges Moraine', Available: http://www.conservation-ontario.on.ca/news/files/ORMEEP_brochure_final.pdf [Accessed: Aug 14, 2009]

Convention on International Trade in Endangered Species (1979), Available: http://www.cites.org/eng/disc/text.shtml#texttop [Accessed: Aug 14, 2009]

De Laporte, A. (2007), 'Tradeoffs between Agricultural Interests and Wetland Ecological Benefits in the Eramosa Watershed', MSc Thesis, University of Guelph.

Drainage Act (2007), *Revised Statutes of Ontario*, Toronto: Queen's Printer for Ontario. Chapter D.17.

Environmental Management and Protection Act (2007), *Revised Statutes of Saskatchewan*. Regina, SK: Queen's Printer for Saskatchewan, Chapter E-10.21.

Environmental Protection Act (2009), *Revised Statutes of Ontario*, Toronto: Queen's Printer for Ontario, Chapter E. 19.

Environmental Protection and Enhancement Act (2008), *Revised Statutes of Alberta*, Edmonton: Queen's Printer for Alberta, Chapter E-12.

Fairchild, G.L., Barry, D.A.J., Gess, M.J., Hamill, A.S., Lafrance, P., Millburn, P.H., Simard, R.R., and B. J. Zebarth (2000), 'The Health of Our Water: Toward Sustainable Agriculture in Canada.' 61-73, Research Branch of Agriculture and Agri-Food Canada, Publication 2020/E.

Food and Drugs Act (2008), *Revised Statutes of Canada*. Ottawa: Queen's Printer for Canada, Chapter F-27.

Fox, G. and J. Kidon (2002), 'Canada'. In: F. Brouwer and D.E. Ervin (Eds), Public Concerns, Environmental Standards and Agricultural Trade. Wallingford; CABI Publishing, pp. 141-179.

Grayson, Rural Municipality of (2004), Basic Planning Statement, Bylaw 2004-01.

Guidelines for the utilization of biosolids and other wastes on agricultural land (1996), Ministry of Environment/Ontario Ministry of Agriculture and Rural Affairs, Publ. No. 3425.

Hager, Heather (2009), 'Payment for ecological goods and services', Available: http://www.topcropmanager.com/content/view/4392/182/, [Accessed: July 29, 2009]

Lakes and Rivers Improvement Act (2006), *Revised Statutes of Ontario*, Toronto: Queen's Printer for Ontario, Chapter L.3.

Lefebvre, A., Eilers, W. and Chunn, B., eds. (2005), 'Environmental Sustainability of Canadian Agriculture: grid-Environmental Indicator Report Series - Report #2.', Agriculture and Agri-Food Canada, Ottawa, ON, page 124-130. (Report)

Lintner, A. M. and A. Weersink (1999), 'Endogenous transport coefficients: Implications for improving water quality from multi-contaminants in an agricultural watershed', *Environmental and Resource Economics* 14(2):269-296.

Livestock Operations Act (1999), Regulation 99-32. Queen's Printer for New Brunswick, Chapter 99-262.

Loreburn, Rural municipality of (1998), Basic Planning Statement, Bylaw No. 01/98.

McKitrick R. (2005), 'Air Pollution, Health and Mortality: Separating Fact from Fiction', Presentation to the Association of Major Power Consumers of Ontario, Toronto, April 30.

Migratory Birds Convention Act (2009), *Revised Statutes of Canada*. Ottawa: Queen's Printer for Canada, Chapter 22.

Montpetit, É. (2002), 'Policy Networks, Federal Arrangements, and the Development of Environmental Regulations: A Comparison of the Canadian and American Agricultural Sectors', *Governance* 15: 1-20.

Moore, J.A. and M.J. Gamroth (1993), 'Calculating the Fertilizer Value of Manure from Livestock Operations', OSU Extension Service, Corvallis, OR.

North Qu'Appelle, Rural Municipality of (2003), Basic Planning Statement, Bylaw 03-02.

Oginskyy, A. (2005), 'Modeling the impacts of nutrient pollution regulations on swine farrow-to-finish farms', MSc Thesis, University of Guelph.

Ontario Ministry of Agriculture and Rural Affairs Online (2009a), 'MDS Implementation Guidelines Factor Tables', Available: http://www.omafra.gov.on.ca/english/landuse/guide_p7.htm [Accessed: Aug 20, 2009]

Ontario Ministry of Agriculture, Food and Rural Affairs Online (2009b), 'Complying with Environmental Legislation On-Farm, including the Nutrient Management Act, 2002', Available: http://www.omafra.gov.on.ca/english/nm/approvals/complyingnma.htm [Accessed: Aug 14, 2009]

Ontario Nutrient Management Act (2009), *Revised Statutes of Ontario*. Toronto: Queen's Printer for Ontario. Chapter 4.

Ontario Soil and Crop Improvement Association (2008), 'Environmental Cost-Share Opportunities for Ontario Farmers Available through the Canada-Ontario Environmental Farm Plan', Brochure.

Ontario Water Resources Act (2009), *Revised Statutes of Ontario*. Toronto: Queen's Printer for Ontario. Chapter O.40.

Pest Control Products Act (2006), *Revised Statutes of Canada*. Ottawa: Queen's Printer for Canada. Chapter 28.

Pesticides Act (2009), *Revised Statutes of Ontario*. Quebec City: Queen's Printer for Ontario. Chapter P-11.

Planning Act (2005), Provincial Land Use Policies Regulation Amendment, Queen's Printer for Manitoba, Chapter P80.

Plant Protection Act (2005), *Revised Statutes of Canada*. Ottawa: Queen's Printer for Canada. Chapter 22.

Prince Edward Island Department of Agriculture Online (2009), 'Guidelines For Manure Management For Prince Edward Island', Available: http://www.gov.pe.ca/af/agweb/index.php3?number=70588 [Accessed: Aug 20, 2009]

Robinson, G.M. (2006), 'Canada's environmental farm plans: transatlantic perspectives on agri-environmental schemes', *Geography Journal*, 172(3): 206–218.

Rudy, Rural Municipality of (1994), Basic Planning Statement, Bylaw No 4-94.

Safe Drinking Water Act (2002), *Revised Statutes of Ontario*, Toronto: Queen's Printer for Ontario. Chapter 32.

Salvano, E., A. N. Rousseau, G. Debailleuil, and J. P. Villeneuve (2006), 'An environmental benefit-cost analysis case study of nutrient management in an agricultural watershed', *Canadian Water Resources Journal,* 31(2): 105-22.

Saskatoon, the City of (1997), The Saskatoon Planning District Development Plan, Bylaw No. 7600.

Smithers, J. and M. Furman (2003), 'Environmental farm planning in Ontario: exploring participation and the endurance of change', *Land Use Policy,* 20(4): 343–56.
Species at Risk Act (2009), *Revised Statutes of Canada*, Ottawa: Queen's Printer for Canada. Chapter 29.
Statistics Canada (2009), 'Agriculture', Available: http://cansim2.statcan.gc.ca/cgi-win/cnsmcgi.pgm?Lang=E&SP_Action=Theme&SP_ID=920&SP_Portal=2 [Accessed: Aug 20, 2009]
Thompson, K. (2009), The ALUS Project, personal communication (August).
Wild Animal and Plant Protection and Regulation of International and Interprovincial Trade Act (2003), *Revised Statutes of Canada*, Ottawa: Queen's Printer for Canada. Chapter 52.
Yang, W., C. Sheng, and P.Voroney (2005), 'Spatial targeting of conservation tillage to improve water quality and carbon retention benefits', *Canadian Journal of Agricultural Economics,* 53: 477-500.

Chapter 5

Decentralized, outcome oriented management of agricultural environmental issues in New Zealand

Anton Meister and Nicole Beechey

Introduction

New Zealand lies in the southwest Pacific Ocean approximately 1600 km south east of Australia and consists of two main islands, the North and South Islands, plus several smaller islands. The combined size of the two main islands is 26.8 million hectares, which is similar in size to Japan or Britain but with a population of only 4.0 million (Ministry of Agriculture and Forestry, 2003). The geography is predominantly hilly and mountainous, with 50% of the land classifiable as steep, 20% as moderately hilly, 30% as rolling or flat, and only 25% below 200 metres above sea level (Ministry of Foreign Affairs and Trade, 2002). Approximately 30% of the land is protected in a network of national parks and reserves, few of which are found on the lowlands. Only 3% of land area is used for urban areas and roads, leaving the majority of the remaining land for agriculture and forestry (Ministry of Foreign Affairs and Trade, 2002).

New Zealand's agriculture

New Zealand's temperate climate and fertile soils support almost every kind of farming which dominates New Zealand's landscape (Parliamentary Commissioner for the Environment, 2004). There are about 70,000 farms in the country and in 2007, 14.7 million hectares were used for agriculture which represents almost half of New Zealand's total land mass (Ministry of Agriculture and Forestry, 2007b; Parliamentary Commissioner for the Environment, 2004). Livestock farming accounts for 83.5% of agricultural land usage while forestry represents a further 12.5% of land usage leaving approximately 4% of land for the production of horticultural and grain crops (Statistics New Zealand, 2007). Agricultural land is predominantly privately owned, and generally farmed by the landowner as a commercial operation.

Farming is a significant part of contemporary New Zealand society. In the international community, New Zealand's image is that of 'a land where there are more sheep than humans' and the 'Clean and Green' image[1] (Meister and Shakur, 2003). Although over 85% of the population now lives in urban areas, the agriculture and forestry sectors continue to play a vital role in New Zealand's economy (Meister, 2002; Parliamentary Commissioner for the Environment, 2004). The New Zealand economy is based on its specialized natural advantage: pastoral farming, horticulture, forestry, seafood, tourism, and niche manufacturing (Ministry of Agriculture and Forestry, 2005).

Sheep and beef farming extend over approximately 10 million hectares and dairy farming another 2 million hectares (Ministry of Agriculture and Forestry, 2007c). Over the 20 year period from 1987 to 2007, the sheep population has declined from 64 million to approximately 38 million. The beef cattle population has seen a smaller decline from 4.8 million in 1987 to 4.3 million in 2007 while the total dairy cattle number has increased over the same period from 3.2 million to its current level of 5.2 million (Ministry of Agriculture and Forestry, 2007c). Total hog, goat and horse populations have declined over the same 20 year period while poultry numbers have remained relatively steady and the deer population has grown (Ministry of Agriculture and Forestry, 2007c).

Horticultural exports in New Zealand represent 15% of New Zealand's food and fibre exports, totalling over $2.6 billion in 2007 (Aitkin *et al.*, 2007). In 2009, New Zealand's horticultural exports grew in value to over $3.4 billion. New Zealand's major fruit crops include kiwifruit, wine grapes[2] and apples which represent 29%, 23% and 13% of horticultural exports respectively (Aitkin *et al.*, 2007). The area planted in arable crops has been declining and the area planted in outdoor vegetable has remained fairly steady while there have been noticeable gains in the production of indoor vegetables (Ministry of Agriculture and Forestry, 2007a).

Agriculture remains New Zealand's dominant export sector. As the trend towards more intensive farming systems in New Zealand continues the impacts on 'natural capital' must be observed and mitigated as necessary (Parliamentary Commissioner for the Environment, 2004). Since agricultural products represent a large percentage of exports, natural resources must be managed carefully to ensure sustainable economic growth for the benefit of present and future generations. The positive image New Zealand has created with regards to environmental issues and green space attracts foreign tourists and provides a differentiated product for the export of meat, wool, timber and fish. To protect this image and the environment, New Zealand must take strong measures to curb pollution, manage the landscape and protect its indigenous flora and fauna which will directly affect the agricultural sector (Organisation for Economic Co-operation and Development, 1996).

Agriculture and environmental issues

A number of environmental issues have been raised over the past century by the development of New Zealand's agriculture. While farming in New Zealand is generally less intensive than in other countries, significant soil erosion problems have resulted from removal of the natural forest cover for pastoral farming in some areas of New Zealand. The removal of indigenous vegetation raises the issue of protecting biodiversity. The resulting sediment along with nutrient run-off and discharge of agricultural wastes has contributed to

[1] 'Clean and Green' is an image New Zealand projects of the environment in which agriculture is conducted which is characterized by abundant supplies of clean water and air, soil protected by low use of fertilizers and pesticides, reliance on all year outside grass feeding and low human population density.
[2] Exported as wine.

water quality concerns in some areas. The conversion of land to horticultural and dairy production has made the issues of water allocation, water quality and minimum stream flow important to New Zealand's environment.

Production rates from pastoral farming have increased in the last decade, mainly due to more intensive farm operations and more productivity per animal (Organisation for Economic Co-operation and Development, 2007). In the dairy sector, farmers responding to economic signals have intensified their operations by converting suitable dry-stock pasture and exotic forestry into dairy farms while existing dairy farms have become more intensive. Examples of increased productivity per animal include higher lambing percentages and greater milk production per cow. The results of more intensive farming in some regions include changes in soil health, reduction of freshwater quality and increased greenhouse gas emissions (Ministry for the Environment, 2007a; Parliamentary Commissioner for the Environment, 2004).

These environmental issues have been recognised and environmental management of land, water, air and biodiversity resources is occurring under the Resource Management Act 1991. To retain the clean and green image New Zealand is proud of, the country must take a serious approach to handling environmental problems like surface and groundwater quality which continue to deteriorate (Ministry for the Environment, 2007a; Parliamentary Commissioner for the Environment, 2004). Although scientists have pointed to the possibilities of nutrients causing eutrophication, reduced clearness of lakes and other nitrate problems, New Zealand has been slow to enact policies to deal with the issue.

Environmental management in New Zealand

Current policy directions reflect the increasing concern about the effects of farming systems on the environment. Directing policy development to deal with environmental concerns provides opportunities for New Zealand to position its products as coming from a high quality, sustainable farming environment. New or existing policies acknowledge the need to protect the land resource which includes biodiversity and landscape preservation with some of the physical threats being soil erosion, weeds, pests, and water pollution. Animal health and food safety associated with current agricultural practices and systems are being dealt with under appropriate legislation. The protection of the consumer and the environment is also reflected in the current moratorium on genetically modified organisms and the establishment of an Environmental Risk Management Authority (ERMA). The ERMA will deal with current and future proposals for the growing or introduction of genetically modified crops or organisms.

Management of New Zealand's natural resources is governed by a number of statutes including the Resource Management Act (1991), the Local Government Act (2002), the Hazardous Substances and New Organisms Act (1996), the Conservation Act (1987) and the Fisheries Act (1996). The Resource Management Act 1991 (RMA) is the overriding statute and has a single purpose: to promote sustainable management of natural and physical resources (New Zealand Government, 2005).

The RMA's central functions are to:

- manage the environmental effects of activities on air, water, and land;

- manage the use of publicly owned or managed natural resources (water, geothermal resources, the surface of lakes and rivers, river beds, and the foreshore and seabed);
- control the discharge of contaminants to land, air, and water (Ministry for the Environment, 2007a, p. 47).

The RMA involves several key concepts (Organisation for Economic Co-operation and Development, 1996, p. 32):

- the development of comprehensive effects-based legislation for environmental management, requiring examination and consistent treatment of all factors in any issue (removing the need for sector-specific responses);
- the desirability of intervening only where required and clearly justified;
- the requirement of clearly focused outcomes (targets) where intervention is justified; and
- the need to use appropriate policy instruments to achieve cost-effective solutions.

The RMA entails a shift away from mandating technologies or discharge standards to protect the environment towards a focus on ambient environmental quality, with flexibility as to how it is to be achieved (Organisation for Economic Co-operation and Development, 1996). The reform moved legislation from a prescriptive method to an enabling mode. As Le Heron and Pawson (1996) state, the planning regime established by the RMA regulates the environmental effects of resource development, rather than directly controlling land usage.

Communities most closely affected by the resource usage are responsible for the implementation of the RMA. Thus, local government is a critical part of environmental management in New Zealand, much more so than in other OECD countries (Ministry for the Environment, 2007a, p. 47). As a result, the majority of resource management programs relevant to agriculture are now carried out at the regional level. These include soil conservation activities, water quality monitoring and control, pest management, etc. Councils develop policies in consultation with their communities to address the issues under the RMA. Under the RMA, central government has the authority to develop policy statements on matters of national significance, national environmental standards, and water conservation orders. However, no set of overriding rules or measures exists as each regional authority is responsible for identifying environmental risks in their respective areas and developing regulations to address these threats.

The Ministry for the Environment (MFE) coordinates the development of environmental standards and guidelines to help local authorities and resource users implement their responsibilities under the RMA. Standards and guidelines help define the 'environmental bottom line' of sustainable management. Standards and guidelines prescribe the minimum amount of regulation to best achieve the desired environmental outcome. Furthermore, national standards should only be developed where the advantages of protecting national values or providing national consistency outweigh the advantages of regional resource management.

In 2003, the Ministry for the Environment and the Ministry of Agriculture and Forestry jointly launched the Sustainable Water Programme of Action (SWPoA) to identify priorities for government action to improve freshwater management in New Zealand. The SWPoA has a particular focus on addressing the pressures on water bodies from land-use change and intensification. The SWPoA is now leading the development of national and regional policies to deal with freshwater demand and water quality management (Ministry for the Environment, 2007a, p. 278).

In 2004, New Zealand's central government introduced 14 national environmental standards for air quality. National policy statements on water management, flood management, electricity transmission and telecommunication infrastructures are also being developed. Regional councils manage weeds and pests under the Biosecurity Act of 1993. The regional councils must identify weed and animal pests, develop a pest management strategy, and then choose the most effective method for pest control.

Implementation of the RMA

Regional councils develop regional plans and policy statements, set environmental standards, and implement the plans by granting permits for resource use (Ministry of Agriculture and Forestry, 2005). The regional councils are responsible for resource use consents which permit the following:

- take, use, dam or divert water. All abstraction of water from any natural source requires permission from a Regional Council;
- discharge contaminants to land, air or water (including marine waters). Point source discharge requires permission from the relevant Regional Council to enable the council to protect native flora and fauna, drinking water supplies, commercial needs as well as cultural and recreational values for clean water;
- occupy the coastal marine area;
- place structures or carry out other activities in river and lake beds;
- control activities on land for the purpose of soil conservation, the maintenance and enhancement of the quality or quantity of water in a water body, the avoidance or mitigation of natural hazards and the prevention or mitigation of any adverse effects from storage, use, disposal or transportation of hazardous substances.

Policies and rules within Regional and District Plans manage the effects of land use. Each rule in Table 5.1 outlines the requirements, if any, applying to a land use activity falling within that rule category (Horizons Regional Council, 2003; Morris and Workman, 1998).

Activities which affect air quality, water quality and quantity, lake and river beds, and the coastal marine area are restricted under the RMA unless allowed by a rule or resource consent. By contrast, land use activities are allowed under the RMA unless restricted by a rule. When permitted, the activity is allowed given adherence to standards and conditions set. When not permitted, a consent must be obtained which in turn has a set of conditions associated with it. Table 5.2 below illustrates the implications of the Resource Management Plan for the Manawatu-Wanganui Region (Horizons Regional Council, 2006).

Water quality management and impacts on farming

The decline in water quality due to nutrient loading has not been effectively reduced through effluent controls because of the increasing intensity of farm operations (Parliamentary Commissioner for the Environment, 2004). Effluent controls ignore the impact of diffuse factors such as run-off from pastoral land, cattle in streams and rivers, over fertilization, stocking density and nutrient leaching. According to the Environmental Performance Review conducted by the OECD for New Zealand, the largest sources of nitrogen leaching in New Zealand are livestock urine and manure (Organisation for Economic Co-operation and Development, 2007).

Table 5.1 Land use rules and description

Land use rule	Description
Permitted activity	- No resource consent required if the activity complies with conditions of regional plan
Controlled activity	- Resource consent required from council and must be granted if applicant demonstrates the activity will comply with standards and terms of regional plan
Restricted discretionary	- Resource consent required from council - can decide not to grant consent if adverse effects level is sufficiently high - Council restricts its discretion to a specified list of matters to turn down consent and impose additional conditions
Discretionary activity	- Resource consent required from council - can decide not to grant consent if adverse effects level is sufficiently high - Council can use broad discretion on granting consent
Non-complying activities	- Activities the council would generally not grant a resource consent owing to the potential for very significant adverse effects on the environment
Prohibited activities	- No consent can be granted, evidence exists showing significant adverse effects resulting from the activity, activities prohibited by section 105(2)(b) of the Historic Places Act 1993

Table 5.2 Rules affecting farming in the Manawatu-Wanganui region of New Zealand

Activity	Classification	Implication – conditions, standards and terms for activity
Vegetation clearance and land disturbance involving Hill Country Highly Erodible Land	Discretionary	Excavation, filling, tracking and soil cultivation, shall not be allowed unless the activity: - will lead to an environmental benefit - is in accordance with farm plan - is for establishing a fence - is for harvesting trees with replanting - is for establishing commercial forestry
Significant land disturbance on land that is not Highly Erodible	Discretionary	Activities that cause significant land disturbance will be regulated in terms of area affected (<1 ha/ year) or volume of fill, activities involving subdivision
Production forestry	Controlled	- No removal of significant indigenous vegetation - No disturbance of archaeological sites, waaihi tapu or koiwi remains
Discharging farm effluent onto production land	Controlled	- No discharge into water - All storage and treatment facilities sealed

Table 5.2 (Ctnd)

		- Distance regulations - No ponding of effluent on soil surface - Permitted N loading rate (kg/ha/yr) - Grazed pasture (150) - Cut and carry grass (600) - *Pinus radiata* (150) - Eucalyptus (250) - Maize silage (200)
	Permitted	- Applied in accordance with the Code of Practice for Fertilizer Use - Total N loading rate below permitted rates - Nutrient budget model used to plan and carry out the fertilizer application
Effluent from Piggeries	Discretionary	
Drainage of regionally significant wetlands	Non-complying	
Uses of surface water complying with primary allocation	Controlled	
Minor uses of surface water	Permitted activity	- Less than 15m^3/day per property - Screen must be used, and the intake velocity shall be sufficiently low, to onsure that the intake does not cause juvenile fish or other fauna to be damaged or killed - The water shall be used efficiently - Intake shall not lower water level in natural wetland greater than 0.5 ha in area - Regional Council to be notified of location, maximum intake, and use for water

Fonterra, which is a large dairy cooperative, is leading an effort to control the damaging effects of nutrient loading. The majority of dairy farmers in New Zealand belong to Fonterra. The 'Dairying and Clean Streams Accord' was signed by Fonterra in 2003. Farmers are provided with information on the importance of compliance with the Accord through field days and discussion groups (Ministry for the Environment, 2008). The Accord outlines five targets to reduce effluent pollution as determined by regional councils, the Ministry for the Environment, the Ministry of Agriculture and Forestry in conjunction with Fonterra to be achieved by 2012. The targets are:

- Exclude dairy cattle from 50% of streams, rivers and lakes by 2007, 90% by 2012.
- By 2007, 50% of regular crossing points have bridges or culverts, 90% by 2012.
- Immediate, 100% compliance with resource consents and regional plans.
- A nutrient input and output management system in place by 2007 on 100% of dairy farms.
- By 2005, 50% of regionally significant wetlands are fenced, 90% by 2007.

The Dairying and Clean Streams Accord Snapshot of Progress 2008/2009 outlines the current progress in reaching the above targets (Ministry of Agriculture and Forestry, 2010). The progress report shows that with respect to target one, 80% of dairy cattle are now excluded from streams, rivers and lakes, with two years remaining to meet the 90% target for 2012. Target number two has been achieved and number four is at 99% of dairy farms. Producers are struggling to meet targets three and five. In the case of target three, only 60% of producers are compliant. For target five, only six of 13 regional councils have defined and identified regionally significant wetlands. Only three of these six regions have met the 2007 target.

In some areas of New Zealand water quality has deteriorated to the point where it may be unsafe for livestock to drink or, in the worst cases, lakes have become eutrophied (Parliamentary Commissioner for the Environment, 2004). Regional councils have begun taking further actions to address the water quality issues by enacting policies to limit nitrogen leaching into waterways and groundwater. This move is also supported by the New Zealand government's SWPoA 2003 which is designed to identify priorities for government action to improve freshwater management in New Zealand. In 2007, the government approved the development of a national policy statement on freshwater to safeguard water quality and aquatic ecosystems (Ministry for the Environment, 2007a).

In the Waikato region, one of New Zealand's most well-known lakes, Lake Taupo, is threatened by nutrients. The Waikato Regional Council has decided that the flow of nutrients into the lake must be capped and nitrogen runoff into the lake must be reduced by 20% (Environment Waikato Regional Council, 2010).

Case Study: Nitrogen Management in the Lake Taupo Catchment

The problem

Scientific evidence over the past 30 years shows that the water quality of Lake Taupo has been declining. Historically, the lake had extremely low levels of nitrogen and other nutrients, which limited the growth of nuisance plants in its waters (Yerex, 2009). However, major land developments in the catchment since the 1930s and changes to more intensive land uses around the lake have increased the amount of nutrient and sediment entering the lake (Yerex, 2009). Landowners have worked to protect the lake from sediment, through extensive fencing, tree planting and land retirement under the Taupo Catchment Control Scheme. These actions successfully controlled sediment and phosphorus entering the lake, but have not prevented increasing nitrogen levels. The increase in nitrogen levels creates greater numbers of tiny free-floating algae, and nutrient dependent weeds, slimes and potentially toxic algae near lakeshore settlements (Environment Waikato Regional Council, 2004). Thus, the clarity and quality of Lake Taupo has been significantly reduced as a result of the increased nutrients in the lake. Since pastoral uses of the land contribute almost 93% of the manageable nitrogen entering the lake, nitrogen management focuses on agricultural uses as illustrated through the method Environment Waikato has established for protecting Lake Taupo as outlined below.

Nitrogen management

Applying less than 75 kg of nitrogen fertilizer per hectare for growing crops will be permitted under the nitrogen management rules for Lake Taupo (Environment Waikato Regional Council, 2010). For producers that graze animals, the stocking rates outlined in Table 5.3 shows the maximum number of animals per hectare that is allowable under the new nitrogen management rules so long as nitrogen fertilizer is not used.

If nitrogen fertilizer in excess of 75 kg is applied or the stocking levels exceed their respective maximums consents are required. This involves approximately five steps which are outlined below (Environment Waikato Regional Council, 2010).

- Meet with Environment Waikato to identify and provide farm data for benchmarking.
- Environment Waikato checks information and sends to AgResearch to prepare draft report.
- Review of draft report to check for accuracy.
- Final report prepared, sent with a letter confirming Nitrogen Discharge Allowance (NDA).
- Prepare a Nitrogen Management Plan and apply for consent.

Table 5.3 Maximum stocking levels for grazing animals in Lake Taupo Region

Animal	Animal/ha
Dairy cow	0.55
Beef cattle	0.8
Calves	3.3
Horses	0.8
Sheep	7.7
Deer	3.3
Alpaca/Llama	3
Goats	10
Free range pig	2.5

Source: Environment Waikato Regional Council (2010)

Benchmarking

Essentially the benchmarking process represents a grandfathering scheme, assigning a nitrogen discharge allowance (NDA) based on historical output. To assign a NDA there is a substantial amount of information required by Environment Waikato. The most obvious items revolve around the application of fertilizer including the method of application, total amount of fertilizer utilized as well as the quantities of nitrogen, phosphorus, potassium and sulphur applied including when and where it was applied. Additional information is required regarding the stock grazing the land including detailed records, percentage of male stock, supplements fed to the stock and the number of stock purchased, sold, born and those that died (Environment Waikato Regional Council, 2010).

Obtaining consents

Four major options exist for what to do after the benchmark has been established. In each option a producer must submit Forms A and B as well as a $500 deposit (Environment Waikato Regional Council, 2010). These three submissions are all that is required to maintain the same fertilizer application and stocking rates. In order to alter production practices and retain all of the nitrogen, a Nitrogen Management Plan (NMP) must also be prepared which meets the NDA established through benchmarking (Environment Waikato Regional Council, 2010). If a producer wishes to sell nitrogen to permanently remove it from the Lake Taupo catchment, a NMP is again required and producers are encouraged to communicate with the Lake Taupo Protection Trust. To purchase or sell nitrogen, thus altering nitrogen levels, a producer must ensure the party that is being traded with also applies for consent and prepares a NMP to meet the proposed NDAs (Environment Waikato Regional Council, 2010).

Biodiversity

New Zealand's international identity is a result of the unique indigenous biodiversity and distinctive landscapes. Many indigenous species are found nowhere else on earth as a result of the country's isolated evolution and the diversity of land and seascapes (Ministry of Foreign Affairs and Trade, 2002; Taranaki Regional Council, 2004). New Zealand is home to many extraordinary species, including the world's only flightless parrot, the kakapo, the kiwi, a bird with nostrils at the end of its beak, a primitive frog that lays eggs that hatch into adult frogs *Leiopelma* species and the giant weta, a large insect which fills a role that small rodents play in other countries. Both species of New Zealand bat are endemic, as are all four frog species, more than 90% of insects and marine molluscs, 80% of vascular plant species[3] and 25% of all bird species (Taranaki Regional Council, 2004).

New Zealand was one of the last places on earth to be settled by humans. Since that time, indigenous biodiversity has been in a state of decline. In the last 700-800 years, humans and their accompanying pests have caused the extinction of:

- 32% of native land and freshwater birds;
- 18% of sea birds;
- three of seven frog species;
- at least 12 invertebrate species (snails and insects);
- one fish; one bat and up to three reptile species;
- 11 plant species (Ministry of Foreign Affairs and Trade, 2002; Taranaki Regional Council, 2004).

The 2007 Environment of New Zealand review states 'In 2007, New Zealand's biodiversity faces the same pressures as 10 years ago. Introduced animal and plant pest species remain a serious threat to New Zealand's biodiversity. Native plants and animals that survived the initial habitat modification caused by human settlement continue to be threatened by introduced predators and competitors.' (Ministry for the Environment, 2007a, p. 401).

---

[3] Vascular plants include ferns, flowering plants and trees, but do not include mosses and liverworts (of which New Zealand has over 1,000 species).

New Zealand's biodiversity is important for scientific discovery as well as cultural and economic purposes as these ecosystems perform many important services. The ecosystem provides clean air and water, decomposes wastes and recycles nutrients, maintains healthy soils, aids pollination, regulates local climate, and reduces flooding. The services provided by the diverse New Zealand ecosystem help sustain the country's agricultural and forestry production (Ministry for the Environment, 2007a, p. 351).

In response to New Zealand's decline in indigenous biodiversity, in the 1980s and 1990s the Government placed emphasis on reforming the environmental protection and management legislation as well as administrative structures. Of particular importance were the establishment of the Department of Conservation in 1987 and the enactment of the RMA 1991, which includes the duty of those administering the RMA to protect areas of significant indigenous vegetation and significant habitats of indigenous fauna.

Much of the effort to protect New Zealand's native vegetation occurs on approximately 8.4 million hectares of legally protected land (Ministry for the Environment, 2010). Within the OECD, New Zealand has the greatest amount of land protected for conservation purposes at 33.4% with an average 15.5% of land protected across all OECD countries. The majority of the upland and mountainous land is protected whereas the coastal areas, lowlands, wetlands and dunelands have little protection (Ministry for the Environment, 2010).

New Zealand's biodiversity is also managed through regional and territorial authorities. Indigenous biodiversity is managed through the RMA under section 5(2)(b) which describes 'safeguarding the life-supporting capacity of air, water, soil and ecosystems' (Ministry for the Environment, 2007b). Section 6(c) of the RMA describes the 'protection of areas of significant indigenous vegetation and significant habitats of indigenous fauna as a matter of national importance' (Ministry for the Environment, 2007b). Lastly, section 7(d) and section 2(1) of the RMA state that it has 'regard to the intrinsic value of ecosystems' wherein 'intrinsic values include genetic and biological diversity' (Ministry for the Environment, 2007b).

The Biosecurity Strategy 2003 has three major goals to protect New Zealand's biodiversity and indigenous life forms. They are as follows:

- Prevention and exclusion - preventing the entry and establishment of pests and unwanted organisms capable of causing unacceptable harm to the economy, environment and people's health.
- Surveillance and response - early detection, identification and assessment of pests and unwanted organisms capable of causing unacceptable harm and, where appropriate, deployment of a rapid and effective incursion response that maximizes the likelihood of eradication.
- Pest management: effective management (including eradication, containment and control) of established pests and unwanted organisms capable of causing harm to the economy, environment and people's health (Biosecurity Council, 2003).

Land conservation and management

Conservation of private land mostly occurs through voluntary efforts. The New Zealand Landcare Trust, Queen Elizabeth II National Trust (QEII), Nga Whenua Rahui (NWR) and

Nature Heritage Fund (NHF) are all examples of programmes that individuals can voluntarily participate in to help preserve New Zealand's natural landscape.

The New Zealand Landcare Trust was established in 1996 to promote sustainable land management, including the conservation of biodiversity (New Zealand Landcare Trust, 2010). The Landcare Trust currently funds projects that will improve and maintain indigenous life forms in New Zealand by restoring waterways, inventorying biodiversity and enhancing protection for biodiversity (New Zealand Landcare Trust, 2010). Projects funded through the Landcare Trust include the Waikawa River Restoration Project, Biodiversity Southland Partnership, Biodiversity Restoration Northland and Biodiversity Restoration Southland (New Zealand Landcare Trust, 2010).

The QEII was established in 1977 to protect, preserve and enhance open space in New Zealand (Queen Elizabeth II National Trust, 2010). Through the QEII landowners sign covenants which are legally binding protection agreements registered on the land title. Once this occurs the land is protected for both the current and any future owners. Each QEII covenant is visited every other year to monitor the site (Queen Elizabeth II National Trust, 2010). Additionally, landowners may receive funding to fence the land and for surveying and registering the land. As of June 2009, 3,189 covenants have been registered with an additional 524 approved and waiting to be registered. Over 109,000 hectares have been protected through this programme (Queen Elizabeth II National Trust, 2010).

NWR is a contestable fund that was established in 1990 to assist in the voluntary protection of native forests on Maori-owned land (Department of Conservation, 2010). Since then it has been expanded to include other indigenous ecosystems. It is administered by committee but serviced by the Department of Conservation (DOC) and receives an annual allocation from the government (Department of Conservation, 2010). There are three methods of protecting land under the NWR, covenanting, Maori reservations and physical protection. Covenants under NWR are similar to those under QEII whereas the reservations are a more stringent protection of the land. The fund will also provide funding to help establish fences to protect indigenous land from farmlands (Department of Conservation, 2010).

The NHF is a contestable fund like the NWR but is open to all private landowners not just Maori and was initiated in 1990. The NHF also began specifically to assist in the protection of forests until 1998 when it expanded to non-forest ecosystems (Nature Heritage Fund, 2010). There are two options under the NHF, direct purchase and covenanting. Similarly to QEII, the NHF will provide some funds for fencing, survey and legal costs of covenanting. The criteria to receive funding through the NHF include that it must be representative of the natural ecosystem, sustainable, contribute to the natural integrity of the landscape as well as the physical and spiritual well-being of the people (Nature Heritage Fund, 2010).

Land management and soil erosion control

Many areas in New Zealand are prone to soil erosion due to natural ecological and geological conditions. Soil erosion is a serious environmental issue in New Zealand and can occur in many forms including fluvial, stream bank, surface and mass movement erosion (Ministry of the Environment, 2007c). Fluvial erosion occurs when flowing water cuts channels or gullies into the soil and is common in the Marlborough and Manawatu regions. Stream bank erosion is a particular version of fluvial erosion wherein the banks of waterways have lost their tree cover and are easily worn away. Surface erosion most commonly occurs outside the hill country in New Zealand and is caused by wind and rain washing soil particles off the surface of the land (Ministry for the Environment, 2007c). Mass movement erosion is commonly caused by storms in hill country and can severely damage fences, drains, roadways and buildings. This type of erosion can cause a whole slope to slide, with hills steeper than 28 degrees having the worst possibility of erosion (Ministry for the Environment, 2007c). Regions with land that is considered hill country have over 7 million acres at risk of erosion of which 48% is being farmed, thus increasing the risk of erosion (Ministry for the Environment, 2007c).

Under the RMA, regional and district councils are to ensure that natural resources, including ecosystems and biodiversity, are managed sustainably. While the central government's financial support for farmers has largely ended, landowners have increasingly taken responsibility for soil conservation. Many regional councils operate extension advisory services on resource management issues and some provide financial assistance for farm and riparian erosion control schemes. Voluntary and government-subsidized soil conservation programs, including the East Coast Forestry Project, have achieved localized reduction in soil erosion on farmland in hill country (Ministry for the Environment, 2007a).

Regional and district councils have developed policies and rules for sustainable land use. Some regional councils have approached farmers individually for partnerships and to develop an advisory role in the preparation of farm plans. Some councils subsidize the development of the plan, while others make subsidized tree stock available to farmers. Several councils operate ratepayer funded programs to assist farmers with soil conservation and sustainable land management activities.

The approaches to land disturbance include both regulatory and non-regulatory approaches. Most rules and regulations apply when changes are being made in terms of afforestation, tracking, vegetation removal through burning and uprooting. With grazing being an existing use, farmers are not affected except in areas that have been declared sensitive.

Increasingly, farmers are taking collective action to achieve sustainable land use. Over 250 land-care or community based groups have been formed throughout the country to address local land-management issues, which include soil conservation. Many of these projects work in partnership with local authorities and research agencies. Federated Farmers, farm discussion groups, and producer boards are also working on a variety of issues related to sustainable land care.

In the East Coast region the soil erosion problem is far greater than elsewhere in New Zealand. Because of the magnitude of the problem voluntary approaches are no longer acceptable and a strict regulatory scheme is being put into place (See East-Coast Case below).

Case study: The East-Coast region

The Gisborne region comprises 7.8% of the North Island and 26% of the country's severely erodible land. The impacts of erosion (flooding and sedimentation) in the Gisborne region are greater than other areas of New Zealand, particularly for the southern part of the region, the Waipoa catchment. After a major disaster in 1988 the Government established the East Coast Conservation Forestry Scheme aimed at planting the most severely eroding land in the headwaters of the catchments. The scheme was replaced in 1992 by the East Coast

Forestry Project (ECFP) which aimed to plant 200,000 hectares over 28 years at a rate of 7,000 hectares per year. The project provides a grant for private growers for planting non-forested areas and tending sites (Bayfield and Meister, 2005).

When forestry returns declined, this scheme lost its effectiveness and the erosion problems remained. The Gisborne District Council has implemented regulations on the lands with the worst erosion which fall into Land Overlay 3 A (LO3A). Within the LO3A regulations, activities are permitted provided that effective tree coverage on all treatable land has been established or maintained or a Sustainable Hill Country Project Works Plan is developed and implemented for the treatable land (Gisborne District Council, 2009). Under either method of protecting the land from erosion a District Conservator of the Gisborne District Council must certify the plans and an annual report on the nature, extent, progress and results of the plans must be provided to the District Conservator (Gisborne District Council, 2009). Both methods are contingent upon the ECFP continuing to exist in order to provide some financial compensation to farmers.

Climate change

The government as a signatory to the Kyoto Protocol is committed to returning New Zealand's net emissions to 1990 levels by 2012. Agricultural greenhouse gas emissions in New Zealand account for 48% of total emissions (Climate Change Information - New Zealand, 2010b). The Emissions Trading Scheme (ETS) was created to address the commitments made by New Zealand in the Kyoto Protocol. Under the ETS, New Zealand Units (NZUs) are traded to aid in the reduction of greenhouse gas emissions (Climate Change Information - New Zealand, 2010d). One NZU represents the right to emit one tonne of carbon dioxide or the equivalent of one of the following greenhouse gases covered under the Kyoto Protocol: methane, nitrous oxide, hydrofluorocarbons, perfluorocarbons and sulphur hexafluoride (Climate Change Information - New Zealand, 2010d). A NZU is submitted for each tonne of emissions produced by a firm or individual required to participate in the ETS. Each sector of New Zealand's economy which produces greenhouse gases will be obligated to participate in the ETS, however, different timelines apply as outlined in Table 5.4 below.

Table 5.4 Entry in the ETS of different sectors in New Zealand

Sector	Emission reporting	Surrender NZU	Allocations of NZU
Agriculture	January 2012	January 2015	Output related
Energy	July 2010	July 2010	None
Forestry	January 2008	January 2011	Per hectare
Industry	January 2010	July 2010	Some activities
Liquid fossil fuels	January 2010	July 2010	None
Waste	January 2012	January 2013	None

The period from July 1, 2010 to December 31, 2012 is a transition period for all sectors that are required to surrender NZUs. During this period, participants will only be required to submit one NZU per two tonnes of emissions. Participants will also be able to buy NZUs from the government at a fixed price of $25 per NZU (Climate Change Information - New

Zealand, 2010c). Sectors including energy, liquid fossil fuel and waste sectors will not receive allocations of NZUs as they are able to pass the costs of implementing the ETS on to their customers. In the case of agriculture, the allocation of NZUs will be 90% of the baseline that has yet to be established. The baseline will be the industry average emissions per unit of output and will be phased out by 1.3% per year from 2016 (Climate Change Information - New Zealand, 2010a).

Agricultural processors must participate in the ETS in order to minimize administrative and compliance costs. In future, the participation may be extended to farmers as issues surrounding compliance, verification and administration of a farmer based participation are solved. In the short term, opportunities for emissions abatement in the agricultural sector are limited making strong emissions reductions unlikely. Methane presents a particular challenge and accounts for two-thirds of the emissions produced by agricultural activities. Some opportunities exist for nitrification inhibitors.

In a recent study by the New Zealand Institute of Economic Research (2008) a model was developed to measure the impacts of the proposed ETS. The results of the model show that costs would notably increase in the agricultural sector. In 2025, dairy farming would decline by 12.9%, and sheep and beef farming by 6.6%. The price of land for dairy farming would fall by 40.6% and 23.4% for sheep and beef farming. The study also mentions that the current design of the ETS does not account for domestic and export market exposure to competition from countries that do not regulate their emissions. Additionally, much of the emissions abatement in New Zealand under the ETS would be replaced by other countries increasing production and approximately 25% of the emissions reduction would be nullified by increased production in external markets (New Zealand Institute of Economic Research, 2008).

Conclusion

While in many ways agriculture is less intensive than in other countries, agriculture does impact the environment in New Zealand. Over the last decade, especially, the shift to more intensive pastoral land use has placed greater pressures on the health of soils, water quality and biodiversity.

Under the New Zealand government's guidelines for sustainable land management and the RMA pressure has increased on land-users to use sustainable practices. These pressures have imposed costs on farmers as they operate under effluent discharge control regulations and costs will likely increase under nitrogen leaching regulation.

Voluntary actions and non-regulatory methods are achieving positive results with regards to soil erosion. Cost sharing is used in some regions to encourage farmers to change their land use strategies, however in most cases the costs associated with soil erosion control fall to the farmer.

District plans can limit private property rights in the interest of protecting indigenous biodiversity, especially in wetland areas. In order to protect and enhance waterways, wetlands dunes and native vegetation, some regional councils provide incentives to farmers for biodiversity management. Biodiversity protection on private land is predominantly achieved through voluntary action and covenants. Some of the transaction costs of entering a covenant are covered through funds set up to aid in this process; however, the majority of the cost of not using the land lies with the land owners.

New Zealand is using many of the tools and approaches commonly used overseas to tackle the issue of agricultural sustainability. The growing intensification of agriculture in New Zealand has caused concerns about the current approach to environmental management. Perhaps the non-regulatory emphasis on environmental management is not sufficient to achieve sustainable agriculture. While voluntary action and non-regulatory approaches have made significant strides towards a more sustainable agricultural industry, it is now clear that regulations will form a key part of effective environmental and natural resource management. Throughout the chapter, illustrations have been provided of the growing regulations for the management of water quality, soil erosion, the preservation of biodiversity, and greenhouse gases. While this imposes constraints and costs on agricultural producers, it also offers opportunities for partnerships with government and other organizations. Through these partnerships, more sustainable farming can be achieved through adaptation and cooperation.

References

Aitkin, A. G., Hewett, E., and Wallace, B. (2007). *Fresh Facts: New Zealand Horticulture 2007.* Auckland: HortResearch.

Bayfield, M. A., & Meister, A. D. (2005). *East Coast Forestry Project Review.* Wellington: New Zealand Ministry of Agriculture and Forestry.

Biosecurity Council (2003, August). *Tiakina Aotearoa - Protect New Zealand, The Biosecurity Strategy for New Zealand.* Retrieved July 29, 2010, from Ministry of Agriculture and Forestry Biosecurity New Zealand: http://www.biosecurity.govt.nz/files/biosec/sys/strategy/biosecurity-strategy.pdf

Climate Change Information - New Zealand (2010a, August 2). *Agriculture allocation: NZUs for Agriculture.* Retrieved September 7, 2010, from Climate Change Information - New Zealand: http://www.climatechange.govt.nz/emissions-trading-scheme/participating/agriculture/allocation/

Climate Change Information - New Zealand (2010b, July 5). *Agriculture in the Emissions Trading Scheme.* Retrieved July 30, 2010, from Climate Change Information - New Zealand: http://www.climatechange.govt.nz/emissions-trading-scheme/participating/agriculture/

Climate Change Information - New Zealand (2010c, September 6). *Liquid fossil fuel suppliers' obligations: Reporting emissions and surrendering NZUs.* Retrieved September 7, 2010, from Climate Change Information - New Zealand: http://www.climatechange.govt.nz/emissions-trading-scheme/participating/fossil-fuels/obligations/

Climate Change Information - New Zealand (2010d, June 18). *New Zealand Units (NZUs).* Retrieved July 29, 2010, from Climate Change Information - New Zealand: http://www.climatechange.govt.nz/emissions-trading-scheme/about/emission-units.html

Department of Conservation (2010). *Nga Whenua Rahui Fund.* Retrieved July 29, 2010, from Department of Conservation: http://www.doc.govt.nz/getting-involved/volunteer-join-or-start-a-project/start-or-fund-a-project/funding/for-landowners/nga-whenua-rahui/nga-whenua-rahui-fund/

Environment Waikato Regional Council (2004). *Protecting Lake Taupo: A Long Term Strategic Partnership.* Retrieved August 3, 2010, from Environment Waikato Regional Council: http://www.ew.govt.nz/PageFiles/7058/strategy.PDF

Environment Waikato Regional Council (2010). *Nitrogen management in the Lake Taupo catchment.* Retrieved August 3, 2010, from Environment Waikato Regional Council: http://www.ew.govt.nz/Projects/Lake-Taupo/Nitrogen-management-in-the-Lake-Taupo-catchment/

Gisborne District Council (2009, January 21). *Chapter 6, Land Disturbance and Vegetation Clearance (Proposed).* Retrieved July 12, 2010, from Gisborne District Council District Plan Chapters: http://www.gdc.govt.nz/assets/District-plan-text/Chapters/DOCSn77572v1CRLDPChapter06LandDisturbanceandVegetationClearance.pdf

Horizons Regional Council (2003). *Land and Water Regional Plan.* Palmerston North: Manawatu-Wanganui Regional Council.

Horizons Regional Council (2006). *The ONE Plan: Resource Management Plan for the Manawatu-Wanganui Region, Working Document, Version 4.* Palmerston North: Horizons Regional Council.

Le Heron, R. and E. Pawson (1996). *Changing Places. New Zealand in the Nineties.* Auckland: Longman Paul.

Meister, A. D. (2002). New Zealand. In F. Brouwer, & D. E. Ervin, *Public Concerns, Environmental Standards and Agricultural Trade* (pp. 215-254). New York, USA: CABI Publishing.

Meister, A. and S. Shakur (2003). *The New Zealand agricultural sector: Policy approaches and initiatives used to help farmers adapt.* Massey University, Palmerston North: Applied and International Economics.

Ministry for the Environment (2007a). *Environment New Zealand 2007.* Wellington: New Zealand Ministry for the Environment.

Ministry for the Environment. (2007b, April 3). *Resource Management Act and biodiversity.* Retrieved July 23, 2010, from New Zealand Ministry for the Environment: http://www.mfe.govt.nz/issues/biodiversity/initiatives/private-land/rma.html

Ministry for the Environment (2007c, September 17). *Soil Erosion.* Retrieved July 5, 2010, from New Zealand Ministry for the Environment: http://www.mfe.govt.nz/issues/land/soil/erosion.html

Ministry for the Environment (2008). *The Dairying and Clean Streams Accord. Snapshot of Progress - 2006/2007.* Wellington: New Zealand Ministry for the Environment.

Ministry for the Environment (2010, April). *Legally protected conservation land in New Zealand, Enivronmental Report Card.* Retrieved July 5, 2010, from New Zealand Ministry for the Environment: http://www.mfe.govt.nz/environmental-reporting/report-cards/biodiversity/2010/

Ministry of Agriculture and Forestry (2003, November). *Agriculture and Forestry in New Zealand. An Overview.* Retrieved June 10, 2010, from New Zealand Ministry of Agriculture and Forestry: http://www.maf.govt.nz/mafnet/rural-nz/overview/nzoverview004.htm

Ministry of Agriculture and Forestry (2005). *Situation and Outlook for New Zealand Agriculture and Forestry.* Wellington, New Zealand: New Zealand Ministry of Agriculture and Forestry.

Ministry of Agriculture and Forestry (2007a, June 30). *Horticulture and Arable Statistics.* Retrieved July 2, 2010, from New Zealand Ministry of Agriculture and Forestry: http://www.maf.govt.nz/statistics/horticulture/

Ministry of Agriculture and Forestry (2007b, June 30). *Land Use Statistics.* Retrieved July 7, 2010, from New Zealand Ministry of Agriculture and Forestry: http://www.maf.govt.nz/statistics/land-use/

Ministry of Agriculture and Forestry (2007c, June 30). *Livestock Statistics.* Retrieved July 2, 2010, from New Zealand Ministry of Agriculture and Forestry: http://www.maf.govt.nz/statistics/pastoral/livestock-numbers/

Ministry of Agriculture and Forestry (2010, March 18). *The Dairying and Clean Streams Accord Snapshot of Progress 2008/2009.* Retrieved July 23, 2010, from New Zealand Ministry of Agriculture and Forestry: http://www.maf.govt.nz/mafnet/press/2010/dairy-clean-streams-background.htm

Ministry of Foreign Affairs and Trade (2002). *Towards Sustainable Development in New Zealand: a report on New Zealand's response to Agenda 21, the programme of action adopted at the Rio de Janeiro Earth Summit on environment and development in 1992.* Wellington: New Zealand Ministry of Foreign Affairs and Trade.

Morris, S. and M. Workman (1998). *Regional Council and Unitary Authority Management of Land Disturbance and Vegetation Removal Activities under the Resource Management Act 1991: Ministry for the Environment Working Paper.* Wellington: New Zealand Ministry for the Environment.

Nature Heritage Fund (2010). Retrieved July 23, 2010, from Department of Conservation: http://www.doc.govt.nz/getting-involved/volunteer-join-or-start-a-project/start-or-fund-a-project/funding/for-landowners/nature-heritage-fund/

New Zealand Government (2005). *Resource Management Act 1991 (1991 No. 69).* Wellington: New Zealand Government.

New Zealand Institute of Economic Research (2008). *The impact of the proposed Emissions Trading Scheme on New Zealand's Economy.* Wellington, New Zealand: NZIER.

New Zealand Landcare Trust (2010). *DOC Biodiversity Projects.* Retrieved July 29, 2010, from New Zealand Landcare Trust: http://www.landcare.org.nz/projects-groups/doc-biodiversity/

Organisation for Economic Co-operation and Development (1996). *Environmental Performance Reviews. New Zealand.* Paris: Organisation for Economic Co-operation and Development.

Organisation for Economic Co-operation and Development (2007). *Environmental Performance Reviews. New Zealand.* Paris: Organisation for Economic Co-operation and Development.

Parliamentary Commissioner for the Environment (2004). *Growing for Good: Intensive Farming, Sustainability and New Zealand's Environment.* Wellington: New Zealand Parliamentary Comissioner for the Environment.

Queen Elizabeth II National Trust (2010). *QEII Open Space Covenants.* Retrieved July 29, 2010, from Queen Elizabeth II National Trust: http://www.openspace.org.nz/Site/Covenants/default.aspx#H60213-7

Statistics New Zealand (2007, June 30). *Agricultural Areas in Hectares by Farm Type.* Retrieved July 5, 2010, from Statistics New Zealand: http://www.stats.govt.nz/browse_for_stats/industry_sectors/agriculture-horticulture-forestry/2007-agricultural-census-tables/land-use-farm-counts.aspx

Taranaki Regional Council (2004). *Indigenous Biodiversity for the Taranaki Region: Regional Policy Statement for Taranaki Working Paper.* Stratford: Taranaki Regional Council.

Yerex, S. (2009, July). *Protecting Lake Taupo: The Strategy and the Lessons.* Retrieved August 3, 2010, from Environment Waikato Regional Council: http://www.ew.govt.nz/PageFiles/7058/ProtectingLakeTaupopublication.pdf

Chapter 6

Voluntary certification schemes and legal standards in the EU

Roel Jongeneel and Thomas Herzfeld

Introduction

Farmers in the EU face increasing interest on production processes from two sides. Whereas farmers are increasingly involved in supply-chain initiated certification schemes in order to provide assurance about the quality and the safety of their products and the environmental sustainability of their production techniques, agricultural policy requests the compliance with certain legally enforced minimum standards. Private certification schemes and cross compliance requirements target often related fields, which raises the concern to what extend they could interact.

The development of certification schemes is fuelled by developments in consumer markets and in the food chain. In the modern economy only few market transactions take place on the basis of product appearance and price alone (Keyzer, 2002)[1]. The food chain is recognized more as key to fulfilling consumer expectations for improved quality, provenance with regard to products as well as desired guarantees about the nature friendly and animal welfare securing way of production. Food chains play important communication roles, embodying among others sustainability and food related discourses.

As a consequence downstream industries have a clear interest in schemes enhancing and guaranteeing quality and safety of agricultural products. Several food crises occurring in Europe only increased the urgency and consumer concern. The various schemes existing nowadays are often overlapping and not always consistent with each other. The initiative to certify products and production techniques may come from producers' organizations, from the processing industry, from multiple retailers or from regional public authorities.

The EU's obligatory cross compliance rules are aimed to enforce minimum standards, and (voluntary) certification schemes represent two approaches to validate whether certain minimum standards are met by farmers. On the one hand, there has been growing interest in developing closer links between the two approaches, as demonstrated by the European Commission's March 2007 paper on cross compliance (COM (2007) 147). On the other

[1] From a welfare theoretical perspective certification and legal standards might appear to be peculiar interventions, since they impose restrictions on market exchanges with the purpose of improving social welfare. However, it can be shown quite generally that in the absence of certification and standards the variability and uncertainty with respect to product composition and characteristics may cause market failure, even under perfect competition (see also Keyzer (2002) for some proofs).

hand, blurring the distinction between public and private sector responsibilities has been heavily criticized in the literature (e.g. Busch, 2009). Therefore, this chapter explores the extent to which there are similarities and differences between certification schemes and the EU's (obligatory) cross compliance. The analysis shows that there are potential synergies between the two approaches, but also suggests that there are clear limits to the extent to which these synergies can be developed, if this was to be considered desirable by the actors involved.

This chapter focuses on private certification schemes, and those standards that apply at farm level, although EU or member state driven schemes are acknowledged where this is relevant to the analysis. In the following, private certification schemes and cross compliance criteria will be briefly described. Subsequently, a comparison of the two approaches to guide production processes and product quality follows. Beside a comparison of attributes, the potential synergies between cross compliance and private standards will be discussed along three levels of aggregation: the farmer, the supply chain and the regulator/ legislator. Finally, we will reflect on the interaction between stricter cross compliance and the willingness to adopt private standards as a strategy of statutory over-compliance - the farmer's strategy to keep regulated outcomes far below or above the statutory defined limits.

Basic characteristics of certification schemes and cross compliance

Characteristics of certification schemes

Following Meuwissen *et al.* (2003; p. 172) certification is defined as the (voluntary) assessment and approval by a (accredited) party on a (accredited) standard. Key elements according to this definition are 'assessment' and 'approval of some standard'. The goals/ motivations of the innovator of a specific standard cover a wide range. They can be used to increase transparency and to reach defined performance and practice standards in the supply chain and by means of this improve consumer trust and competitive advantage (i.e. generate a price premium), to reduce risk liability claims, to enhance the control of livestock epidemics, to improve recall efficiency, etc[2]. Certification schemes can be ordered according to the criteria they impose. A brief examination of certification schemes applied in EU's agricultural sector shows that they comprise several criteria, among which:

- Special quality and food safety standards
- Cooperation in the supply chain
- Environmental and nature friendly character of product
- Social and ethical aspects of the production process
- Organic production method
- Small-scale production
- Touristic values
- Products coming from certain origin or region.

Combinations of the different criteria could also be possible. In addition, the requirements of schemes vary. Some schemes involve a farm audit, whilst others may require record

[2] For a theoretical study of costs and benefits see Leland (1979) and for more applied studies discussing (some) costs and benefits see Meuwissen *et al.* (2003), Caswell *et al.* (1998), Unnevehr and Jensen (1999), and Roberts *et al.* (1996).

keeping or the production of farm management plans. Furthermore, attaining certification often allows the producer to take advantage of any associated branding, such as the use of a logo, to help communicate the quality of the product to stakeholders such as processors, retailers and consumers. Some schemes are not visible to consumers because they have a business to business focus.

Alternatively, certification schemes could be classified according to their (potential) geographical spread:

- local: e.g. Protected Designation of Origin (PDO), Protected Geographical Indication (PGI);
- national: e.g. Red tractor label in the United Kingdom (Assured Food Standards, AFS), the Integrated Chain Governance (IKB) in the Netherlands; and
- international: e.g. organic agriculture, GlobalGAP.

The certifiable parties are potentially all parties in the supply chain. Certification schemes can focus on specific groups (e.g. organic farmers, compound feed industry) within the supply chain, as well as encompass several groups at the same time (e.g. traceability systems and meta management schemes). As such the certification schemes can be ranked according to their scope or degree of encompassing supply chain activities. Certification schemes relevant when considering primary agriculture can be focusing on:

- Feed companies (suppliers)
- Farms (primary production)
- Processing companies (downstream industry)
- Chain (any chain participant)

Certification schemes in general differ from the activities by national surveillance and control systems. The main activity of the latter is to check and evaluate whether the systems implemented at the farm (and company) level fulfil the regulatory requirements. The main motivation of public regulation is the creation of a common playground for agricultural producers in the EU with respect to environmental and food safety aspects. The 'approval of good practice' by the certification schemes is likely to go beyond the requirements imposed by the regulatory system, assigning public regulation a role of securing a minimum food safety and product quality level.

Figure 6.1 provides a general overview of the private (or public-private) certification schemes (see left part), the certifiable parties in the supply chain (see middle) and the regulatory standards (see right part). The regulatory standards taken into account are the SMRs and GAECs requirements as they are part of the EU's cross compliance package.

Certification schemes may also be classified according to the certifying parties (see left side of Figure 6.1). In principle schemes may arise due to the initiative of any individual or group of stakeholders. Such an individual or group is the certifying party which ultimately formulates the standards used for certification. The standard's development might be done in more or less close cooperation with the supply chain partners. If an accredited standard is used, the certification procedure needs to be carried out by an accredited party. All other types of (non-accredited) certification can be certified by either accredited parties or other secondary (e.g. interest groups, clients) or third party organizations (product boards, branch organizations, semi-governmental organizations). Secondary and third parties may thus choose for schemes to be inspected and monitored by themselves or own surveillance agencies, or for management of the scheme by an accredited party. The choice for an accredited certification might improve the credibility of the scheme to outsiders. Accordingly, certification schemes can be ordered based on the accreditation procedure. The following three-way classification can be made:

- first degree schemes: accreditation is granted by independent certification agencies, that are acknowledged by a national governmental body or Board of Accreditation (in line with the European Standard for Product Certification EN45011);
- second degree schemes: accreditation is granted by a branch organization;
- third degree schemes: certification and quality schemes of individual producers, who link their label to own products.

Figure 6.1 Supply chain, certification schemes and regulatory standards

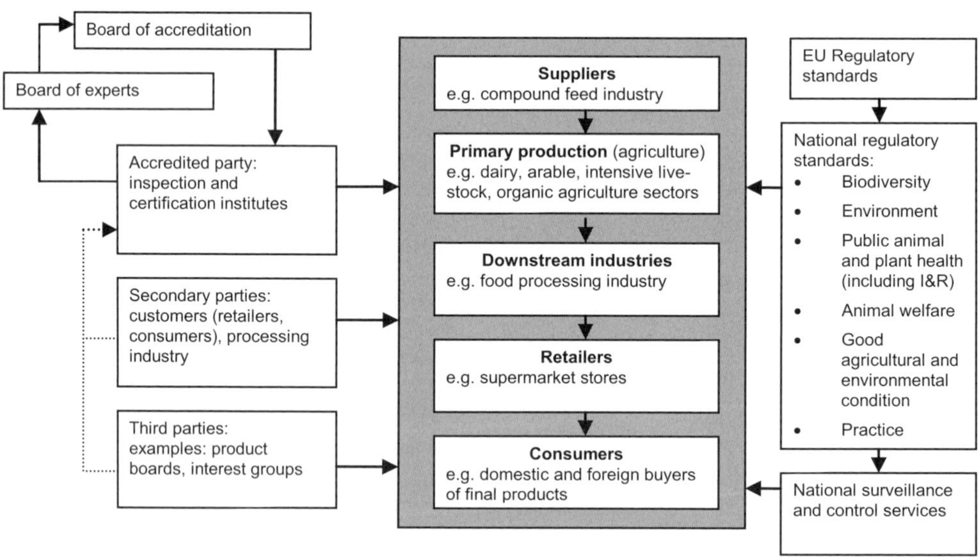

Source: adapted from Meuwissen et al. (2003, pp. 170 and 173)

In the case of accredited standards, chain certification implies that each participant acquires the certification scheme under consideration. For non-accredited standards chain certification generally implies that specific requirements are specified for subsequent stages in the chain.

Finally, the certification schemes can be distinguished with respect to the kind of inspection. As for the accredited standards, schemes linked to ISO39/EN45004, are inspection schemes, based on a checklist, and in principle only valid on the day of inspection. Other schemes, based on ISO65/EN45011 norms, are product certification schemes in which not only processes, but also products are tested on specified standards. They enable visible claims on the end products (e.g. organic product or eco-labels). Lastly, there are certification schemes aimed at evaluating complete management systems, the so-called system certification schemes. The latter ones are based on ISO62/EN45012 (ISO9001:2000).

The certification bodies in charge of controlling compliance with the certification schemes are primarily private. The work of private control bodies is usually surveyed by the organization responsible for the certification scheme, i.e. either by government or by private institutions. In particular since product and system certification schemes do not use straightforward checklists, individual auditors' interpretations of the standards are of utmost importance. For example, an auditor has to judge whether a farmer fulfils the specified requirements. In order to prevent large interpretation differences between auditors, accreditation of the inspection agencies has to be carefully crafted. The national body or Board of Accreditation takes care of this, and certifies the inspection and certification institutes.

The basis for collaboration of farmers with the certification scheme is always voluntary and is considered to be a way to facilitate and improve the marketing of agricultural products. As such, participation in certification schemes not only involves costs but might also generate benefits. Sometimes they might be called quasi-voluntary as the downstream industries make compliance with quality schemes obligatory and don't accept delivery without compliance (Bredahl *et al.*, 2001; Caswell *et al.*, 1998, p. 555).

Characteristics of EU's cross compliance

Cross compliance was introduced as part of the CAP Reform of 2003 with Regulation 1782/2003. Cross compliance involves Member States making receipt of the direct payment aid, called the Single Farm Payment (SFP), conditional on farmers meeting two sets of standards. The first, called Statutory Management Requirements (SMRs), relate to 19 pieces of EU environmental, public, animal and plant health and animal welfare legislation. These are listed in Annex III of the Regulation. The second set, referred to as standards of Good Agricultural and Environmental Condition (GAEC), relate to the appropriate management of soils and the minimum maintenance of agricultural land and features found on that land. The framework for the GAECs (including about 10 separate standards) is outlined by Annex IV of the Regulation (see Table 6.1). These SMRs and GAEC standards apply to all farmers who claim the SFP, and apply to the entire farm holding. Member State authorities must undertake inspections on at least one per cent of farms claiming the SFP to ensure that the standards are being met and those found not to meet the required standards may face reductions or total withdrawal of the SFP. The level of this deduction depends on the severity of the identified infringement.

The GAEC framework and the obligation to preserve the national ratio of permanent pasture at the 2003 level were, at least in principle, new requirements. With the Health Check, the original (obligatory) standards were subdivided into those which are compulsory and those which are optional, though Member States that had already implemented these before 2008 should keep them (see EU Regulation 73/2009 for further details and Table 6.1). Following the Health Check, Member States must introduce a new standard for buffer strips next to watercourses, and another new requirement is compliance with water use authorization standards. Moreover, the retention of landscape elements was revised to refer more specifically to features to be maintained.

Table 6.1 GAEC standards including Health Check adjustments

Issue	Compulsory standard	Optional standard
Soil Erosion Protect soil through appropriate measures	Minimum soil cover	Retain terraces
	Minimum land management reflecting site specific conditions	
Soil Organic Matter Maintain soil organic matter levels through appropriate practices	Arable stubble management	Standard for crop rotations
Soil Structure Maintain soil structure through appropriate measures		Appropriate machinery use
Minimum Level of Maintenance Ensure a minimum level of maintenance and avoid the deterioration of habitats	Retention of landscape features, including, where appropriate, hedges, ponds, ditches, trees in line, in group or isolated and field margins (revised)	Minimum livestock stocking rates or/and appropriate regimes
	Avoid the encroachment of unwanted vegetation on agricultural land	Establishment and/or retention of habitats (new)
	Protection of permanent pasture	Prohibition of the grubbing up of olive trees
		Maintenance of olive groves and vines in good vegetative condition
Protection and Management of Water (new)	Establishment of buffer strips along watercourses (new)	
Protect Water against Pollution and Run-off, and Manage the Use of Water (new)	Where use of water for irrigation is subject to authorization, compliance with authorization procedures (new)	

Source: *Annex III of EU Regulation 73/2009*

Results from a comparative analysis

It is evident that there are a number of differences and similarities between cross compliance and certification schemes. The underlying objectives of certification schemes and cross compliance differ. Whilst the aim of cross compliance is to promote a more sustainable agriculture, enforce compliance with existing legislation, avoid land abandonment and maintain the area of permanent pasture, the aim of certification schemes is typically to set standards that respond to market requirements. As such these standards may or may not correspond to those set by cross compliance. This means that there are limits to any comparison that can be made. The key similarities are that both approaches establish minimum standards, operate an inspection protocol to check compliance with these standards and enforce sanctions if these standards are not met.

An additional qualification before engaging in a comparative analysis relates to the implicit assumptions of the researcher. Economic theories of regulation can be organized into two groups: the so-called public interest and the private interest theories (den Hertog, 2010). Whereas the first assumption departs from a benevolent regulator which aims to pursue the public interest, regulators are assumed to optimize their own individual objective functions within the second theoretical framework. By definition, for all certification schemes the existence of private aims should be assumed. A second difference between the two theoretical categories relates to the amount of information available to the regulator. Public interest theories assign the regulator sufficient information to effectively promote the public interest. Conversely, private interest theories deny the public regulator sufficient information with respect to firm behaviour. In the following, we rely on the body of private interest theories as a foundation of the comparative analysis.

In the remainder of this section, first, the (more theoretical) comparative analysis of cross compliance and certification schemes will be structured along an increasing degree of aggregation from the micro-level (the farm) to the macro-level (the society). Second, an empirical assessment is made in which the overlap between cross compliance and voluntary certification schemes in the Netherlands, the United Kingdom (UK), Germany and Spain are examined.

Micro level

Whereas third-party certification requires a voluntary action of the respective farmer, cross compliance is compulsory for any farmer who claims the SFP. In this respect, each farmer has the freedom to adopt the system which serves best his or her goals. However, obtaining a certification might act as a de facto market entry requirement, at least for the commercially dominating supply chains. Such a characteristic stands in contrast to cross compliance which should secure an equal market access given the compliance with legal (minimum) food safety requirements. At the same time, the farmer is free to dis-adopt the certification scheme. That is, not to renew the certificate after its term expires.

Furthermore, cross compliance criteria cover the whole farm business. In this respect they are similar to those third-party certification schemes which follow a systemic approach like organic agriculture. Alternative, non-systemic, standards provide repetitive control points for different modules for specific products or groups of products and, therefore, disregard other agricultural activities. A mixed dairy farmer with grain production might choose to comply with just one of the GlobalGAP modules – crop or beef – or all of them.

Another difference relates to the inspection rate and definition of inspection sample. The minimum inspection rate for cross compliance, as specified in Regulation 1782/2003 is one per cent of farmers claiming the Single Farm Payment, meaning that, in theory, farmers may only be inspected once every one hundred years. However, agencies apply a risk based sampling which results in significantly higher probabilities of inspection for farms situated in environmentally sensitive areas or those suspected not to comply with a higher number of SMRs. Additionally, the risk based sampling relies on issues like detected non-compliant behaviour in earlier inspections. For many certification schemes *all* members are inspected much more frequently, often on an annual basis. Whereas public sanctions following a detection of non-compliance display a high degree of differentiability ranging from different fines through prison sentences to closure of business, certification schemes stick mainly to warnings or the withdrawal of the certificate. Indirectly, farmers could face the exclusion from marketing channels depending on supply chain characteristics.

Meso level

From the perspective of the supply chain the compliance with public rules guarantees a common ground. Chain partners complement these rules with tailor made own standards and agreements given. As mentioned above supply chain characteristics might yield the voluntary certification schemes quasi-mandatory. Increasing pressure to adopt schemes is expected with a diminishing number of marketing channels or the existence of bottlenecks in supply chains. High adoption rates in pig, cattle and dairy farming serve as an illustration. For example, 90% of the pig meat traded in the Netherlands satisfies the IKB-Pig standard. Meat not having this standard will face obstacles hindering smooth trade. At the same time, lack of alternatives might create lock-in situations where farmers stick with specific marketing channels and processors/ demanders of their products. The anti-trust agency might have an interest in following the increasing role and pseudo-obligatory character of certification schemes because nation- or EU-wide quality assurance schemes can violate competitiveness laws.

Macro level

As regards cross compliance, standards are part of national legislation ratifying EU regulation. With respect to these obligatory standards implicit in the EU's cross compliance package, it turned out that not all Member States were at macro level compliant. This is reflected in several infringement cases on the lack of properly implementing the underlying EU cross compliance regulations. Whereas proper implementation is one thing, adequate monitoring, inspection and sanctioning is another thing, which is left with the authority of Member States. A non-negligible side effect of cross compliance is that the European Commission contributed to a more homogeneous approach to the sanctioning of farmers when being non-compliant, irrespective in which member state they live (Jongeneel and Brouwer, 2010, 194).

As compared to the cross compliance standards, voluntary certification schemes are relatively independent (e.g. standard choice, conditions and target levels). It remains safe to assume a more coherent implementation of certification schemes and their monitoring, inspection and sanctioning across nation states compared to cross compliance. However, these schemes can be stated still to be subject to national legislation. For example, certifying bodies and auditors need to be accredited. Moreover, when these schemes get a pseudo obligatory character they might conflict with EU and/or national competition law. Since a number of these certification schemes are not or no longer tied to national domains the need for EU-wide harmonization needs attention. Whereas national and European Parliament elections secure checks and balances of EU regulation, society's influence on private certification schemes is very unclear.

Empirical assessment of overlap between certification schemes and cross compliance[3]

Below we give more details of the standards set by selected certification schemes as they relate to the requirements of cross compliance. These standards are discussed in relation to the four headings under which the Statutory Management Requirements (SMRs) in Annex

---

[3] This part relies on Farmer *et al.* (2007).

III of Regulation 1782/2003 are listed: environmental; identification and registration of animals; public, animal and plant health; and animal welfare. The linkages between certification schemes and standards set to achieve GAEC are also discussed. A comparison, albeit based on a small number of schemes, allows an evaluation to be made as to whether certification schemes include standards that farmers are legally obliged to meet for cross compliance. If common ground is identified, potential synergies between the two approaches could be explored further, if this is considered desirable to the stakeholders that might be involved. An alternative and equally acceptable scenario is that the two approaches are mutually exclusive and deal with separate standards in order to meet differing objectives. Certification schemes may include standards relevant to cross compliance as well as other standards not covered by the framework of cross compliance. These other standards are beyond the scope of this comparison.

The overview provided attempts a first qualitative assessment of the current level of synergy between SMRs and GAEC standards and the standards set by certification schemes. This analysis concentrates on seventeen schemes in the Netherlands, ten schemes in the United Kingdom (England) and four schemes in Spain. An indication of the degree of overlap is also provided for Germany. This analysis shows where certification schemes operate below the baseline standards set by cross compliance and where certification schemes operate above cross compliance standards. Part of the schemes that were considered, appeared to be not applicable to any of the cross compliance requirements and are omitted from the tables provided below. It should also be noted that the SMRs and GAEC standards in the three countries examined in this section are substantially different, and therefore the comparison is made in terms of national implementation of the SMRs and GAEC standards. This analysis relies on expert judgement because in many cases the standards set for certification schemes do not explicitly refer to cross compliance standards, or in the case of SMRs, the legislation on which the SMRs are based. Identifying the equivalence of private and public standards is not a straightforward task given the use of different wording.

The overlaps between certification schemes and the environmental SMRs in England, the Netherlands and Spain are summarized in Table 6.2. The environmental SMRs are for the Birds Directive, the Habitats Directive, the Nitrates Directive, the Sewage Sludge Directive and the Groundwater Directive. Some schemes, such as Assured British Meat and Assured Chicken Production in England, include a general statement that members must comply with all current legislation. Assured Chicken Production goes further, as do some other schemes such as Assured Combinable Crop Schemes, by stating that producers should notify their certification body of any prosecutions brought or likely to be brought against them with respect to animal welfare, animal movements, food safety or environmental legislation (including cross compliance requirements). In a similar vein, the Soil Association standards state that members' agricultural activities should comply with all relevant cross compliance requirements.

The review summarized in Table 6.2 shows that many schemes set standards that fulfil the requirements of the SMRs relating to the Sewage Sludge Directive, the Nitrates Directive, and to a smaller extent, the Groundwater Directive. Most of the schemes reviewed are much weaker in setting standards that meet the requirements of the SMRs for the Birds and Habitats Directives. None of the Spanish schemes reviewed for this exercise include equivalent standards. The main exceptions are what might be referred to as the higher level environmental schemes, which are EKO, Demeter, LEAF and Soil Association

Table 6.2 Certification scheme standards compared to environmental SMRs covered

Certification scheme	Wild birds	Ground-water	Sewage sludge	Nitrates	Habitats
Netherlands					
GlobalGAP base certificate arable sector	1	1	3	2	1
GlobalGAP combinable crops	1	1	3	3	1
GlobalGAP base certificate animal sector	1	1	3	2	1
GlobalGAP cattle and sheep	1	1	3	2	1
GlobalGAP dairy	1	1	3	2	1
GlobalGAP pigs	0	1	3	2	0
IKB pigs	0	0	0	0	0
KKM milk	0	0	0	0	0
PROduCERT free range cattle	0	0	0	0	0
PROduCERT free range pigs	0	0	0	0	0
PROduCERT free range laying hens	0	0	0	0	0
KPA base certificate	0	0	0	0	0
KPA product specific product Environmental label (Milieukeur product)	0	1	0	2	0
KPA farm Environmental label (Milieukeur bedrijf)	0	1	0	2	0
EKO label arable production	0	3	3	3	2
EKO label animal production	0	3	3	3	3
Demeter-label	3	3	3	4	3
England					
Assured British Meat	0	3	3	3	0
Assured British Pigs[a]	0	0	0	0	0
Assured Chicken Production	1[b]	0	0	0	1
Assured Combinable Crops Scheme	0	3	3	2	0
Assured produce scheme	1[b]	3	3	3	1
National Dairy Farm Assured Scheme	0	1	2	3	0
Nature's Choice	1	3	3	3	1
Linking Environment and Food (LEAF)	3	3	3	3	3
Soil Association Organic	3	3	3	3	3
Spain					
Denominación de Origen	0	0	0	0	0
Producción integrada of each Autonomous region	0	3	0	2	0
COVAP	0	2	0	1	0
CERTIFOOD S.L.	0	2	2	2	0

Notes: a) States that all relevant legislation must be complied with but makes no specific reference to the cross compliance Directives and Regulations; b) Limited references to environmental management as desirable but no specific reference to cross compliance requirements.
Keys: 0 - SMRs or GAEC standards not covered by certification scheme; 1 - Minor coverage of SMRs or GAEC standards by certification scheme; 2 - Substantial coverage of SMRs or GAEC standards by certification scheme; 3 - Full coverage of SMRs or GAEC standards by certification scheme; 4 - Certification scheme standards go beyond cross compliance requirements.

organic and apply also to other Member States. Some of the other schemes reviewed loosely address these SMRs through a requirement to introduce an environmental plan. For example, Assured Chicken Production requires free range poultry producers to setup either an environmental plan which covers the conservation of wildlife or a Farm Biodiversity Action Plan. With GlobalGAP, for example, there are recommendations with respect to environmental issues meaning that the farmer does not need to meet them in order to be awarded certification. The LEAF Marque standards make no specific reference to the five environmental Directives included in the cross compliance SMRs. However, the LEAF Marque requires farmers to maintain a nutrient management plan, a manure management plan and a whole farm conservation plan.

In Germany, not included in the table, the only scheme that might deal with the requirements of the Birds and Habitats Directives is the EU eco-management and audit scheme (EMAS). This is because although EMAS has no defined set of standards, it has the scope to include relevant requirements. EMAS may therefore also include requirements for the Groundwater, Sewage Sludge and Nitrates Directive SMRs. GlobalGAP and some of the Qualität und Sicherheit schemes also include standards which match the environmental SMRs implemented in Germany.

The analysis of Dutch, English and Spanish certification schemes shows that where the identification and registration of animals SMRs are applicable, standards equivalent to the SMRs have either not been implemented by the certification scheme, or at the other extreme, have been fully implemented (see Table 6.3). The GlobalGAP dairy scheme, the Soil Association scheme and the Assured British Meat scheme all fully cover the requirements of the animal identification and registration SMRs. However, in England only the Assured British Pigs scheme lists the specific legislation relating to the relevant SMRs. Assured British Meat has specific standards relating to the identification and traceability of cattle and sheep and refers to the need for correct tagging and movement records as required by the legislation. The EKO label in the Netherlands only addresses the requirements in vague terms. Three of the four Spanish schemes reviewed have good coverage of the animal identification and registration SMRs. The PROduCERT and Freedom Food schemes do not include identification and registration standards. In the case of Freedom Food, this may be because the standards are largely focused around animal welfare issues. In Germany, the Qualität und Sicherheit schemes for cattle and pigs, and the Qualitätsmanagement Milch scheme fully match the national SMRs for animal identification and registration.

Table 6.4 indicates the degree of overlap between a number of Dutch and English certification schemes, and the nationally derived SMRs for EU plant, animal and public health legislation. There appears to be a strong overlap between the certification schemes and the SMRs for food safety given that a key objective of many certification schemes is to ensure food safety by undertaking checks along the food chain. In the Netherlands GlobalGAP dairy and the KKM milk scheme are considered to exceed the SMRs. In comparison the PROduCERT schemes do not cover the food safety SMRs.

The arable and higher level environmental schemes tend to cover the statutory requirements for plant protection products. With GlobalGAP, the SMRs are only met if the producer obtains both the arable sector base certificate and the combinable crop certificate. With the Assured Combinable Crops Scheme there is a direct reference to the need to comply with the national legislation for plant protection products and codes of practice.

Table 6.3 Certification scheme standards compared to SMRs related to the identification and registration of animals

Certification scheme	Identification and registration, ear tags and passports	Identification and registration of bovine animals and beef products	Identification and registration of ovine and caprine animals
Netherlands			
GlobalGAP base certificate arable sector	n.a	n.a.	n.a.
GlobalGAP combinable crops	n.a.	n.a.	n.a.
GlobalGAP base certificate animal sector	3	0	0
GlobalGAP cattle and sheep	3	3	3
GlobalGAP dairy	3	3	3
GlobalGAP pigs	3	0	3
IKB pigs	3	0	3
KKM milk	3	3	0
PROduCERT free range cattle	0	0	0
PROduCERT free range pigs	0	0	0
PROduCERT free range laying hens	0	0	0
KPA base certificate	0	0	0
KPA product specific product Environmental label (Milieukeur product)	0	0	0
KPA farm Environmental label (Milieukeur bedrijf)	0	0	0
EKO label arable production	n.a.	n.a.	n.a.
EKO label animal production	2	2	2
Demeter-label	3	3	3
England			
Assured British Meat	3	3	3
Assured British Pigs	3	n.a.	n.a.
National Dairy Farm Assured Scheme	3	3	n.a.
Freedom Food	0	0	0
Soil Association Organic	3	3	3
Spain			
Denominación de Origen	-	1	1
Producción integrada of each Autonomous region	-	3	3
COVAP	-	3	3
CERTIFOOD S.L.	-	3	3

Keys: - = comparison not made; n.a. – not available; 0 - SMRs or GAEC standards not covered by certification scheme; 1 - Minor coverage of SMRs or GAEC standards by certification scheme; 2 - Substantial coverage of SMRs or GAEC standards by certification scheme; 3 - Full coverage of SMRs or GAEC standards by certification scheme; 4 - Certification scheme standards go beyond cross compliance requirements.

Table 6.4 Certification scheme standards compared to plant, animal and public health SMRs

Certification scheme	Plant protection products	Hormonal, thyrostatic substances and beta-agonists	Food law and safety	Transmissible spongiform encephalopathies	Foot and mouth disease	Swine vesicular disease	Blue-tongue
Netherlands							
GlobalGAP base certificate arable sector	2	n.a	2	n.a..	n.a.	n.a.	n.a.
GlobalGAP combinable crops	3	n.a.	3	n.a.	n.a.	n.a.	n.a.
GlobalGAP base certificate animal sector	0	3	2	2	2	2	0
GlobalGAP cattle and sheep	0	0	3	2	2	2	0
GlobalGAP dairy	0	0	4	2	2	0	0
GlobalGAP pigs	0	3	3	0	0	2	0
IKB-pigs	0	3	2	0	0	3	0
KKM-milk	0	3	4	2	2	0	0
PROduCERT free range cattle	0	0	0	0	0	0	0
PROduCERT free range pigs	0	0	0	0	0	0	0
PROduCERT free range laying hens	0	0	0	0	0	0	0
KPA-base certificate	2	0	2	0	0	0	0
KPA-product specific product Environmental label (Milieukeur product)	3	0	2	0	0	0	0
KPA-farm Environmental label (Milieukeur bedrijf)	3	0	2	0	0	0	0
EKO label arable production	4	n.a.	2	n.a.	n.a.	n.a.	n.a.
EKO label animal production	4	4	2	0	0	0	0
Demeter-label	4	4	2	0	0	0	0
England							
Assured British Meat	n.a.	3	3	2			
Assured British Pigs	n.a.	0	3	0			
Assured Chicken Production	n.a.	n.a.	3	n.a.			
Assured Combinable Crops Scheme	3	n.a.	3	n.a.			
Assured Produce Scheme	3	n.a.	3	n.a.			
National Dairy Farm Assured Scheme	n.a.	3	3	2			
Nature's Choice	3	n.a.	3	n.a.			
Linking Environment and Food (LEAF)	3	n.a.	3	n.a.			

Table 6.4 (Ctnd)

Soil Association Organic	3	3	3	3
Spain				
Denominación de Origen	0	2	2	-
Producción integrada of each Autonomous region	3	3	3	-
COVAP	0	3	3	-
CERTIFOOD S.L.	3	3	3	-

Keys: - = comparison not made; n.a. – not available; 0 - SMRs or GAEC standards not covered by certification scheme; 1 - Minor coverage of SMRs or GAEC standards by certification scheme; 2 - Substantial coverage of SMRs or GAEC standards by certification scheme; 3 - Full coverage of SMRs or GAEC standards by certification scheme; 4 - Certification scheme standards go beyond cross compliance requirements.

EKO, Demeter and Soil Association organic have a strong focus on pesticide and herbicide application, and thus their requirements match or exceed the national plant protection SMRs. In Germany, the Qualität und Sicherheit schemes for fruit and vegetables and potatoes as well as GlobalGAP also match the SMRs for the use of plant protection products. In comparison with the apparently wide coverage of the food safety SMRs in England and the Netherlands, only the Qualität und Sicherheit scheme satisfies the German requirements, although GlobalGAP and Qualitätsmanagement Milch partly satisfy the requirements.

A number of the meat schemes do not include standards which cover the animal disease SMRs, including Assured British Meat and Assured British Pigs in England. Similarly in the Netherlands a direct reference to the need to make immediate notification of the occurrence of contagious diseases is not mentioned in the literature for the relevant schemes.

As Table 6.5 shows there is a substantial overlap between the requirements of certification schemes and the likely animal welfare rules that will be implemented in the Netherlands and England. All the schemes examined have full or substantial coverage of EU requirements, and in some cases exceed them. With respect to the standards for the housing of calves, the organic labels EKO and Demeter (both for the Netherlands) include requirements which exceed statutory requirements. All the schemes reviewed in England appear to comply with the national requirements. The certification scheme standards for the housing of pigs also exceed the legislative requirements in the Netherlands. In England, the Freedom Food and Soil Association Organic standards go beyond legislative requirements, whilst in Germany, the Qualität und Sicherheit scheme meets the national statutory requirements for the housing of calves.

Discussion of results

The synergy between cross compliance and certification scheme raises some interesting questions about the benefits of one approach when compared to the other.

The benefits at least partly depend on the ability to set appropriate standards given the objectives set, enforce those standards and ensure appropriate sanctions are applied. This implies that compliance with standards must be able to be checked during an inspection visit, and that these inspection visits must occur on a frequent basis in order to verify continued compliance.

It may also be noted that if a certification scheme has been in existence for longer than cross compliance, membership of that particular certification scheme may enhance the likelihood of the farmer meeting cross compliance standards if at least some of the scheme's standards match those required by cross compliance.

Table 6.5 Certification scheme standards compared to animal welfare rules

Certification scheme	Housing of calves	Housing of pigs
Netherlands		
GlobalGAP base certificate animal sector	2	2
GlobalGAP cattle and sheep	2	2
GlobalGAP dairy	2	2
GlobalGAP pigs	2	2
IKB-pigs	n.a.	4
KKM-milk	2	0
PROduCERT free range cattle	0	n.a.
PROduCERT free range pigs	n.a.	4
EKO label animal production	4	4
Demeter label	4	4
England		
Assured British Meat	3	n.a.
Assured British Pigs	n.a.	3
National Dairy Farm Assured Scheme	3	n.a.
Freedom Food	3	4
Soil Association Organic	3	4
Spain		
Denominación de Origen	2	2
Producción integrada of each Autonomous region	3	3
COVAP	3	3
CERTIFOOD S.L.	3	3

Keys: n.a. – not available; 0 - SMRs or GAEC standards not covered by certification scheme; 1 - Minor coverage of SMRs or GAEC standards by certification scheme; 2 - Substantial coverage of SMRs or GAEC standards by certification scheme; 3 - Full coverage of SMRs or GAEC standards by certification scheme; 4 - Certification scheme standards go beyond cross compliance requirements.

Cross compliance and certification schemes are similar to the extent that they are both systems that seek to ensure compliance with a set of standards. Their approach to ensuring compliance is also similar in that they both establish inspection protocols and enforce sanctions.

There are also some clear differences between private certification schemes and cross compliance:

- One difference is in terms of the issues covered by the schemes. Private certification schemes do not tend to include standards which farmers are legally obliged to meet.

- There appears to be an assumption that the legal baseline is already met. Among the countries examined, there appear to be very few, if any, schemes that incorporate all of the cross compliance standards that apply in the Member State concerned.
- Certification schemes rather focus on a specific commodity sector, a farming system, or the quality of an end product. Very few schemes take a whole farm approach by seeking compliance with a range of standards that apply to all land on the holding, as is the case with cross compliance.
- In a few cases, certification scheme standards exceed national cross compliance standards. This occurs for those standards which the scheme specializes in, particularly animal welfare requirements for schemes with an animal welfare focus and environmental requirements for higher level environmental schemes.
- Cross compliance applies to the majority of EU farm holdings, whilst sector specific private certification schemes often apply to only a minority of holdings, although some schemes do have wide coverage. Thus certification schemes, at least among those in the Member States reviewed, do not set a horizontal, uniform baseline of minimum standards in the way cross compliance does.
- All producers who seek certification tend to be inspected every year, whilst just one per cent of farmers claiming the Single Payment are inspected for cross compliance in any year. The higher frequency of certification scheme inspections underlines their credibility, so long as standards are meaningful and well defined, compliance is verifiable during an inspection and appropriate sanctions are rigorously enforced. The difference in inspection rates needs to be contextualized by the respective scale of public and private approaches, with cross compliance applying to a much larger number of farmers than private certification schemes do.
- The difference in inspection rates is somewhat influenced by the accompanying sanctions, with the financial sanctions applied through cross compliance potentially more severe than those applied by most certification schemes, where farmers are generally provided with time to rectify a breach before certification is withdrawn. In the event a farmer's membership of a certification scheme is cancelled, the farmer may lose access to the market and encounter financial consequences as a result.
- Also, despite the fact that many schemes do not comprehensively cover all the legal requirements designated by cross compliance, the branded logos provide evidence that certain standards have been met and these logos are generally visible and meaningful to the consumer.

There is also an argument that membership of a certification scheme may ease compliance with cross compliance standards. However, as the evidence in this analysis shows, compliance with certification scheme standards cannot act as a proxy for compliance with EU legal standards.

Potential benefits and limitations to developing synergies between certification schemes and cross compliance

To explore the apparent similarities and differences between certification schemes and cross compliance a subset of certification schemes needs to be excluded. First of all, certification schemes must be accessible for all farmers within the EU. The development of these synergies could have some advantages, as well as disadvantages. The arguments presented here very much depend on the intentions of public and private bodies and whether there is a common interest, or not, in facilitating a collaborative approach to either setting standards or enforcing those standards.

Firstly, the identification of synergies could lead to the harmonization of standards between certification schemes and those found in EU legislation, thereby reducing potential consumer confusion in differentiating product labels and determining the sustainability of farm produce. Harmonization would mean that the standards in private certification schemes include within them the standards for cross compliance, where these are relevant. Secondly, harmonization could be of benefit to the cross compliance inspection regime and result in increased administrative efficiencies. For example, it may be the case that farmers who are certified for meeting specific standards are less likely not to meet cross compliance standards. Membership of certain certification schemes could be a factor in the risk sample that Member States use to target farms for cross compliance inspections. Such an approach would require some confidence that certification schemes rigorously enforce standards that closely match those set for cross compliance. Harmonization may therefore be to the advantage of certification schemes if such schemes can credibly demonstrate to potential members that the standards set by the scheme meet the requirements of cross compliance. This could be an enabling factor in helping farmers to meet the EU legal standards. It may also be a selling point of a scheme if scheme membership is taken into account in the risk sample for cross compliance inspections, meaning that members will be less likely to receive a cross compliance inspection.

The case against harmonization can also be argued on the grounds of duplication of standards. This may not be desirable if a certification scheme sets a standard that the farmer is legally obliged to meet through cross compliance. It may also be inefficient if the farmer is inspected for the same standard during a cross compliance inspection and an inspection required for certification scheme membership.

Harmonization could, of course, raise fundamental questions about the mutual role of government and private bodies in ensuring legal standards are met and in encouraging farmers to meet standards that exceed the legal minima. It may be argued that it is the role of government to ensure that farmers meet minimum regulatory standards and that certification schemes have a complementary role to play in ensuring additional standards not provided for by existing legislation are met by farmers in order to satisfy the demands of the market.

It may be inappropriate and not in the public interest for certification schemes to inspect mandatory standards such as those for animal disease control. National administrations may rightly be reluctant to entrust the private sector with the control of mandatory standards, even if the certification bodies are accredited by state-run accreditation services. Member State administrations would also need to have the ability to sanction private bodies if they fail to adequately inspect and control legal standards. It may also be argued that certification schemes may not want to set overly ambitious standards, as this could dissuade farmers from applying for membership. Certification schemes may have an inherent conflict of interest in this respect.

Voluntary overcompliance – upgrading

Taking into account a further development of cross compliance it might be beneficial for farmers to comply with more demanding standards now. That is, farmers' uncertainty about future minimum requirements will be a determinant of adoption of certification schemes which are closest to cross compliance but go beyond cross compliance requirements. From a purely economic point of view such a behaviour seems to be irrational. However, there are reasons for farmers to do more than required. Given the size of most European farms we rule out the motive to use overcompliance as a marketing tool, e.g. by developing an own standard, and focus on the two aspects risk aversion and expectation of future regulatory changes.[4]

Risk aversion: Product quality and agricultural process characteristics are mostly mediated through natural processes. By definition, farmers lack means of overly precise management of clearly defined outcomes. However, in many cases standards and cross compliance focus on precise levels neglecting variation induced by nature. Risk averse producers act rationally in using less of the regulated input like fertilizers or pesticides. Similarly, the use of more labour input than required adds to this behaviour.

Expectation of future regulatory changes: Any investment within European agriculture is highly influenced by local, national or EU regulations. Forming expectations on future regulatory changes and taking them into account in planning investments might deliver comparative advantages in the future. The longer the investment horizon the more beneficial overcompliance can be. However, uncertainty with respect to future statutory changes increases too.

Conclusion

The evidence presented in this chapter shows that there are some fundamental similarities in the objectives of certification schemes and cross compliance, as well as the way they establish an operational framework. However, there are also a number of clear differences in terms of the scope of the schemes and the way in which they function which may impose limits to any collaborative approach. Examples are differences in coverage, domain, monitoring and inspection, punishment, communication to consumers, etc. Overall, it seems that the extent to which cross compliance and private certification schemes are mutually compatible approaches to meet standards deserves further consideration. Our analysis demonstrates that there is an opportunity for the mutual exchange of best practice between certification schemes and cross compliance. Although limitations exist, there is scope to further explore the synergy between public and private approaches.

The European Commission and members states' governments explore the framework of the future agricultural policy. Proposals include a further greening of the CAP by increasing conditionality on direct payments eligible to farmers. In case more targeted payments will be introduced, the synergies between public and private requirements might get even more attention. Furthermore, a consolidation of certification schemes within the private sector could not be excluded. In the case that smaller regional or national certification schemes disappear or merge into larger international ones, synergies might become more obvious.

---

[4] For a more theoretic discussion see Cavaliere (2000) using a game theoretic approach. Shimshack and Ward (2008) present an empirical explanation of overcompliance by paper mills.

References

Bredahl, M.E., Northen, J.R., Böcker, A., Normile, M.A. (2001). Consumer Demand Sparks the Growth of Quality Assurance Schemes in the European Food Sector, in: Regmi, A. (Ed.), Changing Structure of Global Food Consumption and Trade. US Department of Agriculture, Washington D.C., pp. 90-102.

Busch, L. (2009). The Private Governance of Food: Equitable Exchange or Bizzare Bazaar? *Agriculture and Human Values,* 36 (3), 1-8.

Caswell, J.A., Bredahl, M.E., Hooker, N.H. (1998). How Quality Management Metasystems Are Affecting the Food Industry, *Review of Agricultural Economics,* 20, 547-57.

Cavaliere, A. (2000). Overcompliance and Voluntary Agreements, *Environmental and Resource Economics*, 17, 195-202.

den Hertog, J. (2010). Review of Economic Theories of Regulation, in: Tjalling C. Koopmans Research Institute (Ed.), Tjalling C. Koopmans Discussion Paper Utrecht University, Utrecht, NL, p. 59.

Farmer, M., Swales, V., Jongeneel, R., Karaczun, Z., Müssner, R., Leipprand, A., Schlegel, S., Poux, X., Ramain, B., De Roest, K., Varela Ortega, C., Simó, A., (2007). Exploring the Synergies between Cross Compliance and Certification Schemes, in: LEI (Ed.), Report. Wageningen UR, The Hague, p. 82.

Jongeneel, R., Brouwer, F. (2010). Compliance with agri-environmental standards in Europe and the US. In: Goetz, S.J. and Brouwer, F. (eds.) New Perspectives of Agri-environmental Policies: A multidisciplinary and transatlantic approach. Abingdon, Routledge, 177-198.

Keyzer, M. (2002). Labeling and the realization of cultural values. *De Economist*, 150(4), 487-511.

Leland, H.E. (1979). Quacks, Lemons, and Licensing: A Theory of Minimum Quality Standards, *Journal of Political Economy,* 87, 1328-1346.

Meuwissen, M.P.M., Velthuis, A.G.J., Hogeveen, H., Huirne, R.B.M. (2003). Traceability and Certification in Meat Supply Chains, *Journal of Agribusiness,* 21, 167-81.

Roberts, T., Buzby, J.C., Ollinger, M. (1996). Using benefit and cost information to evaluate a food safety regulation: HACCP for meat and poultry, *American Journal of Agricultural Economics,* 78(5), 1297-1301.

Shimshack, J.P., Ward, M.B. (2008). Enforcement and Over-compliance, *Journal of Environmental Economics and Management*, 55, 90-105.

Unnevehr, L.J., Jensen, H.H. (1999). The economic implications of using HACCP as a food safety regulatory standard, *Food Policy,* 24, 625-35.

Chapter 7

Environmental and health standards in the dairy sectors of the EU, the US and New Zealand

Irina Bezlepkina, Roel Jongeneel, Floor Brouwer, Koen Dillen, Anton Meister, Kees de Roest and Matty Demont

Introduction

With the 2003 Mid-Term Review (MTR) of the EU Common Agricultural Policy (CAP) a number of adjustments to agricultural support were introduced. One of the most substantive changes was the introduction of a system of decoupled payments per farm (known as the Single Farm Payment). These decoupled payments were made conditional on recipients meeting environmental, food safety, animal and plant health, and animal welfare requirements as well as standards of good agricultural and environmental conditions (obligatory cross compliance). Also outside the EU measures are taken to address environmental harm, with the US Concentrated Animal Feeding Operations (CAFOs) legislation as a notable example. Compliance with standards potentially might affect production costs and thus competitiveness.

This chapter focuses on environmental compliance costs issues in the dairy sector of the EU mainly and the implications of those issues for farm level and external competitiveness between the EU and two other main players: the US and New Zealand. The EU's presence on the world market is strong for all the major dairy products. In 2007 the EU-27 shares of world trade are 24% for butter, 18% for skimmed milk powder, 20% for whole milk powder and 37% for cheese (DairyCo, 2010). New Zealand exports about 33% of global dairy products (Blayney *et al.*, 2006). With the enlargement of the EU in 2007, the dairy industry in the EU-27 became very large with 24.2 million dairy cows in comparison with 9.2 and 4.2 million dairy cows in the US and New Zealand respectively (USDA, 2010). Milk production in the EU-27 in 2007, where dairy represents the largest agricultural sector (European Commission, 2006), is 1.6 times larger than that of the US, where dairy is the second largest sector in terms of cash receipts at the farm level (Miller and Blayney, 2006) and about nine times larger than in New Zealand (USDA, 2010).

Would the competitiveness of the EU dairy sector that deals with perishable products and has to comply with environmental standards be influenced? Not much evidence is available to assess the costs of compliance with regulations. An EU study assessing the administrative burden on farms arising from the CAP identifies that cross compliance controls represent 1% of the total administrative costs (DG-AGRI, 2007). However, costs could rise due to changes in production strategies in complying with regulations. Kuik

(2006) reviewed studies that estimated farm level costs of measures to reduce nitrogen pollution in agriculture across Europe. He found that the differences in costs, ranging from €6 to €236 per hectare, can be explained by the structure of the sector (dairy, beef, pigs and poultry, crops, mixed), livestock intensity, historical rates of fertilizer application and the vulnerability of soils to nitrate leaching. In terms of the trade implications of environmental regulations, research has mainly investigated the effects within the manufacturing sector (see e.g. Xu, 2000; Le Roux *et al.*, 2008 for an overview of studies). For the agricultural sector, the work of Cassells and Meister (2001), who simulate the effects of compliance to water quality standards in the New Zealand dairy sector, is most closely related to the objective of this chapter. The contribution of their paper is that it analyses the extent to which the imposition of farming standards gives rise to on-farm costs. The cost increase due to compliance with regulations at the farm level needs to be further scaled up to the sector level in order to be integrated into the analysis of trade thereby assessing external competitiveness.

As presented in Chapter 2 for Europe and in Chapters 3 and 5 for the US and New Zealand, various agricultural environmental policy initiatives have been considered before a selection of comparable standards influential for the competitiveness of the dairy sector was made. The next section describes the data employed in the analytical approach. Thereafter the impacts of the Nitrates Directive and the Identification & Registration Directive at the sector level in the EU (Germany, France, Italy, the Netherlands and the UK) and its key competitors (the US and New Zealand) are presented. Representative farm studies were used as a basis for the cost calculations. Best-estimates of compliance are borrowed from the literature and expert opinions. A computable general equilibrium model which is presented in Chapter 1 is used to assess the impact of compliance costs on competitiveness of the three trading nations in global trade. The negative impact of the analysed measures on EU imports and exports is less than 1%. If a smaller increase in compliance takes place, trade impacts will be even smaller. When the standards for nitrate pollution taken by the US and New Zealand are taken into account along with a full compliance assumption in all countries analysed, this would only marginally improve EU exports. The trade impacts obtained under the assumption of no changes in key competitor countries can, hence, be interpreted as upper bounds of the likely trade impacts.

Analytical approach and data

The concept of competitiveness has no unique definition and is used in a broad set of contexts and levels of aggregation (firm, product, country, industry). As presented in Chapter 1, the analytical approach and design of this study is based on qualitative as well as quantitative modelling assessments, differentiating between farm, sector and global levels of competitiveness.

Firm level competitiveness addresses the comparison of firms in a market. A firm is more competitive compared to the other if it can supply a product at lower costs, without affecting its overall profitability in the long term. In the case of farms, the profitability can therefore be approached by the farm's gross income (revenues and farm support payments) minus the costs of variable and (quasi-)fixed factors.

If compliance with standards affects production costs at the farm level, aggregate agricultural production is affected resulting in a shift of the supply curve, i.e. a supply response. Other well-known mechanisms shifting the supply curve are the introduction of milk quotas (introduced in 1984), reductions in support prices for butter and skimmed milk powder (effectuated with the Luxembourg Reform in 2003), introduction of milk premiums as part of the decoupled payments (Bouamra-Mechemache *et al.*, 2009). The supply response, in turn, displaces the equilibrium between supply, demand and international trade. Differing standards and degrees of compliance with standards, and heterogeneity of farming conditions between trading partners within the sector can affect the market share of those trading in this sector. Market share between sectors is also affected, due to substitution and complementarity and spill-over effects between produced commodities in the economy.

In this chapter the assessment of potential cost implications due to the compliance with standards is made for selected Member States to demonstrate the effects of considerably different variation in the implementation of standards. At the global level, the analysis for the dairy sector is carried on with the GTAP model ensuring for comparability of competitiveness between EU and non-EU countries by including in the analysis comparable standards and potential cost implications for New Zealand and the US.

The standard specification of GTAP model as presented in Chapter 1 has some properties that do not align with the situation in the 2005 European dairy market. First of all, an alternative closure is specified making the output level exogenous and thus fixed. This ensures for implementing the bounding quota constraint in the European milk production. Secondly, to account for the 2003 Mid-Term Reform in the dairy sector as specified above, the base year of the model which refers to 2001 is adjusted. While decoupled payments are assumed not to affect production decisions, price reductions are implemented in the model uniformly by a change in taxes. In this way a new base scenario for the competitiveness analysis in the dairy sector is created. Finally, the improvement in compliance is estimated as the change in compliance as compared to 2005, the year for which best estimates of compliance could be obtained. This is done by shocking the model by the percentage cost increase (evaluated at sector level) relative to the base year due to the simulated increase in compliance. The results of the GTAP model are then interpreted as estimates of improvement in compliance since 2005, i.e. they are evaluated as if they were happening in 2001 (the GTAP base year data). For this reason, percentage changes as compared to the baseline scenario are analysed and not absolute numbers. For the EU-15, we impose compliance only on Member States which represent a significant share (5%) of EU exports. We assume that other countries are too small to significantly affect European competitiveness on the world market. The remaining countries are aggregated into: (i) the rest of OECD (ROECD) countries and (ii) the rest of the world (ROW). The analysis is performed for selected EU-countries, the EU-15 as a whole and two selected non-EU countries namely the US and New Zealand.

As elaborated in Chapter 1, the additional costs associated with compliance to environmental regulations had to be assessed at the farm level and consequently scaled up to national level. The necessary data assessments are the initial and target level of compliance with the standard, percentage of affected farms, per unit cost estimates arising due to compliance with the standard. In the case of the EU dairy these costs are derived for representative specialized dairy farms as recorded in the Farm Accountancy Data Network (FADN) in 2003. For non-EU countries, the season 2005/2006 farm as well as regional data are used from the US Department of Agriculture (USDA), the Ministry of Agriculture for New Zealand, and regional authorities. The representative farm approach is also used for the US and New Zealand dairy where farms from different regions are distinguished for illustrative purposes.

Since the observed costs include the already implicit costs due to current level of compliance, for calculating the percentage of additional costs, the total costs are thereby corrected for the level of compliance. At the sector level both affected and non-affected farms count through the share of specialised farms in total milk production. The assessment of costs at the sector level in the US and New Zealand is also done using representative farm approach, distinguishing farms in different regions for illustrative purposes.

The analysis of competitiveness at the farm level by means of profitability indicator is done using the 2006 data from the International Farm Comparison Network (IFCN, see Hemme, 2007). The advantage of these data is that they have been harmonized across EU as well as non-EU countries and distinguish family labour costs (see Table 7.2), thereby ensuring comparability and allowing for focusing on the net results (gross margin and profitability).

Standards, farm and sector cost implications

For our detailed cost analysis, we selected the expected three most influential standards relevant for the EU: Nitrates Directive, Identification and Registration of farm animals and Food safety and its competitors. The US and New Zealand generally face less regulation. Although the policy-approaches in these countries currently rely more on voluntary actions, more stringent forms of regulation are being implemented or might be introduced in the future. Therefore scenarios simulating the effect of the Nitrates Directive will consider changes in regulations for key competitor countries as well. For the US case, we analyse the requirement to develop and follow a Comprehensive Nutrient Management Plan (CNMP) in order to comply with the Environmental Protection Agency's Clean Water Act (CWA) regulatory requirements. For the case of New Zealand, the impact of the water quality regulations is assessed.

Nitrates Directive

The Nitrates Directive (91/676/EEC) is perceived by EU dairy farmers as one of the most restrictive standards (Jongeneel and Brand, 2010, p. 202). The Directive was adopted in 1991 aiming at the protection of waters against pollution by nitrate from agricultural sources. If land is located within a Nitrate Vulnerable Zone (NVZ), the farmer must comply with NVZ Action Programme Measures which limit the organic manure loading per hectare at 170 kg of nitrate per ha (with exception for grassland in Germany and The Netherlands allowing for 250 kg/ha). Such limits on the disposal of manure by animals while grazing require sufficient slurry storage facilities (or alternative arrangements) to cater for the closed period.

In the US, the approach toward managing the environmental impact from farming has been largely voluntary or with compliance being a condition for cost-sharing assistance with best management practices. The CWA, which was originally focused on point sources of pollution, has been expanded to non-point pollution, with agriculture identified as one of the key sectors. The farms that feed animals primarily in confinement are designated by Clean Water Act rules as Concentrated Animal Feeding Operations (CAFOs) and are the focus of water quality regulations at the federal and state levels. However, it was not until 2002 that the federal government issued specific rules governing CAFOs, requiring the design and implementation of a CNMP. CAFOs are treated as a regulation 'similar' to the EU Nitrates

Directive. Since only 2% of all confined livestock operations are CAFOs, this policy currently only affects a small proportion (0.04%) of US dairy farms.

In New Zealand, effluent from the farm is to be disposed on the land (after ponding) and fertilizer applications should follow a Code of Good Practice. The latter states manure discharge to be a controlled activity, where the rate of application may not exceed 150 kg N per ha annually (and no more than 50 kg per ha within a period of 24 hours), buffer zones should be respected and runoff and ponding of effluent should be avoided. Because the issue of nutrients derived from intensive farming has become a big concern, a private-public partnership tries to come up with self-regulation (the partnership includes the nationwide dairy cooperative Fonterra as well as regional councils).

Table 7.1 reflects the calculation of costs at the farm and sector level for selected countries for the situation when all affected farms improve their degree of compliance in such a way that they all achieve full compliance. Dairy farming in the US does not rely on pasture system; therefore the acreage per farm is not mentioned (row c in Table 7.1). A slightly different procedure was followed for the US and New Zealand cases which are presented below therefore leaving some places in Table 1 non-filled for these countries.

For selected EU Member States in Table 7.1, the number of affected and non-affected farms is assessed through farming intensity and location of farms in NVZs of Europe. Farms with lower densities than 2 LU/ha ('extensive farms') are assumed to face no significant cost increases. Having studied the available literature (Andrews *et al.*, 2000; Kuik, 2006), we rely on per hectare cost estimates from Kuik (2006), adjusted for inflation (row d). Total costs (FADN, 2003) per an average size dairy farm are corrected for record keeping costs of manure applications (in row e) since Kuik's estimates did not include these costs (approximated at €150 per dairy farm in a NVZ-area in all selected countries). For calculating the percentage of cost increase at the farm level (row j), the total costs (in row g) are corrected for the initial level of compliance (row a). At the sector level both affected and non-affected farms count through the share of specialized farms in total milk production (row i).

For the full-compliance situation, the percentage increase in total sector costs varies between 0.10% and 4.07% (row k). The absolute cost increases for this group ranges from €400 to €8,800 per farm (row e minus row f). In reality this effect might be somewhat lower, since (temporary) derogation provisions are not taken into account. Moreover, it should be noted that many of the less intensive specialized dairy farms, as well as the non-specialized dairy farms are assumed to have zero compliance costs. For Europe, the highest percentage cost increase was estimated for Italy based on a case study carried out for dairy farms of different size and structure exercising various manure management systems as presented in De Roest *et al.* (2007).

Formulae mentioned in Table 7.1 are also not applicable to the last two columns. For these non-EU countries separate calculations of percentage cost increase were carried out. Further details on cost assessment are summarized in Jongeneel *et al.* (2008). For New Zealand, the farm monitoring data on 21 dairy regions (MAF, 2005) are used to assess the costs associated with similar standards in this country. Compliance costs relating to water quality regulations are calculated at the farm level for the land-based disposal of dairy shed effluent. The percentage of farms that still had to comply for the period 2005/2006 is assessed to be 15%. Regional differences in average herd sizes and in consent and monitoring costs are incorporated into the analysis. Compliance costs at the farm level based on the data from 2003 are approximated at 10,447 NZD (or €5,886). Farm costs are then aggregated to obtain a total cost for the New Zealand dairy industry to comply with water quality regulations. From this total cost the compliance costs per kilogram of Milk Solids produced are estimated. The national dairy farm budget is used to calculate the percentage cost increase at the farm level which amounts to 3.2% (Table 7.1, row j). For the industry as a whole, the percentage cost increase is 0.48%, given that 85% of farms are already compliant.

Table 7.1 Calculated production cost increases in selected countries due to compliance with environmental standards (Nitrates Directive for the EU countries)

		France	**Germany**	**Italy**[4]	**Nether-lands**	**United Kingdom**	**United States**[4]	**New Zealand**[4]
a	Initial level of compliance (%)	40	60	20	75	95	44	85
b	Targeted level of compliance (%)	100	100	100	100	100	100	100
c	Average specialized dairy farm size (ha)	124	39	28	43	75		123
d	Nitrates Directive costs per ha (€)	50	50	50	105	110		
e	Nitrates Directive costs per farm, including record keeping costs (€1000) (=c*d+€150)	6.3	2.1	11.0	4.7	8.4	8.7[1]	5.9[2]
f	Nitrates Directive costs per farm already implicit in total costs (€1000) (=a*e/100)	2.5	1.3	2.2	3.5	8.0		
g	Total costs (per farm[3]) as in FADN database 2003 (€1000)	121.2	112.6	184.1	177.8	219.5		
h	'Corrected' total costs per farm (excluding implicit Nitrate costs, €1000) (=g-f)	118.7	111.3	183.8	174.3	211.5		
i	Percentage of specialized affected dairy farms in the sector (%)	5.0	23.0	85.0	79.0	52.0	2.0	15.0
j	Percentage increase farm costs (%) (=(b-a)*e/h*100)	3.18	0.75	4.78	0.67	0.20	1.30-4.20	3.20
k	Percentage increase sector cost (%) (=i*j/100)	0.17	0.17	4.07	0.53	0.10	0.02	0.48

Notes: [1] *The value of 9,788 USD in 2003 is converted at the rate 1 USD = 0.89 Euro (www.oanda.com);* [2] *The value of 10,447 NZD in 2005/6 is converted at the rate 1 NZD = 0.56 Euro (www.oanda.com);* [3] *Here it is assumed that for specialized dairy farms all production costs can be related to dairy production.* [4] *Formulae is not applicable to Italy, US and New Zealand. See text below for more explanations.*

Source: own calculations based on data from EU-FADN, DG AGRI G3 and De Roest et al. (2007)

Two typical dairy farms from Wisconsin and California were chosen for the cost analysis in the US not only because they are the top two dairy states, but also because they represent two distinct milk production systems (see also Chapter 3). Wisconsin dairy farming is represented by a typical traditional smaller dairy farm with 92 cows, whereas Californian dairy farming is represented by a modern farm with 980 cows (based on the agricultural season 2006). To calculate costs per farm, costs of nutrient management, keeping costs, off-farm transportation costs, land treatment and manure costs are included, amounting to an average of €7,308 per farm or €37 per animal on annual basis. The cost of compliance with the CNMP requirements on the typical Wisconsin dairy farm amounts to €1.37 per 100 kg of milk produced (€0.31 per 100 kg in California). For an average US dairy farm this translates into an increase of 2.0% (Table 7.1, row j). Given the low percentage of farms which are subject to CAFOs and the fact that about 44% of farms are already compliant, the percentage cost increase at the national level averages to 0.02%.

The above example for two US farms with different size, cost level and structure provides a range of farm level cost increases from 1.3% to 4.2% and clearly illustrates that compliance can have different effects on individual farm performance and, hence, internal competitiveness. Table 7.2 illustrates competitiveness analysis with respect to the Nitrates Directive for the EU and comparable standards in non-EU selected farm types. The percentage rates of cost increase as presented in Table 7.1 are applied to the data from Hemme *et al.* (2007) on typical dairy farms in selected countries. The cost data from these farms have been harmonized and recalculate per energy corrected milk (ECM) which is adjusted to 4% fat, and 3.3% protein. These typical farms are of quite different size but with rather high yields (6,500- 9,000 kg per cow per year) with one exception. Grazing dairy systems in New Zealand produced 4,361 kg milk per cow per year since they are mainly producing from grass, grass silage or hay with no concentrates or only a small amount.

Among all 38 countries, reported in Hemme *et al.* (2007), all studied EU countries group under the same high costs category per 100 kg of ECM (> €30). Costs are significantly higher due to decoupling of direct payments which has altered the non-milk returns of the dairy enterprise. Variable costs include costs for purchased feed, fertilizer, seeds, fuel, maintenance, land rents, interest on liabilities, wages paid, veterinary services and medicine, water, insurance, etc. Variable costs of milk production are quite comparable over countries with the exception of New Zealand. Here the costs are quite low, which compensates for rather unfavourably low milk price in this country. The difference in total costs per 100 kg ECM between countries is mainly due to fixed costs, which include depreciation on purchase prices for buildings and machinery and costs for using own production factors (land owned, family labour input, and equity including quota). Labour costs per 100 kg ECM are rather different reflecting differences in scale of farming and is also an important factor determining long run competitiveness.

In Table 7.2, farm financial results are presented for the situation when full compliance with the Nitrates Directive (or similar regulations outside the EU) is assumed. As Table 7.2 shows, relatively small (0.2-4.8%) changes in costs can significantly reduce gross margin (0.4-8.4%) and even more so profitability (4.1-91.7%).

Identification and registration of farm animals

The EU Directive on Identification and Registration (I&R) of animals (92/102/EEC, and Regulations 911/2004, 1760/2000, and 21/2004) is one of the most important burdens according to the farmers (DG-AGRI, 2007). Most time consuming is the check of animal identification especially in extensive farms or in cattle breeding, as animals are often outside and in different fields and sometimes difficult to approach, compared to dairy cows kept indoors. I&R of farm animals is characterized by a significant degree of non-compliance, with 30% non-compliance being no exception. Besides, the inclusion of animal I&R results is very demanding for controlling agencies (about 36 hours per farm in England or 40 hours in the Netherlands).

Table 7.2 Farm structure and level of competitiveness in selected countries in 2006

Codes as in Hemme *et al.* (2007)	**France FR-38**	**Germany DE-85BW**	**Italy IT-133**	**Nether-lands NL-96**	**United Kingdom UK-157NW**	**United States US-80WI**	**New Zealand NZ-294**
Number of dairy cows, LU	38	85	133	96	195	80	294
Milk production per cow, kg ECM	6541	8216	8259	9089	7366	8703	4361
Costs per 100 kg FCM, €							
costs for means of production	25.2	22.4	25.1	20.7	21.5	19.6	10.3
in that costs of feed	7.1	7.2	10.6	7.7	9.1	4.5	4.5
land costs	1.6	2.6	2.7	2.8	2.0	1.9	4.0
labour costs	16.4	12.2	9.3	10.8	6.7	7.8	3.5
capital costs	3.2	2.3	1.2	2.6	1.5	1.3	3.6
total costs	46.8	40.9	40.3	42.6	31.9	30.6	21.4
variable costs	21.5	23.0	25.7	22.1	22.4	22.8	14.4
fixed costs	25.3	17.9	14.6	20.5	9.5	7.8	7.0
Revenue per 100 kg ECM, €							
milk revenue	28.5	28.5	31.9	27.4	27.2	24.3	18.1
cattle revenue	5.3	5.6	2.0	3.9	2.2	5.7	1.8
direct payments	8.6	5.6	5.0	4.0	3.4	1.1	0.0
other revenue	0.0	0.3	1.5	0.5	0.0	0.1	0.0
revenue	42.4	40.0	40.3	35.8	32.7	31.3	19.9
Results per 100 kg ECM, €							
gross margin (revenue – var.costs)	20.9	16.9	14.6	13.7	10.3	8.5	5.5
profitability (gross margin – fix. costs)	-4.4	-1.0	0.1	-6.7	0.9	0.7	-1.5
Results per 100 kg ECM under full compliance to Nitrates Directive, €							
% additional costs associated with full compliance to Nitrate Directive	3.2	0.8	4.8	0.7	0.2	2.0	3.2
gross margin at full compliance	20.2	16.8	13.4	13.6	10.3	8.0	5.0
profitability at full compliance	-5.9	-1.3	-1.9	-7.0	0.8	0.1	-2.2
% change in gross margin due to full compliance	-3.3	-1.0	-8.4	-1.1	-0.4	-5.4	-8.4
% change in profitability due to full compliance	-25.1	-24.2	-97.1	-4.1	-7.4	-91.7	-31.7

Source: based on data from Hemme et al. (2007) and Table 7.1

The results of the cost estimates at farm level for five selected Member States are presented in Table 7.3. It is assumed that about 0.8 cow units of other animals (heifers/bulls) per dairy cow is present at the farm and that 95% of the dairy cows give birth to a calf, which has to be registered (row c). In addition it is assumed that there is an eartag loss rate of 15%, which requires a proper and timely eartag replacement (row e). The costs due to requirements of the Identification and Registration standards consist of the eartag costs and the labour (per animal) required for registration (row h).

The procedure of approximating the additional costs is similar to that for the Nitrates Directive. The percentage cost increase at the farm and sector level is the same since all farms in the sector are affected. In selected Member States the estimated additional costs are below 0.14%, which is relatively marginal.

Table 7.3 Calculated cost increase in selected EU Member States due to compliance with the Identification and Registration standards

	Unit	France	Germany	Italy	Nether-lands	United Kingdom
a	Initial level of compliance (%)	90	65	70	90	70
b	Targeted level of compliance (%)	100	100	100	100	100
c	Average number of animals per specialised dairy farm	76.5	77.7	81.8	114.2	158.6
d	Estimated number of calves born (=a*0.95/1.8)	40.4	41.0	43.2	60.3	83.7
e	Estimated number of lost eartags(=15%*c)	11.5	11.7	12.3	17.1	23.8
f	Labour costs per animal (€)	1.75	1.75	12.00	1.75	[2]
g	Costs tags per animal (€)	1.80	2.92	3.00	2.75	4.20
h	Total I&R costs per farm (€) (=(d+e)*(f+g))	193.0[1]	246.0	831.8	348.3	451.4
i	I&R costs per farm already implicit in total costs (€) (=a*h/100)	173.7	159.9	582.3	313.5	316.0
j	Total costs (per farm) as in FADN database 2003, €	121197	112586	184097	177805	219474
k	'Corrected' total costs per farm (excluding implicit I&R costs, €) (=g-h*f/100%)	121023	112426	183515	177492	219158
l	Percentage increase farm cost (%) (=(b-a)*h/k)	0.02	0.08	0.14	0.02	0.06
m	Percentage increase sector cost (%) (=l)	0.02	0.08	0.14	0.02	0.06

Notes: [1]Fixed costs per farm in France at the level of €9 are added. [2] For the UK no specific labour costs were distinguished. They are included in the costs per animal.

Source: own calculations based on data from EU-FADN, DG AGRI G3

Food safety (hormone use)

Consumer concerns as regards food safety have led to hormone use prohibition in the EU. In the US, recombinant bovine somatotropin (rBST), a growth hormone that stimulates milk production, has been approved for use in dairy cows since 1994. Alongside the US, BST (or rBST) is used by at least 16 other dairy producing countries (Jarvis, 2002). The EU, Canada and Japan rejected legal BST use, and also within the US the technique was (and still is) controversial, at least within certain groups and regions. See Jarvis (2002, p. 103) for further details.

Monsanto – the monopolistic supplier of the product (brand name: POSILAC) – reports that in 1999 about 13,000 US dairy farms were using BST, applying it to 9 million dairy cows (approximately 30% of the cows in supplemented herds). The use of this hormone leads to production increases. Tauer (2002) shows that application of rBST in the first year may increase an average herd's milk yield by 419-575 kg, whereas late adopters are able to directly realize an annual milk yield increase of about 480 kg per animal. Jarvis (2002), who analyzed the potential effects of BST on world dairy markets, estimates that the total US milk supply increased by 3% due to the (pure) BST application[1].

Whereas BST will increase yields and thus revenues, it simultaneously increases input use (feed) and thus costs (including the technology fee farmers have to pay for BST use). In an *ex ante* analysis, Perrin (1991) estimated per unit cost savings varying from 0.5% to 4.4%, although milk yields were increasing by approximately 15% (note the simultaneous cost increase effect). Jarvis (2002, p. 109), using a slightly different approach, provides a maximum per unit cost decline of 5% (per unit average). When simulating the trade impact of hormone use ban in the US (see next section), we use the latter estimate and interprete the cost advantage as foregone which translates into a 5% cost increase for US farmers.

Standards and external competitiveness

External competitiveness is analysed along six scenarios. The first three scenarios in Table 7.4 simulate compliance to the Nitrates Directive in EU Member States at various levels of compliance. Additionally to the full-compliance calculations presented in Table 7.1, the cost increases were assessed for scenarios 2 and 3 as follows: 'compliance improved by by 20%' when all affected farms improve their degree of compliance by 20% (to a maximum of 100%) as compared to the base year level and 'minimum of 75% compliance' when all affected farms improve their degree of compliance by 10% to a minimum of 75% as compared to the base year level. In all three scenarios it is assumed the level of compliance in non-EU trading partners remained at the initial level. Scenario 4 simulates full compliance to the Nitrates Directive within the EU and similar regulations in the US and New Zealand.

For the case of full compliance to Nitrates Directive in EU countries (scenario 1), the European dairy sector loses 0.71% of its export mainly to the rest of OECD and the US. Moreover, it increases its imports by 0.80%. As was already noted in the introduction, EU imports are playing a very limited role, and the main impact will thus be on exports. Since the GTAP model does not distinguish disaggregated dairy product markets, it is impossible to indicate how various product markets (e.g. butter, skimmed milk powder, whole milk powder, hard and soft cheeses, casein, etc.) will be affected. For example, the EU is known to export various speciality cheeses to the US market. The exports of such high value-added products are likely to be less affected than the 'average' dairy export product simulated within GTAP. The predicted export reduction of 0.70% of the EU to the US is therefore likely to be an upper bound. The extra imports may be an overestimation due to the way tariff rate quotas (TRQ) are modelled in GTAP. All TRQs in the dairy sector are modelled as bilateral TRQs and an arrhythmic mean tariff is calculated from the in- and out-quota tariff. However, for the dairy sector this representation is acceptable because of the fixed nature of the global TRQ in the sector (based on historical trade volumes). The other countries will increase their export in order to fill the gap the EU leaves. However, the total traded volume decreases by 0.05%, which is quite small.

---

[1] For comparative reasons note that the autonomous (genetic) milk yield increase is about 2% per annum. Whereas bST application creates a one-shot increase, genetic progress creates a continuous milk yield increase.

Table 7.4 Percentage changes in trade due to compliance to various standards

	Scenario	EU import: total	EU export: total	Exports of which to...			Total trade
				USA	Rest of OECD	Rest of World	
1	Nitrate EU: 100%	0.80	-0.71	-0.70	-0.77	-0.70	-0.05
2	Nitrate EU: +20%	0.51	-0.42	-0.45	-0.46	-0.40	-0.02
3	Nitrate EU: minimum 75%	0.40	-0.33	-0.37	-0.37	-0.31	-0.02
4	Nitrate EU and non-EU: 100%	0.30	-0.41	-0.63	-0.44	-0.38	-1.07
5	I&R EU : 100%	0.21	-0.15	-0.18	-0.18	-0.14	-0.01
6	I&R EU : 100% and Nitrate EU : 100%	1.01	-0.87	-0.89	-0.95	-0.85	-0.06
7	Ban on hormone use in US: 100%	-0.25	2.44	27.57	2.52	-0.80	0.24

Notes: Regional impacts are presented for the situation of the EU being a net exporter.

Source: GTAP calculations

Because of the lower costs for European farmers in Scenarios 2 and 3 compared to the first scenario (see also Table 7.1) there is a smaller effect on the competitiveness of the European dairy sector. The decrease in export volume is 0.42% and 0.33% respectively. Because of the smaller price effect on the world market the total traded volume only reduces by 0.02%. The rest of the OECD and the rest of the world fill most of the reduced European export. In scenario 4, when next to the EU countries also their competitors fully comply with nitrate measures, the results do not change substantially when compared to scenario 1; total trade declines at a higher rate (1.07%) whereas the total export of the EU still decreases (by 0.41%). This is not surprising given quite low percentage cost increase at sector level in the non-EU countries.

Scenarios 5 and 6 report the simulation results of compliance of EU countries to the Identification & Registration standards. There are no costs for non-EU countries and slightly increased costs for EU countries resulting in the smallest trade decline (0.01%) among all scenarios. A combined effect of full compliance of EU countries to the Nitrates Directive and I&R leads to an EU export loss of 0.87%. Scenario 7 finally simulates the impact of the hormone ban in the US thereby affecting mainly US trade and benefiting the EU dairy sector with an export volume increase of 2.44%.

Conclusions

In this chapter the impact of compliance to standards on production costs in the dairy sector is estimated using a farm level analysis and taking into account actual farm accountancy data. Representative farm studies have been conducted and used as a basis for the cost calculations. Best-estimates of compliance are used, but these still contain a certain degree of uncertainty. In a number of cases alternative approaches and alternative sources were used to cross-check both production costs and degree of compliance in order to test for the robustness of our results. As regards the impact of the Nitrates Directive and Identification and Registration standards on production, clearly the nitrates standard causes the most significant impact. At farm level, the percentage cost increase varies from 0.19% to 6.8% within the selected Member States, from 2% to 4.2% for US farms and is about 2% in New Zealand. Profitability and gross-margin of farms with a lower degree of compliance (France, for example) are substantially affected. At sector level, full compliance to the nitrates standard for all affected farmers would involve an increase in production costs between 0.1% and 0.6%, with rates varying between Member States. The estimated percentage cost increase associated with full compliance with the Identification and Registration standard by all affected farmers would be less than 0.15% and, hence, rather marginal. As shown in Table 7.4 the negative impact of these measures on EU imports and exports are less than 1%. If a smaller increase in compliance takes place, these trade impacts will be even smaller. When the standards for nitrate pollution taken by the US and New Zealand are taken into account along with a full compliance assumption in all countries analysed, this would impact the trade balance by 1.07% (the highest rate of all the scenarios) and will result in a lower decline of EU exports (about 0.41%). This suggests that although approaches and measures are formally different there seems to be a non-neglectible degree of functional equivalence between the measures addressing a similar class of problems in different jurisdictions. The EU's international competitiveness does not increase substantially in this situation and therefore improvements in internal farm competitiveness (higher gross margin and profitability) should remain a high priority in farm management. The trade impacts obtained when no changes are assumed to happen in key competitor countries can thus be argued as providing the upper bound of the likely trade impacts. Moreover, as became clear from analysing the dairy farm costs and revenue structure over countries, cost differences induced by environmental and animal registration regulations are only one factor explaining the dairy farm sector's competitiveness. Differences in labour costs (fixed costs) per 100 kg of milk between the considered countries are generally speaking much more important. These latter differences reflect differences in labour productivity, which seem to be related to the farm size structure (economies of scale).

References

Andrews, K., F. M. Anderson, A. Bartonova, H. Horth, N. Kilde, H. V. Larsen and T. Zabel (2000). Study on investment and employment related to EU policy on air, water and waste. Final Report Annex I; Best estimates of costs. WRc Medmenham, Report No. EC 4739/A.1.

Blayney, D., M. Gehlhar, C.H. Bolling, K. Jones, S. Langley, M.A. Normile, and A. Somwaru (2006), 'US Dairy at a Global Crossroads', Economic Research Report, 52p.

Bouamra-Mechemache, Z., R. Jongeneel and V. Réquillart (2009), 'EU Dairy Policy Reforms: Luxembourg Reform, WTO Negotiations and the Quota Regime', *EuroChoices*, 8(1): 13-22.

Cassells, S.M. and A.D. Meister (2001), 'Cost and trade impacts of environmental regulations: effluent control and the New Zealand dairy sector', *The Australian Journal of Agricultural and Resource Economics*, 45 (2): 257-74.

DairyCo (2010), The Dairy Farming Information Centre. www.dairyco.org.uk/datum/dairy-processing--trade.aspx (accessed March 6, 2010).

De Roest, K., C. Montanari, and E. Corradini (2007), 'Il costo della Direttiva Nitrati', BIT SpA - Cassa Padana, Leno (Brescia). CRPA.

DG-AGRI (2007), 'Study to assess the administrative burden on farms arising from the CAP', Final report 2006-G4-03, Directorate General for Agriculture and Rural Development.

European Commission (2006), 'Milk and milk products in the EU - August 2006', in DG for Agriculture and Rural Development. Luxembourg: European Comission, p. 27.

Hemme, T. (Ed.) (2007), 'IFCN Dairy Report 2007, International Farm Comparison Network', IFCN Dairy Research Center, Kiel, Germany, p. 214.

Jarvis, L.S. (2002), 'The potential effect of recombinant bovine somatotropin on world dairying', In V. Santaniello, R.E. Evenson and D. Zilberman (eds), Market development for genetically modified foods, Wallingford: CABI Publishing, p. 101-11.

Jongeneel, R.A, I. Bezlepkina, F.M. Brouwer, L.H. Aramyan, K. Dillen, M. Farmer, V. Swales, S. Schlegel, R. Landgrebe, K. de Roest, C. V. Ortéga, P. Esteve, X. Poux, G. Fox, P. Rajsic, A. Meister, Z. Karaczun, J. Winsten (2008). CROSS-COMPLIANCE Facilitating the CAP reform: Compliance and competitiveness of European agriculture Specific Targeted Research or Innovation Project (STREP) Integrating and Strengthening the European Research Area: Deliverable 13 : Product-based assessments to link compliance to standards at farm level to competitiveness. http://edepot.wur.nl/3579.

Jongeneel, R. and H. Brand (2010), Direct income support and cross compliance. In: Oskam, A.J., Meeester, G. and Silvis, H. (eds.) EU policy for agriculture, food and rural areas. Wageningen, Wageningen Academic Press, pp.191-205.

Kuik, O. (2006), 'Ex ante and ex post costs of implementing the Nitrates Directive', Study for DG Environment, Contract No. ENV.G.1/FRA/2004/0081, Free University, Institute for Environmental Studies, Amsterdam.

Le Roux, J., E. Williams, A. Staines and A. Bergmann (2008), 'Environmental Quality and the Cost of Environmental Regulation: a Comparison of Scotland with the International Community', *European Environment*, 18: 45-54.

MAF (2005), '2005 Dairy Monitoring Report Ministry of Agriculture and Forestry', (http://www.maf.govt.nz/statistics/primaryindustries/farm-monitoring/dairy.htm accessed: 15 December 2007).

Miller, J.J. and D.P. Blayney (2006), 'Dairy Backgrounder', Outlook report: LDP-M-145-01, Economic Research Service/USDA, p. 25.

Perrin, R.K. (1991), Ex-Ante Evaluation of Experimental Technology. Unpublished Manuscript, Department of Agricultural and Resource Economics, North Carolina State University, Raleigh.

Tauer, L.W. (2002), 'The impact of bovine somatotropin on farm profits', in V. Santaniellon, R. E. Evenson and D. Zilberman (Eds.), Market development for genetically modified foods. Wallingford: CABI Publishing, p. 81-90.

USDA (2010), Production, Supply and Distribution Online. www.fas.usda.gov/psdonline/psdQuery.aspx (accessed March 5, 2010).

Xu, X. (2000), 'International trade and environmental regulation: time series evidence and cross-sectional test', *Environmental and Resource Economics*, 17, 233-57.

Chapter 8

An EU/US perspective on nutrient management policies and growth hormone bans in the beef sector

Kees de Roest, Jonathan Winsten and Koen Dillen

Introduction

The beef sector in the EU and in the US is subject to several environmental and food safety regulations, which may generate extra costs for beef farmers. In other terms, the negative externalities related to water pollution and to human health uncertainty become internalized into the beef farm cost structure. The enforcement of environmental and food safety policies may have implications for the competitiveness of the EU and US beef sectors in the world market, in particular when decisions are taken unilaterally, but also when significant policy differences exist between the EU and the US.

This chapter presents an analysis of the compliance costs for beef farmers in the EU and the US for specific environmental and food safety regulations and the impact this cost increase may have on the competitiveness of beef production on the world market. The compliance costs will be assessed using typical beef farms in the EU and in the US, whereas the impact on competitiveness will be calculated using the GTAP model as has been described in Chapter 1. The simulation will be presented first under the hypothesis that the EU is implementing unilaterally its own legislation followed by an assessment of the EU-US competitiveness in beef when both jurisdictions enforce their own environmental and food safety regulations. Other commercial blocs such as South America and Australia enter into the GTAP simulations but no calculations have been carried out about their eventual compliance costs with local environmental legislation, as this was beyond the scope of this research.

A myriad of environmental regulations have been issued in the EU and the US. It is beyond the scope of this chapter to undertake a comprehensive and exhaustive examination of the full set of such regulations. The analysis will focus on what are arguably the most topical examples of environmental and food safety legislation: the Nitrates Directive and the Identification and Registration of bovines in the EU and the Clean Water Act in the US. A comparison will be made between the regulatory tools of the Clean Water Act of the US, which aims to improve and protect water quality, and the mandatory standards established by the Nitrates Directive in the EU. An effort will be made to compare the impact on competitiveness of the EU and US beef sectors in case of full compliance with the respective environmental regulations.

Additionally attention is paid to the possible cost and trade consequences of a prohibition on the use of growth hormones in the US. As far as this last issue is concerned, we will assess the cost advantage of US beef producers with the use of growth hormones. Of particular interest will be the examination of the effect that a prohibition on the use of growth hormones in the US might have on the international trade of beef.

An important consideration is the fact that the EU has started to become a net importer of beef in 2003, after several decades of having been a significant beef exporter due to the market regime for beef which allocated abundant export refunds. The gradual break down of this policy has opened the EU market to imports, particularly from South America. The EU beef market is still protected by import tariffs. But it is possible to foresee a further deterioration of the competitive position of the EU beef industry in the future, as low cost producers of South America (Brazil, Argentina and Uruguay) may gain a larger access to the EU market.

Cost implications of the Nitrates Directive

The main requirement under the Nitrates Directive is the maximum limit of 170 kg N/ha manure (or 250 kg N/ha of manure on grassland in Germany and the Netherlands) application rate, which may generate extra costs for a manure disposal. In Nitrate Vulnerable Zones (NVZ), farms exceeding the maximum limit have to find extra land to spread excess manure, either by buying or renting land or paying a licence to spread manure on land of nearby farmers. Moreover, they are obliged to invest in manure storage facilities due to the prohibition on spreading manure in winter time.

Affected beef production

Out of the 491,000 farms engaged in beef production in the EU, about half of them are cow-calf farms, over one quarter (27%) are specialized beef finishing farms and 20% have a coupled production of dairy and beef (Sarzeaud *et al.*, 2007).

Striking to note is that 44% of beef production originates on farms where dairy is the main livestock activity, another 14% is produced on fattening farms and 35% on cow-calf farms. Extensive production systems tend to predominate in the EU as on 47% of the cattle farms the stocking rate does not exceed 1.4 livestock units (LU) per ha. Typically, low stocking rates characterize cow-calf farms, whereas the pure fattening farms often exceed a stocking rate of 1.8 LU per ha.

From Table 8.1, it is clear that many types of beef production systems will not be affected by the Nitrates Directive, since they have a stocking rate of less than 1.8 LU per ha. In terms of farm numbers this amounts to 72% of the beef farmers in the EU. For the remaining 28%, some of the beef finishing farms and a small minority of cow-calf farms will eventually face costs of compliance.

In order to assess the share of farms affected, the limit of 1.8 LU per ha has been compared to the maximum stocking rate allowed within the Nitrate Vulnerable Zones (NVZs). For beef finishing farms, the 1.8 LU per ha would constrain the stocking rate in NVZ by 13% and in cow-calf farms by 38% (Table 8.2).

Table 8.1 Breakdown of cattle farms in beef farming systems in the EU

Production	**Cow-calf**				**Fattening**		**Dairy**		**Small farms**	**Total**
Beef farming systems	**Cow-calf and sheep**	**Cow-calf and fattening**	**Pure cow calf**	**Fattening and sheep**	**Pure fattening**	**Small fattening**	**Dairy and beef**	**Pure dairy**		
Number of farms	49,889	65,100	171,388	9,202	12,656	58,763	123,788	338,725	91,404	920,916
Farm size (ha)	151.9	73.3	69.7	87.9	94.0	45.9	81.5	56.9	19.8	63
Farm size (livestock unit per holding, LU)	52.5	61.4	47.4	39.5	127.2	31.6	93.6	66.7	3.8	58.1
Share of total number of beef farming systems (%)	5	7	19	1	1	6	14	37	10	100
Share of total beef production value (%)	6	11	24	1	6	7	20	24	1	100
Number of farms by intensity (LU/ha)										
<1.4	32,428	32,550	123,399	2,669	1,898	32,907	38,374	15,167	54,842	432,831
1.4 – 1.8	11,474	17,577	25,708	4,417	3,544	11,165	39,612	98,230	11,883	230,229
>1.8	5,987	14,973	22,280	2,116	7,214	14,691	42,088	125,328	24,679	257,856
Share of number of farms by intensity in LU/ha (%)										
<1.4	65	50	72	29	15	56	31	34	60	47
1.4 – 1.8	23	27	15	48	28	19	32	29	13	25
>1.8	12	23	13	23	57	25	34	37	27	28

Source: Sarzeaud et al. (2007)

According to Table 8.1, there are 22,280 cow-calf farms and 7,214 pure fattening farms which exceed the limit of 1.8 LU per ha. If we presume a proportional distribution of these intensive beef farms above and below the maximum allowed stocking rate we can conclude that 87% of the *intensive* beef finishing farms are exceeding the limit imposed by the Nitrates Directive in NVZs and 62% of the *intensive* cow-calf farms are in this condition for a total of 40,580 farms.

Table 8.2 Comparison of 1.8 LU per hectare with maximum stocking rate allowed in Nitrate Vulnerable Zones

Farm type	Number of animals per ha equal to 1.8 LU	N in manure (kg per animal per year)	Maximum stocking rate (head per ha) in NVZ	Share of farms exceeding the 1.8 LU limit (%)
Beef finishing farms, with beef bullocks (1 – 2 years)	2.6	58	2.93	13%
Cow-calf farms, with suckler cows	1.8	68.5	2.48	38%

Source: Elaborated by CRPA on ERM/AB-DLO (1999)

Table 8.3 Estimate of suckler cows and fattening calves in Nitrate Vulnerable Zones in the EU-15

	Suckler cows		Male fattening calves	
	(1,000 animals)	(share in NVZ, %)	(1,000 animals)	(share in NVZ, %)
France	4,148	23.5	1,082	53.5
Spain	1,563	25.8	134	16.9
United Kingdom	1,765	20.3	1,120	24.9
Ireland	1,168	100.0	813	100.0
Germany	747	100.0	1,242	100.0
Belgium	534	37.6	175	59.2
Italy	609	25.8	690	53.5
Other EU countries	1,273	47.5	762	58.3
EU-15	11,808	39.1	6,017	64.0

Source: Elaborated by CRPA on Eurostat- Regio databank

However this number of farms likely overestimates the number of affected farms, because we have assumed that the whole territory of the EU is declared as a Nitrate Vulnerable Zone, and the application rate of 170 kg N/ha for disposal of excess manure would therefore be applicable throughout the EU. In Table 8.3 we estimate the percentage of suckler cows and of male fattening calves in the NVZs of the EU. The estimate has been obtained by using the REGIO databank of Eurostat by overlaying the map of Nitrate Vulnerable Zones on the NUTS III regions of the EU.

Table 8.4 translates these figures into number of farms. The outcome of this analysis reveals that 24,842 beef farms (6.7% of the total number of cattle farms) in the EU-15 will face a cost increase due to the application of the Nitrates Directive.

Impact of improved compliance with Nitrates Directive on beef farm costs

The calculation of the cost of compliance with the Nitrates Directive was carried out considering a typical beef finishing farm producing young bulls in the North East of Italy (De Roest *et al.*, 2007). A significant share of the beef cattle fattening farms with high stocking rates is located in this part of the EU and can be considered typical for this type of activity. Similar fattening units exist in the North West of Italy and in the North of Spain. Weaner calves are primarily imported from France.

Table 8.4 Share of beef farms in the EU affected by the Nitrates Directive

	Cow-calf farms		Beef fattening farms		EU
	Number of farms	**%**	**Number of farms**	**% affected**	**Number of farms**
Farms >1.8 LU/ha	43,240	100.0	24,021	100.0	67,261
Farms exceeding 170 N kg/ha	29,403	68.0	20,898	87.0	50,301
Exceeding farms in NVZ	11,467	39.1	13,375	64.0	24,842
Total EU farms	286,377		80,621		366,998
Total farms exceeding 170 N kg/ha in NVZ	11,467		13,375		24,842
Share of farms exceeding 170 N kg/ha in NVZ (%)	4.0		16.5		6.7

Source: Elaborated by CRPA on Sarzeaud et al. (2007)

Table 8.5 Sample beef finishing farm characteristics

Technical data elements	**Value**
Number of young bulls present	1,200
Average starting weight (kg/head)	370
Average final weight (kg/head)	640
Finishing period (days/head)	210
Net production (kg)	538,000
Agricultural land area (ha)	151

Source: ISMEA/CPRA, 2006

Table 8.6 Slurry production and land required for spreading

Technical elements	**Value**
Slurry production after storage (m^3/year)	15,756
Nitrogen production (kg N)	50,904
Agricultural land available for spreading of manure (ha)	151
Agricultural land needed for disposal of effluents (ha)	226

Source: ISMEA/CPRA, 2006

The related technical and economic data have been drawn from a sample of beef farms maintained by CRPA (Centro Ricerche Produzioni Animali) for annual monitoring of beef farm production costs in Italy (ISMEA/CRPA, 2006) (Tables 8.5 and 8.6).

The yearly slurry production is equal to 15,756 m^3 corresponding to a nitrogen content of 50,904 kg. Currently the average farm consists of 151 ha of agricultural land available for cattle slurry spreading.

If this farm is located in an ordinary zone, the only cost deriving from the current situation is given by storage and slurry distribution on owned farmland. If the area where the farm is located is declared a NVZ, then it would be necessary to add farmland (75 ha) in order to respect the maximum allowed limit of 170 kg N manure per ha. This entails an extra cost for acquiring the right of spreading on lands owned by other farmers and to transport the excess manure to these farmlands that are supposed to be situated 15 kilometres from the beef farm. Furthermore, the size of basins for storing slurry must be increased up to the minimum capacity required, corresponding to 180 storage days.

Table 8.7 compares costs for slurry management in an ordinary zone and a vulnerable zone under the hypothesis that compliance with the Nitrates Directive is achieved through increasing the land base. The total average slurry management cost is expressed per kg of live weight.

Table 8.7 Slurry storage and distribution costs in ordinary zone and in NVZ (increase of land availability)

Costs		In Ordinary Zone	In Vulnerable Zone
Tank size	m^3	5,698	8,547
Tank cost (31 €/m^3)	€	199,432	299,148
Depreciation and maintenance (5% of tank cost)	€	9,972	14,957
Tank load (0.11 €/m^3 of slurry produced)	€	1,800	1,800
Investment interest (3.5% on tank cost)	€	6.980	10,470
Operating interest (2.9%)	€	97	133
Average storage cost	€/kg	0.035	0.051
Average distance for distribution	m	700	7.800
Cask wagon (capacity of 15 m^3))	€	23,000	23,000
Depreciation and maintenance(12% of cask wagon cost)	€	2,760	2,760
Distribution cost on the farmland (0.55 €/m^3)	€	8,618	4,331
Distribution cost on land far from the farm (7.00 €/m^3)	€	–	54,869
Investment interest (3% on cask wagon cost)	€	805	805
Operating interest (2.9%)	€	143	869
Rights of spreading (130 €/ha)	€	–	9,731
Crop yield loss	€	–	13,718
Average spreading cost	€/kg	0.023	0.162
Total average cost for slurry management	€/kg	0.058	0.213

Source: ISMEA/CPRA, 2006

The option to spread manure on neighbouring land may not be feasible when in the surrounding area the land supply is scarce and livestock farming density is high. In this case, manure treatment systems would be needed. The treatment equipment is designed for reducing nitrogen content in the animal slurry (centrifuge for the solid fraction separation and aerator for the liquid fraction), reducing the need for land outside the farm from 75 to 21 ha. In addition, the sale of compost is an extra revenue source that partially offsets the operating costs of the treatment process.

This latter option does not involve significant differences with respect to the first (acquire spreading rights from other farmers). The average costs entailed by the two solutions can be compared to the total production costs as calculated by CRPA (ISMEA/CRPA, 2006) based on a sample of typical Italian beef farms. Considering that the total cost in 2006 was equal to 2.57 €/kg live weight (including the purchasing cost of weaners), the percentage cost increase entailed by both options can be estimated at 5.8% (Table 8.8).

Table 8.8 Slurry storage, treatment and distribution costs in NVZ (investment in treatment equipment)

Cost category		**In Nitrate Vulnerable Zone**
Tank size	m^3	7,265
Tank cost (31 €/m^3)	€	254,276
Depreciation and maintenance (5% of tank cost)	€	12,714
Tank load (0.11 €/m^3 of slurry produced)	€	1,740
Investment interest (3.5% on tank cost)	€	8,900
Operating interest (2.9%)	€	116
Average storage cost	€/kg	0.044
Centrifuge cost	€	75,000
Storage platform cost (650 m^3 for separated solid fraction)	€	97,490
Airing basin cost	€	31,740
Aerator cost	€	16,454
Shovel tractor cost	€	63,000
Total investment cost for treatment equipment	€	283,684
Depreciation and maintenance (8% of equipment cost)	€	22,695
Operating costs (composting, electricity, etc.)	€	36,239
Investment interest (3% on cask wagon cost)	€	9,929
Operating interest (2.9%)	€	683
Average effluents treatment cost	€/kg	0.129
Average distance for distribution	m	741
Cask wagon (capacity of 15 m^3)	€	23,000
Depreciation and maintenance (12% of cask wagon cost)	€	2,760
Cost for liquid fraction distribution (0.55 €/m^3)	€	7,293
Cost for solid fraction distribution (1.70 €/m^3)	€	4,018
Investment interest (3% on cask wagon cost)	€	805
Operating interest (2.9%)	€	182
Rights of spreading (160 €/ha)	€	3,387
Crop yield loss	€	13,718
Average spreading cost	€/kg	0.060
Total average cost for slurry management	€/kg	0.233
Compost sale	€/kg	0.018
Net average cost for slurry management	€/kg	0.215

Source: ISMEA/CPRA, 2006

Impact of improved compliance at sector level

According to the analysis of the structure and the regional distribution of cow-calf and beef fattening farms, 3.0% of beef produced in cow-calf farms and 4.2% of beef cattle in the finishing farms would be affected by Nitrates Directive and subjected to an increase of production costs. If we assume that 50% of the beef farms are already complying with the Nitrates Directive then 1.5% of beef produced in cow-calf farms and 2.1% of beef produced in beef fattening farms will face increased costs in order to attain a 100% compliance at EU level.

The cost increase for affected fattening farms is 0.155€ per kg beef. In order to estimate the impact on EU beef sector, the production cost analysis of the Agribenchmark, coordinated by the Federal Institute for Agriculture (FAL), has been used. This is a data base representing 10 EU countries and 90% of EU beef production. A weighted average of Agribenchmark beef farms generates an average production cost of beef of €2.67/kg in the EU. The increase for those farms located in NVZ which exceed the limit of 170 kg N per ha would then be equal to 5.8%. In the case of 25% compliance 0.83 % of beef production would be affected and the sector cost increase would be 0.048%. If we assume 50% compliance the sector cost increase equals 0.095% and in the case of 75% compliance the cost increase will be 0.143%. This limited cost increase related to compliance with the Nitrates Directive can be primarily attributed to extensive nature of the majority of beef production in the EU (Table 8.9).

Table 8.9 Cost increase for EU beef sector due to compliance with Nitrates Directive

	Cow-calf	Beef fattening	Total	Cost increase (%)
Number of farms exceeding 170 N kg/ha in NVZ	11,467	13,375	24,842	
Share of beef value (%)[1]	74.5	25.5	100	
Share of farms affected by Nitrates Directive (%)	4.0	16.5	6.7	
Share of affected beef (0% compliance) (%)	3.0	4.2	3.3	
Share of affected beef (25% compliance) (%)	0.75	1.05	0.83	0.048
Share of affected beef (50% compliance) (%)	1.50	2.10	1.7	0.095
Share of affected beef (75% compliance) (%)	2.25	3.15	2.48	0.143

[1] *Excluding beef value produced in dairy farms*

Source: ISMEA/CPRA, 2006

Impact of water pollution measures in the US

In the US the Clean Water Act (CWA) is the primary federal law for the protection of the surface water quality in the US. Among the other mandatory standards imposed upon US farmers, the regulatory tools employed by the CWA can be compared to the requirements established by Nitrates Directive in the EU.

Initially the primary focus of the CWA was on point sources of pollution from industry and wastewater plants, but in recent years it has been expanded to include farming activities. Until December 15, 2002, the Federal Government issued specific rules governing Concentrated Animal Feeding Operations (CAFOs) defined as animal feeding operations with more than 1,000 animal units (700 dairy cows, 1,000 beef cows, 2,500 pig or 100,000 broilers)[1]. Under the current rule, all farms designated as CAFOs are required to obtain a permit under the National Pollution Discharge Elimination System (NPDES) that entails the implementation of a comprehensive nutrient management plan (CNMP). The CNMP must be specific for the operation and detail the proper management of all manure produced. It must address the assimilative capacity of the farmland for the manure and other nutrients applied to the land. The implementation of CNMPs often result in operations seeking

[1] Now the EPA has abandoned the Animal Units to define CAFOs and is expressing it in number of heads per species.

additional land on which to spread manure and/or the use of alternative nutrient control strategies for manure.

Feedlot beef farming has been considered in order to estimate the costs of compliance with US regulations concerning water quality (Winsten, 2007). Cost of compliance is related to the implementation of the CNMP that is required for all livestock farms recognized as CAFO. In the case of beef farms, CAFOs are feeding operations with more than 1,000 beef cows.

The US Department of Agriculture (USDA) has estimated the average costs for developing and implementing a successful CNMP in specialized feedlots farms, as $14 per head, for an average size of 1,300 head per operation. These figures are the basis for calculating the average costs of CNMP implementation in a typical cattle feedlot in Texas, which is the leading beef producing state in the US accounting for 27% of total beef production (see Table 8.10).

Table 8.10 Estimated expenses and CWA compliance costs for Texas feedlot, 2007

Expenses	$ per head	%
750 lb. Feeder steer	814.73	72.0
Total feed, handling, and management charge	257.52	22.8
Interest on feeder and 1/2 feed	44.63	3.9
Death loss (1% of purchase)	8.15	0.7
Total expenses	1,125.03	99.5
CAFO CNMP costs	6.21	0.5
Total costs	1,131.24	100.0

Source: USDA (2003)

For this analysis, the CNMP compliance cost per head was constructed for the average size of Texas feedlots with more than 1,000 head, equal to 22,462 head per farm. Because the average CNMP costs decrease as farm size increases, the USDA estimated costs per head (14$) for all US feedlots would overestimate the total costs for the larger feedlots. To account for the lower CNMP costs associated with larger feedlots, an adjustment factor was calculated. When applied to the $14 per head cost for a typical US fattened cattle operation, the resulting CNMP compliance cost is estimated to be $6.21 per head. Based on the average feeder cattle production estimates on Texas feedlots (50,000 head, assumed 450 pound increase in finishing size), the $6.21 CNMP compliance costs per head result in an estimated 0.55% increase in total feedlot production costs, as compared to a 5.8% cost increase in a large feedlot in Italy which has to comply with the Nitrates Directive. It has not been possible to estimate a sector cost increase, because it has been extremely difficult to estimate in a reliable way the number of feedlots and related beef cattle which will be affected by the Clean Water Act. At the same time we presume the sector cost increase will be very small as already the compliance costs at farm level are negligible.

Growth promoters use prohibition: evaluated standards and cost implications

Council Directive 96/22/EC of 1996 concerns the prohibition on the use in livestock farming of substances having a hormonal or thyrostatic action and of beta-agonists. This section of the chapter investigates the extent of the cost advantage that the US beef industry enjoys using these substances, relative to the EU.

The use of hormones in beef fattening cattle increases average daily gain and improves feed efficiency. There is also evidence of drawbacks which hormones may cause to the carcass and eating quality, but there are growth implant strategies available which can considerably alleviate these concerns (Montgomery *et al.*, 2001). Hence, the cost advantage of the use of growth hormones more than offsets any negative effect on carcass quality.

Impact of the prohibition on the use of hormones

In a meta-analysis combining information from more than 170 research trials carried out in the last forty years, the economic impacts of pharmaceutical technologies in cow-calf and beef fattening farms have been assessed (Lawrence and Ibarburu, 2006). The average effects on critical production parameters obtained from this meta-analysis are reported in Table 8.11.

Table 8.11 Potential effects of a ban on growth promotants on the technical efficiency in US beef sector (%)

	Beef cow herds		Feedlot
Wean rate (%)	-2.54	Average daily gain (%)	-14.13
Wean weight (%)	-3.07	Feed to gain (lbs feed/ lbs weight gain) (%)	-8.79

Source: Lawrence and Ibarburu (2006)

Taking into account the adoption rate of growth promotants, which in the US is high, an overall impact on production costs of a ban on growth promotants has been estimated as 7.14%. In other words, a ban on the use of growth promotants would increase the production costs of beef in the US by this percentage, as the wean rate would decrease by 2.54%, the average daily gain would deteriorate by 14.13% and the feed conversion rate by 8.79% (Lawrence and Ibarburu, 2006).

An increase in production costs of this magnitude would have significant consequences for the international competitiveness of US beef production. The GTAP simulations concerning this issue are presented at the end of this chapter.

Identification and registration of animals: evaluated standards and cost implications

Important food safety considerations lay behind the regulations related to the identification of animals. In order to comply with Regulation 1760/2000, beef farms will have to update registers and eartags continuously. The costs generated by this regulation are related to the time necessary to update the registers and to the purchase of eartags for newborn calves, for imported calves and to replace eartags lost during the lifetime.

Impact of compliance with identification and registration of animal requirements

The costs per head, including costs for the eartags accidentally lost, range from a minimum of €1.80 per head in France up to €5.00 per head in the Netherlands (see also Chapter 7).

Table 8.12 Production costs increase per kg beef related to the identification and registration of beef cattle

	France	Germany	Italy	Netherlands	United Kingdom	Spain	Ireland
Cost of identification and registration (per head)	1.80	2.92	2.20	5.00	4.20	2.20	4.20
Production cost (US$ per kg)	6.17	5.59	4.83	6.40	8.54	5.21	6.80
Production cost (€/kg carcass weight)	4.47	4.05	3.50	4.64	6.19	3.78	4.93
Production cost (€/kg live weight)	2.63	2.27	2.17	2.60	3.40	2.07	2.66
Number of cattle sold per year	75	394	1,825	50	48	2,901	80
Kg live weight produced/head	248	263	177	143	143	188	96
Carcass yield (%)	59	56	62	56	55	55	54
Beef production farm	18,500	103,592	323,362	7,150	6,887	545,373	7,690
Percentage loss (%)	5	5	5	5	5	5	5
Cost of eartags (US $ per farm)	134.1	1149.02	4013.90	250.00	201.60	6381.47	336.00
Cost of I&R per kg beef (US $ per kg beef)	0.007	0.011	0.012	0.035	0.029	0.012	0.044
Share cost of I&R (%)	0.16	0.27	0.35	0.75	0.47	0.31	0.89
Beef cattle (males more than 1 year) (1,000 animals)	1,397	1,151	845	93	1,583	340	1,604

Source: LEI for identification and registration costs per head and CRPA for sector costs assessments

In order to estimate the impact on the EU beef sector the production cost analysis of the Agribenchmark coordinated by the Federal Institute for Agriculture (FAL) has been used (Agribenchmark, 2007). Considering the total production costs per farm it is possible to calculate the percentage cost increase per kg beef to be attributed to the identification and registration of beef cattle. The seven countries listed in Table 8.12 represent almost 90% of beef production of the EU-15.

Assuming 100% compliance we obtain an average weighted cost increase of 0.455% for the EU-15. The cost impact of identification and registration varies with the cost per head of registration and with herd size. In countries with a large herd size some economies of scale can be exploited, whereas small herds, like for example in Ireland, are facing more significant effects.

Standards and external competitiveness

Expected impact of Nitrates Directive and identification and registration of animals on beef trade flows

The impact of full compliance with Nitrates Directive was simulated with the GTAP model in order to assess the effects of this policy on the competitiveness of EU beef production in the world market. The previous analysis has found a limited impact on EU beef farms (an increase of costs 0.095% at a 50% compliance rate), primarily due to the low share of farms that would be affected. The effects on EU beef exports could be a 0.68% decrease in quantity, while imports would rise by 0.51%. The main advantage would be taken by Brazil with an increase of 0.28%. For Australia, the US and Canada no significant changes would occur.

Table 8.13 Effect of Nitrates Directive on global bilateral beef export (variation on quantities, in %)

	Importers						
Exporter	**EU-15**	**Japan**	**USA**	**ROECD**	**ROW**	**Total**	**Share export**
Australia	0.53	0.00	0.01	0.03	0.08	0.04	0.03
New Zealand	0.51	-0.02	-0.02	-0.01	0.04	0.18	0.16
EU-15	0.00	-0.75	-0.77	-0.71	-0.65	-0.67	-0.68
Canada	0.61	0.01	0.01	0.02	0.09	0.04	0.02
Brazil	0.52	0.00	0.00	0.00	0.09	0.30	0.28
USA	0.54	0.01	0.00	0.03	0.06	0.04	0.03
Rest of World	0.55	0.00	0.01	0.02	0.05	0.30	0.28
Total	0.53	-0.01	-0.03	-0.03	-0.18	0.01	0.00
Share import	0.51	-0.02	-0.04	-0.04	-0.20	0.00	

According to the previous calculations, the EU-15 is facing a cost increase of 0.455% per kg beef for maintaining a system of identification and registration of beef cattle. Between EU Member States it varies from 0.16% in France up to 0.89% in Ireland.

Table 8.14 shows that cost increases due to I&R regulations would increase the EU beef imports by 2.2% and decrease exports by a similar amount. The EU share of the world export market would decline by 2.23%, while a 2.09% increase of the share on global import is foreseen.

Negative impact on EU beef competitiveness entails positive changes in the market share by some of the main beef exporting competitors. In particular Brazil would expand its exports by 1.1%. The performances of the other competitors on the world market would be affected to a smaller extent. The share gains of Canada, the US and Australia are expected to be negligible. The increase of beef traded by these countries should be limited to less than 0.2%, due to the fact they compete on different markets than those where EU is engaged. Benefits would be higher for the export from the rest of the world, that on the whole is projected to rise by 1.21%.

Combining the effects of both standards (Nitrates Directive and I&R), an overall decrease of 2.73% would occur to EU exports. The reduction in exports to Japan would equal 2.67% while exports to other OECD would decrease by 2.54%. On the whole, Brazil would gain 1.37% of global market share, while changes in the global market shares of Australia, Canada and the US would be negligible.

Table 8.14 Effect of Identification and Registration of beef cattle on global bilateral beef export (variation on quantities, in %)

	Importers						
Exporter	**EU-15**	**Japan**	**USA**	**ROECD**	**ROW**	**Total**	**Share export**
Australia	2.44	0.00	0.03	0.06	0.25	0.16	0.04
New Zealand	2.17	-0.11	-0.09	-0.06	0.11	0.74	0.62
EU-15	0.00	-2.00	-2.01	-1.86	-2.15	-2.12	-2.23
Canada	2.28	0.01	0.04	0.0	0.31	0.13	0.01
Brazil	2.18	0.00	0.01	0.04	0.28	1.22	1.10
USA	2.14	0.02	0.00	0.07	0.21	0.14	0.02
Rest of World	2.25	0.01	0.03	0.05	0.16	1.21	1.09
Total	2.21	-0.03	-0.08	-0.07	-0.61	0.12	0.00
Share import	2.09	-0.14	-0.19	-0.19	-0.73	0.00	

Table 8.15 Effect of Nitrates Directive and I&R of beef cattle on global bilateral beef export (variation on quantities, in %)

	Importers						
Exporter	**EU15**	**Japan**	**USA**	**ROECD**	**ROW**	**Total**	**Share export**
Australia	2.95	0.01	0.04	0.08	0.33	0.20	0.07
New Zealand	2.67	-0.14	-0.11	-0.07	0.14	0.91	0.77
EU-15	0.00	-2.67	-2.73	-2.54	-2.74	-2.73	-2.86
Canada	2.81	0.02	0.05	0.09	0.39	0.17	0.03
Brazil	2.67	-0.02	0.02	0.07	0.36	1.51	1.37
USA	2.66	0.02	0.00	0.09	0.26	0.18	0.04
Rest of World	2.77	0.01	0.04	0.08	0.20	1.50	1.36
Total	2.72	-0.03	-0.10	-0.10	-0.78	0.13	
Share import	2.58	-0.17	-0.24	-0.23	-0.91		

Expected impact of a hormone ban in the US on beef trade flows

Earlier, we presented evidence that a US ban on the use of growth hormones would increase production costs by 7.14%. According to the simulations carried out with the GTAP model, a ban on growth promotants would cause an increase in US beef imports of 32%, whereas exports would decline by 36% and beef production in the US would decline by 5%. In particular, the gap left by the US in the Japanese import market will be primarily filled by Australia.

The decline of American beef production will stimulate a growth of the imports from Australia and Canada. Also the rest of the OECD countries will import more as the US will lose about 41% of its exports on these markets.

In this new situation the EU-15 will remain a net importer of beef but will slightly improve its competitive position. EU exports are expected to rise by nearly 6%.

Table 8.16 Effect of hormone ban in the US on global bilateral beef export (variation on quantities, in %)

	Importers						
Exporter	**EU15**	**Japan**	**USA**	**ROECD**	**ROW**	**Total**	**Share export**
Australia	-0.7	19.8	32.1	22.8	4.0	19.2	16.8
New Zealand	-1.5	18.8	31.0	21.8	4.9	12.1	9.9
EU-15	0.0	21.6	34.1	24.7	3.6	7.9	5.8
Canada	-0.4	19.9	32.2	22.9	2.0	27.1	24.6
Brazil	0.7	21.4	33.9	24.5	3.3	4.9	2.9
USA	-45.2	-33.1	0.0	-32.2	-40.7	-34.7	-36.0
Rest of World	0.5	21.3	33.7	24.5	7.6	10.6	8.4
Total	-2.7	-10.0	32.2	-11.9	-3.4	2.0	0.0
Share import	-4.6	-11.8	29.6	-13.6	-5.3	0.0	

Conclusion

Within the EU, beef is produced in a wide range of farming systems, ranging from the extensive cow-calf farms in Ireland, the UK and the centre of France down to the very intensive beef fattening systems located in Italy and Spain. The Nitrates Directive affects 4.2% of beef cattle raised in intensive finishing farms and 3.0% of beef produced on cow-calf farms. This low percentage of farms affected by the Nitrates Directive explains the limited sector cost increase, which has been estimated at 0.095%. This relatively low cost impact does not have significant consequences for the competitive position of the EU beef production on the world market. The EU trade deficit in beef would increase, as exports would fall by 0.68% and imports would rise by 0.51%. The country which would benefit most from this situation is Brazil increasing its global exports by 0.28% and in particular to the EU (+0.52%).

More important for the beef farms are the regulations concerning the identification and registration of beef cattle. Implemented as a reaction to the BSE crisis, regulations require beef farmers to register all cattle movements and make sure that all animals are correctly identified from birth up to the slaughterhouse. According to our estimates, these measures generate a 0.454% average cost increase for the beef farms in the EU. This cost effect has a significant effect on EU beef trade. Beef imports are higher by 2.2% and exports are lower, by the same percentage under these regulations. Again Brazil's exports to the EU are 2.18% higher and its global exports are 1.1% higher as a result of these EU regulations. The other competitors on the world market are affected to a lesser extent.

Finally, a ban on the use of growth hormones in the US would have a strong effect on beef production costs, with an average cost increase of 7.14%. Such a production cost increase would have a significant effect on the competitiveness of US beef production in the world market. US exports would decline by 36% to those countries which already now import American beef.

References

Agribenchmark (2007) Beef report 2007, *Benchmarking Farming Systems Worldwide,* FAL, Braunschweig.

De Roest, K., C. Montanari, E. Corradini (2007) Il costo della Direttiva Nitrati, BIT SpA - Cassa Padana, Leno (Brescia).

ISMEA-CRPA (2006) Analisi dei costi di allevamento nel comparto bovino da carne, Ismea, Quaderni di filiera n.15, Roma.

Lawrence, J.D. and M.A. Ibarburu (2006) Economic Analysis of Pharmaceutical Technologies in Modern Beef Production, Research Report, Iowa State University.

Montgomery, T.H., P.F. Dew, M.S. Brown (2001) Optimizing carcass value and the use of anabolic implants in beef cattle, *Journal of Animal Science*, 79 (E.Suppl) 296-306.

Sarzeaud P., F. Becherel, C. Perrot (2007) *EU beef farming systems and CAP regulations.* EAAP Technical Series, no.9.

USDA (2003) Costs associated with development and implementation of comprehensive nutrient management plans. Part 1 – nutrient management, land treatment, manure and wastewater handling and storage, and recordkeeping. USDA NRCS, table 38, pp. 105.

Winsten J. (2007) US Research report Cross-Compliance, Winrock International, Research Report of EU Cross Compliance project 'Facilitating the CAP reform: Compliance and competitiveness of European agriculture'.

Chapter 9

Good agricultural and environmental conditions in the EU and their implications for international trade in cereals

Roel Jongeneel, Xavier Poux and Glenn Fox

Introduction

In the EU, cereals are the most widely produced arable crop. European statistics on cereals comprise wheat, barley, maize, triticale, rye, oats, mixed grain, rice, sorghum, buckwheat, millet and canary seed. Cereals account for a third of the EU's utilized agricultural area (UAA). Wheat is the crop that forms the highest proportion in European agriculture accounting for 46% of the total area used for the production of cereals. Wheat is used in animal feed, human diet and also for starch and bioethanol. In 2009, a total area of 25.6 million hectares was cultivated to produce 138.7 million tonnes of wheat. Grain maize constituted 19% of EU 2009 cereal production, the production being 57.7 million tonnes.

When considering the relative share of production traded on the world market, the cereals sector is one of the agricultural sectors that is most opened up, with around 11-15% of the total production being traded each year, and trade depending on weather conditions and stocks. Wheat and maize dominate world cereals trade, (representing 45% and 35% of total EU cereals trade respectively), while rice and barley jointly represent 18% of EU cereals trade (leaving 2% for other cereals such as sorghum, millets and oats) (FAO, 2007-2008 data). Relative to cereals, other agricultural commodities such as meat or dairy products are less exported/imported. As such the competitiveness issue is particularly crucial in the cereals sector. Moreover, in a sense for several countries the price for cereals can be considered of utter importance for the whole economy, as shown in the recent years with 'hunger fights' in 2007/2008 for instance (Dufumier, 2009). As shown in the next section, the number of countries exporting cereals is quite limited, with a huge domination of the USA alone.

Cereals are produced in a great variety of farming systems across the world, mainly in temperate areas. They can be cultivated in small mixed low input farms as well as in large specialized and high input farms, concentrated in the most productive land forming various 'grain belts'. These differences in farming systems entail a great range of environmental impacts of cereal systems. In some cases, extensive cereal systems can be beneficial for biodiversity, while in other cases the use of nutrients, pesticides, inappropriate irrigation practices and heavy tillage will lead to severe negative impacts on natural resources (Bignal and McCracken, 1996; IEEP, 2000).

This chapter addresses the impact of compliance and costs of compliance with environmental standards on the competitiveness of farms producing cereals. In a context where the EU is developing environmental conditions on its agriculture, the overarching question for policy makers is to find a balance between a better integration of environmental concerns in cereal production - in order to meet a strong need at EU level - while not imposing constraints which would threaten the competitiveness of EU cereal producers. The thesis of the chapter is that the system of Good Agricultural and Environmental Conditions (GAECs), which was adopted by the EU as part of cross compliance, prevents a significant unilateral loss of competitiveness. The measurable impact on production costs due to the measures taken is thus found to be limited.

The next section discusses some sustainability issues with respect to cereals production. The following section provides an overview of cereal production and trade aimed at getting further insight into the particular structure of cereal markets and trade. It also serves to motivate the choice of the countries that will be further analysed in the subsequent part of the chapter. The following section focuses on the structure and composition of costs of cereals production in the EU and in main regions of the world. This provides further insight into the relative share of different cost items that can be attributed to some environmental constraints. We will then address the environmental standards and their estimated impact on production costs in cereals. Some comparisons are made with cereals production in Canada. The potential impact of compliance with environmental standards on world trade are assessed. This is achieved by scaling up cost impact at farm level to sector level. Subsequently the change in costs at sector level are imported into the GTAP model, which includes world trade in cereals. Finally, the conclusion puts the findings in the broader perspective of a better integration of environmental concerns, and of other factors influencing the competitiveness of cereal production.

Cereals and sustainability

To start with, we will briefly elaborate on some key environmental issues associated with cereal production. Cereal yields in the EU have increased since the mid-1970s at an average rate of 2.2% per annum. The overall use of pesticides has decreased since the early 1990s (Eurostat, 1997). Arable farming in Europe has impacts on wildlife and on the sustainability of arable systems. Some of the other environmental impacts associated with cereal production include removal of soil, pollution of water, deterioration in ecosystems reflected in aesthetic quality of arable landscape.

Arable soils are vulnerable to erosion by wind and water, soil compaction through the use of heavy machinery, and organic matter reduction through frequent crop removal. Organic matter may erode to water bodies reducing its availability on arable soils. Legume based leys and integrating livestock into arable farming has been found to improve organic matter level in comparison to conventional systems (e.g. Drinkwater *et al.*, 1998). As Evans (1996) shows factors such as the increased area of autumn cultivation, increases in field size and associated loss of hedges, and permanent arable cropping increase levels of erosion. Erosion is clearly slope-related. Skinner *et al.* (1997), for example, reported that 60% of erosion occurs on slopes of less than 7°. Moreover, they argued that hedge removal accelerates erosion, therewith pointing to a linkage between sustainability and landscape preservation. Moreover on flat land and light soils, wind erosion is of importance. The use of heavy machinery and frequent passing with farm equipment contributes to soil compaction, which accelerates runoff. This might also lead to poor soil drainage, cause waterlogging and make crops susceptible to drought. Erosion can also cause loss of nutrients. The risk of soil erosion by wheat cultivation has been found to be low, but is generally higher with maize than with other cereal crops. Available studies in Europe suggest that lack of crop cover increases erosion rates on arable fields (Boatman *et al.*, 1999). Brouwer and Ervin (2002) have found that late harvested spring sown maize increases the exposure to erosion. Erosion has agronomic as well as economic consequences. García-Torres *et al.* (2003), for example, has shown annual economic losses of erosion to be around €53 per ha and €32 per ha for off-site effects.

A relatively recent phenomenon is the production of biomass for bioenergy. This needs to comply with the same environmental standards as food production. As a result of the Biofuels Directive of 2003, both the use and the production of bioethanol have been rising steadily in the EU. In 2004, approximately 528 thousand m^3 of bioethanol was produced in Europe (involving 0.7 million tons of cereals). By 2009, the production of bioethanol for fuel applications had already reached 3.6 million m^3. With the further implementation of the Renewable Energies Directive, which provides for a mandatory biofuel quota of 10% for the year 2020, the demand for bioethanol is expected to rise strongly in the EU. It is estimated that the biofuel requirement will imply a demand for ethanol of about 20 million m^3 in the year 2020 (equivalent to about 50 million tonnes of cereals or about 36% of the EU's wheat production in 2009). Grain used for bioenergy production currently constitutes about 1% of total EU cereal production (2009/10). The use of crop residues, e.g. straw for bioenergy might become more important in the future as so-called second generation technologies become available. Collecting this material from the field, however, creates the risk of depleting soil organic matter content because farmers currently (partly) use it as mulch or soil cover. This will consequently then reduce water retention and increase the risk of soil erosion. In good quality soils, taking straw from fields is allowed in North Western Europe. However there is resistance by many farmers (European Commission, 2006). Bioenergy production from cereal straw can help to mitigate climate change because of reduced emissions, but this is likely to depend on the conditions under which production will take place. The production of bioenergy from cereal straw could also lead to the competition for agricultural raw materials with food production. Cereal straw as raw material for biofuel also serves as raw material for the production of pulp and paper, animal bedding and feed. The use of straw as bedding and feed for animals may contribute to the reduced amount of straw for bioenergy.

Production, trade and consumption of cereals

Three countries (China, US and India) dominate world cereals production and trade (Table 9.1). France and Germany are key producers. The US is the world's dominant net exporter of cereals, followed at a distance by countries like France, Canada, Argentina, and the Russian Federation. As regards the EU, France is the most important net exporter of cereals.

Table 9.1 Production and trade in cereals (million tonnes)

Country	Production	Import	Export	Net exports
China	457.81	7.69	10.14	2.45
US	415.13	6.15	99.98	93.83
India	260.49	1.80	9.72	7.92
Russian Federation	80.21	1.10	16.94	15.83
Indonesia	70.44	5.54	0.20	-5.34
Brazil	69.44	9.61	11.47	1.86
France	59.33	2.28	25.59	23.31
Canada	48.11	3.25	22.78	19.53
Bangladesh	44.71	3.25	0.02	-3.24
Argentina	44.19	0.02	27.93	27.91
Germany	40.63	6.77	9.69	2.92

Source: FAOSTAT

As regards trade, the US dominates the market for wheat and maize. The US has a share of more than half of global production of maize. The EU is a key player only for the wheat market (with current exports about 16 million tonnes), while it is quite marginal for the maize market (it is even a net importer). Table 9.1 shows that the main exporters are not necessarily the main producers, with the exception of the US.

The main importers include three categories:

- Developed countries with a relatively low share of arable land (e.g. Japan, South Korea, Spain, Italy). The rationale to import cereals is simply to compensate for an absence of natural assets with other traded products outside the agricultural sector.
- Developing countries with population density that is simply too high compared to the producing capacity. Egypt is the most obvious example in this case, with only 3% of its territory being used for arable production. Algeria also falls in this category.
- Intermediate countries in the process of development are adopting a comparative advantage strategy. Although they would be able to produce cereals, they have chosen to allocate their agricultural production factors to other commodities.

Four Member States dominate cereals production in the EU (Table 9.2). In addition to France and Germany, also Poland and Italy are large producers and in total they represent 54% of production in the EU. France alone has a share of about 22%.

Figures 9.1, 9.2 and 9.3 provide an overview of the evolution of cereals production at EU Member State level. Production tends to fluctuate over time (notably for Hungary), but the figures show no clear trends.

Cereals are the basis of the human diet in many ways: directly consumed (e.g. products like bread, pastas and processed food based on starch) or indirectly consumed through livestock products (i.e. humans eating animals fed under a cereal diet i.e. granivore animals such as pigs, poultry). About two thirds of EU cereals production is used for feed, be it by farmers (on farm feed mixing, 36%) or indirectly via the compound feed sector. About one quarter of EU cereal production is used for human consumption.

Table 9.2 Composition of EU cereals production in 2009 (1,000 tonnes)

Country	Wheat	Barley	Maize	Triticale	Rye	Oats	Mixed grain	Rice	Other	Total
France	38,324.7	12,879.6	15,299.9	2,015.5	129.8	572.8	215.2	138.1	464.4	70,040.0
Germany	25,190.3	12,288.1	4,527.2	2,514.4	4,270.0	825.6	132.5			49,748.1
Poland	9,789.6	3,983.9	1,706.6	5,234.0	3,712.9	1,415.4	3,884.1		99.9	29,826.4
United Kingdom	14,379.0	6,969.0		66.0	36.0	757.0	30.0			22,237.0
Spain	4,796.8	7,399.7	3,478.7	140.6	181.4	906.2	38.0	899.4	30.4	17,871.2
Italy	6,341.0	1,049.2	7,877.7		11.7	315.1		*1,500.0*	296.9	*17,391.6*
Hungary	4,419.2	1,063.9	7,528.4	360.7	72.5	111.1		11.7	22.8	13,590.3
Denmark	5,996.0	3,421.0		233.1	244.6	267.0	37.9			10,199.6
Belgium	1,978.1	452.6	808.1	45.1	2.7	34.8				3,322.8
Netherlands	1,402.0	309.6	244.9	17.0	11.2	9.6				994.3
Semi-total	112,616.7	49,816.6	41,471.5	10,626.4	8,672.8	5,214.6	4,337.7	2,549.2	914.4	236,220.0
(in % of semi-total)	47.7	21.1	17.6	4.5	3.7	2.2	1.8	1.1	0.4	100.0
Other countries in EU-27	26,108.4	12,582.3	16,306.6	1,434.8	1,205.2	3,196.0	208.9	479.9	408.8	61,930.9
Total EU-27	138,725.1	62,398.9	57,778.1	12,061.2	9,878.0	8,410.6	4,546.6	3,029.1	1,323.2	298,150.8
(in % of total)	46.5	20.9	19.4	4.0	3.3	2.8	1.5	1.0	0.4	100.0

Note: Italic numbers represent best estimates by the authors.

Source: FAO-STAT

Figure 9.1 Cereal production in France, Germany, Poland and the United Kingdom

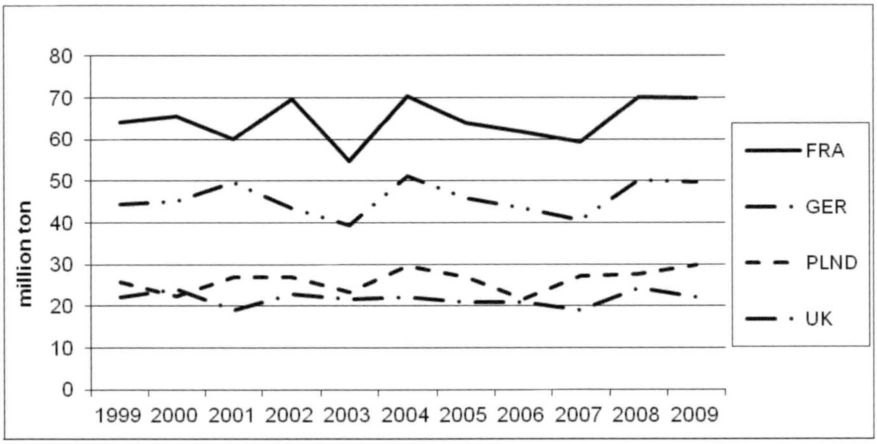

Source: Eurostat (various years)

Figure 9.2 Cereal production in Spain, Italy and Hungary

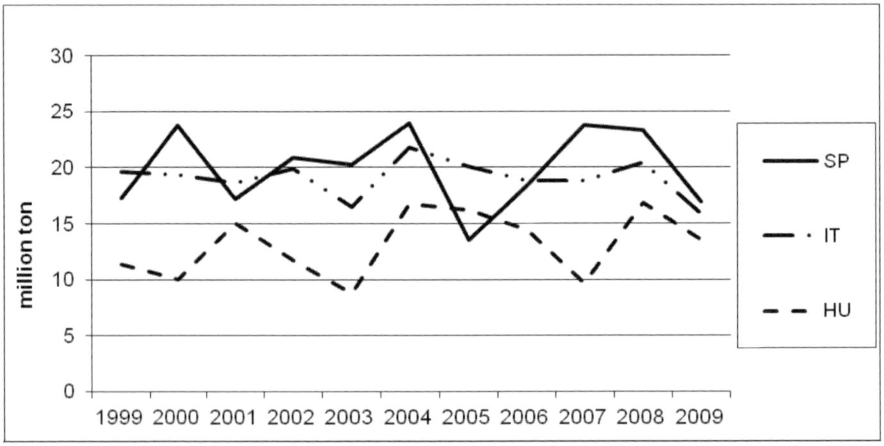

Source: Eurostat (various years)

In the EU, about 65% of the cereals is used for animal food. Cereals are a dry commodity, and are rather easily transported across Europe. Thus, the industries of cereals are often based near the main consumption areas, which are the industrialized livestock areas in Europe. This explains why the main importers are also large livestock producers (i.e. Spain, Italy and the Netherlands).

Figure 9.3 Cereal production in Denmark, Belgium and the Netherlands

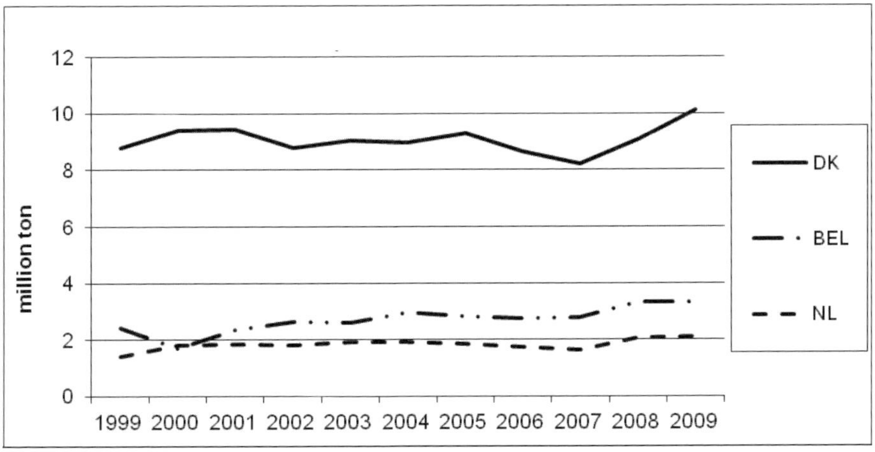

Source: Eurostat (various years)

Arable farming and wheat production in the EU

Cereals are part of a wider mix of crops produced by EU arable farms. Table 9.3 provides an overview of the hectares of wheat grown per specialized arable farm in the EU-27. There is considerable variation across Member States, which reflects among others differences in scale of farming. Large scale cereal farms can be found in Czech Republic, Slovakia, the UK, Germany and France. A significant increase in the hectares of cereals is observed over time in most countries. As can be understood when combining information from Table 9.4 with Table 9.3 this also reflects farm scale increase.

Table 9.4 represents the share of cereals in the total utilizable agricultural area of specialized arable farms in the EU. As the table shows these shares tend to be rather stable over time. Table 9.4 provides an indication of the degree of specialization in cereals in the specialized arable farm sector by Member State. Note that for most Member States the share of cereals is higher than 50%, which indicates the key importance of cereals for the EU arable sector as well as for nearly all EU Member States. Member States showing a relatively low share of cereals in their crop mix are Belgium, The Netherlands, and Portugal.

Figure 9.4 provides the crop specific costs (€/ha) for selected Member States. Costs are related to the whole crop mix, including non-cereal crops. Here, specific costs for crops distinguish between:

- seeds and seedlings, purchased and produced on the farm;
- fertilizers, soil improvers (lime, compost, peat, manure) and crop protection products; and
- other specific crop costs that are the general costs directly connected with crop production (packing and binding materials, soil analysis, etc.).

Specific costs tend to vary between €200 and €400 per hectare. Belgium, and even more so The Netherlands have very high specific costs per hectare and are mainly suitable to grow arable crops with a high gross margin and high yields.

Table 9.3 Hectares of cereals grown per specialized arable farm in EU-27

	1990	1995	2000	2005	2006	2007	2008
Austria		18.85	24.79	29.05	28.33	30.82	32.15
Belgium	19.62	17.74	19.87	23.34	23.71	25.26	29.89
Denmark	22.19	28.64	34.87	38.37	45.21	43.23	45.17
Finland		26.89	34.93	34.10	34.92	36.92	39.41
France	36.26	41.87	53.55	56.70	57.18	56.36	60.30
Germany		46.36	62.38	73.98	80.28	80.71	84.66
Greece	4.66	4.09	4.65	4.81	5.92	5.99	6.42
Ireland	37.32	31.63	33.10	26.41	30.78	33.85	37.96
Italy	5.57	6.87	7.75	10.52	11.00	11.39	
Luxembourg							
Netherlands	12.57	13.17	11.85	15.40	16.79	16.70	19.20
Portugal	3.54	3.06	2.63	3.99	4.32	3.93	5.67
Spain	22.13	27.96	29.38	29.41	30.78	31.33	39.60
Sweden	0.00	32.94	40.81	46.60	52.53	56.47	61.63
UK	88.46	78.27	88.25	90.89	83.45	84.29	92.86
EU-15	13.98	18.12	21.37	25.45	26.56	26.91	36.49
Cyprus				6.66	8.10	6.48	12.28
Czech Republic				151.58	149.57	150.70	137.67
Estonia				54.87	59.61	58.80	62.61
Hungary				40.78	43.71	42.95	42.66
Latvia				44.05	48.21	52.48	54.34
Lithuania				38.79	43.44	45.16	47.65
Malta							
Poland				15.09	14.27	14.25	15.49
Slovakia				203.49	188.15	227.41	226.82
Slovenia				3.94	4.07	4.48	4.66
EU-10				26.03	25.21	25.35	26.15
Bulgaria						48.08	41.87
Romania						11.11	14.52
EU-02						14.55	17.17

Note: A blank-value indicates that no information was available. Note that the EU-10 Member States joined the EU in 2004, whereas Bulgaria and Romania joined the EU in 2009.

Source: EU-FADN

Farm profitability of specialized arable farms, as measured by the Farm Net Value Added per Annual Work Unit-indicator (FNVA/AWU), is presented in Table 9.5. Differences in remuneration per work unit reflect differences in farm scale and intensity of production. It should be also noted that also outside agriculture income per worker varies substantially over Member States. The observed differences across countries reflect such general differences, although remuneration in agriculture is lagging behind the general income evolution. The FNVA/AWU indicator includes the impact of coupled and decoupled direct payments received by specialized arable farmers (e.g. the Single Farm Payment or Single Area Payment, as well as national aids and payments received for delivery of agri-environmental services). Due to the extreme high prices in 2007-2008 period, FNVA/AWU showed a strong increase in 2008.

Table 9.4 Share of cereals in total utilizable agricultural area per specialized arable farm

	1990	1995	2000	2005	2006	2007	2008
Austria		0.51	0.58	0.60	0.58	0.62	0.63
Belgium	0.49	0.40	0.39	0.42	0.43	0.45	0.50
Denmark	0.65	0.61	0.68	0.66	0.66	0.64	0.66
Finland		0.58	0.66	0.62	0.60	0.63	0.66
France	0.57	0.51	0.57	0.56	0.56	0.56	0.59
Germany		0.54	0.57	0.53	0.55	0.55	0.58
Greece	0.59	0.52	0.53	0.54	0.55	0.56	0.59
Ireland	0.63	0.54	0.57	0.54	0.61	0.63	0.70
Italy	0.52	0.52	0.52	0.54	0.51	0.54	
Luxembourg							
Netherlands	0.28	0.27	0.23	0.28	0.30	0.30	0.34
Portugal	0.33	0.28	0.30	0.29	0.29	0.27	0.35
Spain	0.63	0.56	0.59	0.57	0.58	0.60	0.61
Sweden		0.54	0.57	0.53	0.54	0.56	0.59
UK	0.60	0.53	0.54	0.49	0.48	0.48	0.54
EU 15	0.57	0.52	0.55	0.54	0.54	0.55	0.59
Cyprus				0.57	0.61	0.49	0.75
Czech Republic				0.60	0.60	0.62	0.62
Estonia				0.55	0.54	0.50	0.51
Hungary				0.65	0.63	0.64	0.64
Latvia				0.53	0.55	0.56	0.58
Lithuania				0.58	0.62	0.62	0.64
Malta							
Poland				0.64	0.62	0.61	0.64
Slovakia				0.55	0.52	0.55	0.55
Slovenia				0.38	0.46	0.49	0.51
EU-10				0.62	0.61	0.61	0.62
Bulgaria						0.64	0.60
Romania						0.65	0.68
EU-02						0.65	0.66

Note: A blank-value indicates that no information was available. Note that the EU-10 Member States accessed the EU in 2004, whereas Bulgaria and Romania entered the EU in 2009.

Source: EU-FADN

Production costs in the EU and other exporting countries

Figure 9.5 provides the cost structure for wheat production for selected EU Member States as a percentage of the output value (Kleinhanss *et al.*, 2011). The costs of soft wheat production (durum wheat for Italy) are calculated for three periods, i.e. 1999-2003, 2004-2006 and 2007 (conditional on a sufficient number of reliable observations being available). Distinguished cost categories are crop specific costs, energy, other specific costs, overheads, and depreciation and other fixed costs. When looking to the absolute levels, Spain, Poland and Hungary turn out to be relative low cost producers. For the other Member States the costs of soft wheat production vary between €600-€900 per hectare, with the UK being a high-cost producer. Costs of energy vary between €50 and €90 per hectare. Costs of depreciation amount to about €150 per hectare, and are as such an important cost item.

Figure 9.4 Crop specific costs in € per hectare for selected EU Member States (2003-2006 average)

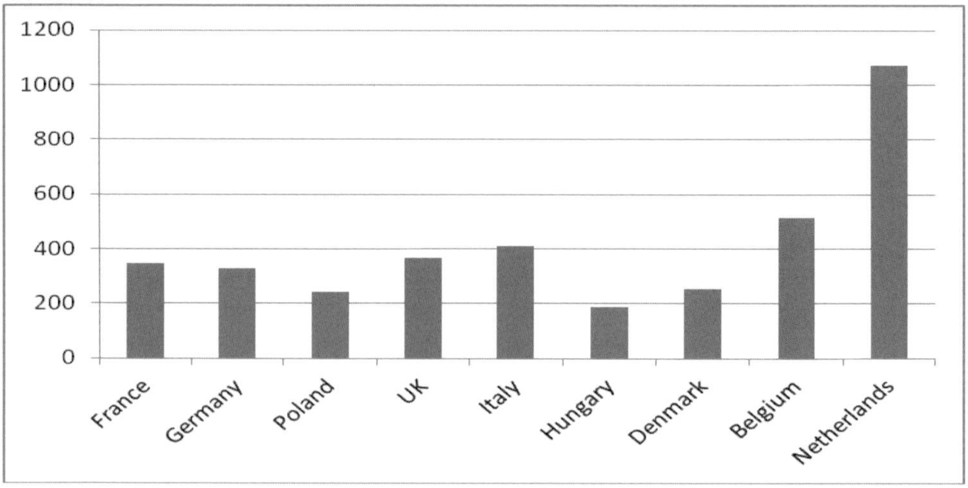

Source: EU-FADN

Yields show wide variation across Member States, and costs per unit of production are more appropriate for an EU-wide comparison than costs per hectare. Production costs vary between €80/ton and €130/ton, with the average being close to €110/ton. Year 2007 is a special year, with a strong increase in the output price of wheat, as well as strong increases in costs (Kleinhanss *et al*., 2011, p. 11).

France is a key producer of cereals with surplus production. Moreover, Kleinhanss *et al*. (2011) and Thorne (2005) indicate that costs of production in France are comparable to other major cereal producing countries in the EU. It was therefore decided to focus on France as the representative case for the competitiveness analysis of the environmental standards. The results are used to generalize them to other Member States (taking into account their specific compliance rates and cost-revenue structure).

The main producing and exporting countries have quite different yields (Table 9.6). This is largely due to natural and socio-economic conditions. The EU and its competitors seem to opt for opposite strategies. While the EU's competitiveness is based on maximizing production per hectare – due to a relative scarcity of agricultural land per inhabitant – the Australian, Argentine, US and Canadian competitiveness is rather based on minimizing costs per hectare.

As these countries have different yields per hectare, the relevant data is not the cost per hectare, but rather the costs per unit of product (€/quintal), which are provided in Figure 9.6. The costs are calculated using a standardized procedure in order to ensure comparability over countries (as compared to Figure 9.5, in Figure 9.6 the costs of land, capital and labour are accounted for).

Table 9.5 FNVA/AWU for specialized arable farms in EU-27 (in €)

Member State	1990	1995	2000	2005	2008
Austria		24,343	22,307	27,810	42,612
Belgium	31,818	39,865	40,368	41,731	42,426
Denmark	19,664	32,605	38,266	50,550	48,020
Finland		20,796	23,844	18,422	24,323
France	19,758	30,069	28,786	25,985	37,970
Germany		27,308	28,504	31,656	40,345
Greece	5,161	5,990	9,415	10,493	12,829
Ireland	21,846	23,689	31,053	33,437	36,200
Italy	7,176	9,873	11,742	18,480	
Luxembourg					
Netherlands	39,943	50,992	34,248	43,194	54,745
Portugal	2,423	2,593	3,839	3,945	9,324
Spain	8,179	15,271	21,710	15,894	29,013
Sweden		20,824	15,220	21,277	47,145
UK	26,655	42,498	34,127	36,727	52,006
Cyprus				5,641	10,067
Czech Rep.				10,223	16,475
Estonia				8,930	8,700
Hungary				10,982	19,553
Latvia				4,861	10,486
Lithuania				7,105	14,819
Malta					
Poland				4,184	6,343
Slovakia				5,864	10,157
Slovenia				1,459	3,760
Bulgaria					7,269
Romania					17,630

Source: EU-FADN

Figure 9.5 Cost shares of wheat referring to output value for three sub periods for selected EU Member States (1=1999-2003; 2=2004-06; 3=2007)

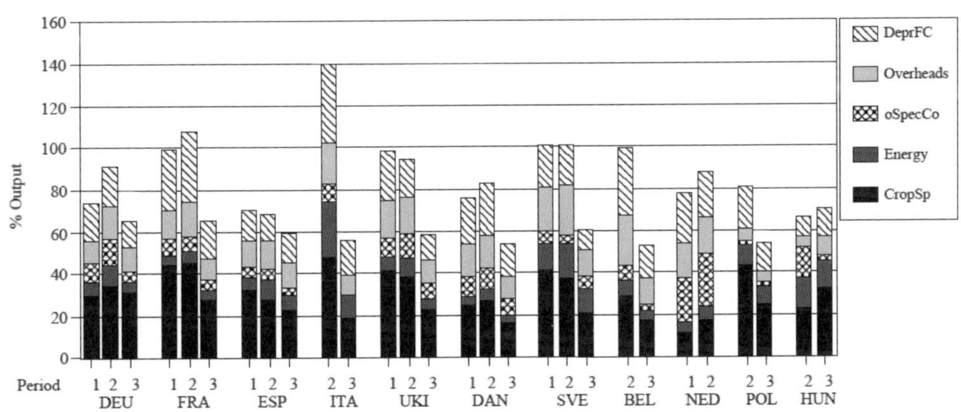

Note: For Italy estimates are for durum wheat.

Source: Kleinhanss et al. (2011, p. 12).

Table 9.6 Comparison of yields and costs/ha for the main exporting countries

	Yield (q/ha)	total costs €/ha
France	83	1,183
Argentina	34	443
Australia	26	275
Canada	26	342
USA	26	391

Source: AGPB; 1998-1999 crop year

Figure 9.6 Cost structure of cereals production inside and outside the EU

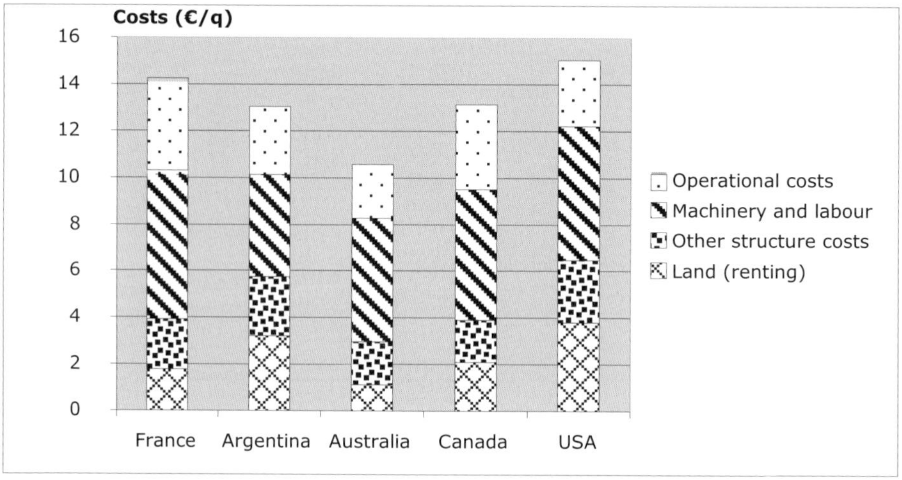

Source: AGPB, 1998/99 data

At lower end, Australia is the most competitive country (for the year indicated), with 10.6 €/q while the other countries range between 13 and 15 €/q. The figure shows that the relatively similar costs are reached through different costs structures. In the US, for example, the land costs are relatively high (circa 25% of all costs), due to a high 'consumption' of land per quintal produced (another way to label the low yields) while, symmetrically, they are relatively low in France (12%). Reciprocally, the operational costs are higher in France (±30% of all costs) than in the US (< 20%).

Meanwhile, the machinery and labour costs levels are rather similar across the countries studied, expressed in €/quintal. While the breakdown between the two items is not available, one can only make assumptions on how this result is obtained. While the French cereals will stand on more intensified cropping patterns, the higher costs are in general compensated by higher yields. Said in other words, the machinery and labour costs are more or less fully proportionate to the yield. Differences in cost structure makes the different systems differently sensitive to variations in yields and prices.

Environmental standards and their costs

This section starts with a short discussion of which standards were finally selected for a quantitative trade-impact analysis. Thereafter it provides background about the calculated sector percentage cost of production increases. Moreover some cases are mentioned to provide some further farm level insights. The farms involved in cereal production will be potentially affected by the following standards or directives:

- 'general': groundwater, sewage sludge, placing of plant protection products, food law, & procedures in matters of food safety; and
- 'zoned': Nitrates Directive (NVZ), Birds Directive (SPA) and Habitat Directive (SCI); the latter two are in the N2000 network.

Livestock production is more affected by environmental standards than cereal production. Cereal production is potentially affected by the Nitrates Directive, in particular the requirement it imposes on achieving a certain degree of soil cover during winter time. More important for the arable sector are the requirements for Good Agricultural and Environmental Conditions (GAECs) (see Table 9.7). GAEC standards principally concern limitation of soil erosion, buffer strips, diversity of cropping pattern, the ban on straw burning, regulation of irrigation. Cereals and crop systems are, in principle, mainly affected by the GAECs although some dispensations may weaken the requirements because individual Member States had some freedom in determining the GAEC standards and could tailor them to their national situation.

In order to evaluate the impact on competitiveness an estimate was made about the sectoral percentage cost increase of the GAEC regulations on cereals. Based on previous country studies the costs of the GAECs for cereals are estimated to be €35/ha, which includes the estimated revenue-reduction due to the need to retain more straw on the fields (organic matter) and the additional costs (tillage, ploughing, etc.). A much higher cost is taken into account for the proper management of idled land, which is estimated to create a cost of €400/ha. This latter estimate includes the costs of adequate soil cover, tillage, fuel and labour costs.

Since the estimated costs associated with the GAECs for cultivated and idled land are different, the use of set-aside land (e.g. idled or used for energy crops) and the rate of set-aside (10% for year 2005, but currently reduced to 0%) are likely to affect the cost increase associated with the GAEC standard. For this reason it was decided to simulate two scenarios: One scenario assumes the set-aside rate to be 10% and all set-aside land to be used as idled land. This scenario can be considered to be a worst case scenario with relatively high costs. The alternative scenario assumes either a 0% set-aside rate, or a situation in which all set-aside land is cultivated with specific crops rather than left idled. Since the relative costs for cultivated land are lower than for idled land, this scenario leads to a relatively low cost of GAEC standard estimate. Together these scenarios provide an idea of the upper and lower bounds of GAEC costs for cereals.

Table 9.8 provides a brief overview of the results, with GAEC-1 the cost increase associated with a 10% set-aside rate being in place, whereas GAEC-2 accounts for the case where the set-aside rate is set to zero. As the relative cost per hectare of land cropped with cereals is much lower than that of set-aside or idled land, the cost increase for GAEC-2 is much (almost 50%) lower than in GAEC-1. For France the buffer strips, which were previously 'included' in the set-aside requirement, in GAEC-2 become an explicit constraint, precluding the French effective set-aside to go to zero, although the formal rate would allow for this.

Table 9.7 GAEC requirements (French example)

Issue	Standards: GAEC
1. Soil erosion: protect soil through appropriate measures	Set-aside of farmland (3%) = buffer strips Obligation of land maintenance (set-aside land, grassland, non-productive land)
2. Soil organic matter: maintain soil organic matter levels through appropriate practices	Ban on burning straw Diversity of cropping pattern
3. Soil structure: maintain soil structure through appropriate measures	Regulation of irrigation Diversity of cropping pattern
4. Minimum level of maintenance: ensure a minimum level of maintenance and avoid deterioration of habitats	Set-aside of farmland (includes buffer strips) General regulations of land maintenance Maintenance regulations of cultivated land Maintenance regulations of set-aside land Maintenance regulations of grassland Maintenance regulations of non-productive land

Source: EU and national French legislation

The final sectoral (additional) costs estimates as presented in Table 9.8 are dependent on the best estimates of compliance (the upper row of the table). These estimated compliance rates take into account the extent to which GAEC requirements reflect pre-existing national legislation (see Chapter 2 for more details).

Table 9.8 draws from FADN data taking into account several farm types (e.g. general field cropping, mixed crops, etc.). Differences in cost increases partly depend on farm structure. At sectoral level, costs increase by less than 1%. However, at farm level costs might increase by up to 4%, depending on the degree of compliance.

Table 9.8 Best-estimates of percentage cost increases for GAECs in cereals

	BE	DK	FR	DE	IT	NL	SP	UK
Best-estimate of degree of compliance	0.90	0.95	0.80	0.95	0.80	0.85	0.90	0.95
GAEC-1 (10% set-aside)	0.43	0.18	0.74	0.15	0.68	0.33	0.82	0.18
GAEC-2 (0% set-aside)	0.21	0.09	0.48	0.07	0.33	0.16	0.40	0.09

Source: own calculations

In Canada, management practices similar to the GAEC standards in the EU are implemented through various cost share programmes. These agricultural practices are known as Beneficial Management Practices (BMPs). Agriculture and Agri-Food Canada (AAFC) (2006) defines BMPs as 'farm management practices that: minimize and mitigate impacts and risks to the environment, by maintaining or improving the quality of soil, water, air and biodiversity; ensure the long term health and sustainability of natural resources used for agricultural production; and, support the long-term economic and environmental viability of the agriculture industry'.

In order to help agricultural producers develop and implement BMPs, the Government of Canada promoted provincially delivered Environmental Farm Plan (EFP) programmes. Such programmes support the agricultural sector to better identify its impacts on the environment; and promote the growth of stewardship activities within the agriculture industry. As part of the programme, Canadian farmers attend an Environmental Farm Plan workshop and complete a workbook designed to assess the current state of the farm and identify areas of concern. Then farmers develop an action plan for addressing the areas of concern. The action plan is confidentially reviewed by a group of locally appointed farmers. Once the Peer Review Committee approves the Action Plan, a farmer can participate in the EFP Cost-Share Program that helps cover a portion of the costs of implementing eligible projects from the action plan (OMAFRA, 2006). Producers are eligible to apply for cost-share incentives through the Canada Farm Stewardship Program, Green-Cover Canada, and the Canada Water Supply Expansion Program.

There are 36 Best Management Practice categories, each containing several practices eligible for funding. Federal government covers up to 60% of the cost of implementing eligible practices. Many practices covered through federal cost share programmes are also eligible for funding under different provincial cost share programmes. As a result up to 90% of the total project cost can be covered by combining federal and provincial funds. However, the coverage varies depending on farmer eligibility, province and type of BMP. In Ontario, funding is available through the Nutrient Management Financial Assistance Program, Wetland Farm Stewardship Incentive Program, Oak Ridges Moraine Environmental Enhancement Program and Greenbelt Farm Stewardship Program.

Although implementation of BMPs is not mandatory, there has been a relatively high degree of participation. For example, between 2005 and 2007, more than 11,000 of 57,211 Ontario farmers implemented or were in the process of implementation of BMPs. Even though their implementation is partly subsidized by the federal and provincial governments, there is a price to the farmers as well. As an example, Ontario farmers bore about a third of the cost of implementation of the management practices eligible for funding. Net costs for participating farmers could amount to about €1000 per farm. Because participation is voluntary, it is assumed that farmers will only participate if the net costs they have to make are offset by gains achieved elsewhere in the farm operation. For this reason the estimated percentage cost increase is assumed to be zero.

Standards and competitiveness

For the GAECs, only scenarios where the EU imposes standards are evaluated. It was estimated that the net implications on costs are effectively zero, or very close to zero.

The simulated trade impacts for both GAEC scenarios are presented in Table 9.9 below. For both scenarios 1 and 2 it is assumed that compliance with the regulation improves from the level as estimated to prevail early 2005 to full compliance. For the GAEC 1 scenario EU exports decline by about 2% whereas imports increase by a similar percentage. In terms of volume (assuming the EU wheat exports are about 16 million tons) this implies a reduction in EU exports of about 300 thousand tons for scenario 1.

For scenario GAEC-2, which has no set-aside obligation, EU exports decline and import increase are respectively about 30% and 40% lower as compared to scenario 1. The impact on total world cereals trade is negligible. Scenario 3 is similar to scenario 2, but presents the impacts, taking into account that in all other sectors (dairy, beef, pigs and poultry) considered, simultaneously the analysed selected standards are fully complied with. As can be seen this only marginally changes results, suggesting a very limited impact of spill-overs and feedback effects from other sectors.

Table 9.9 Percentage change in trade due to compliance with GAECs

	Scenario	EU Import	EU Export	Exports of which to...			Total world trade
				Japan	South Korea	Rest of OECD	
1	GAEC-1 (10% set-aside)	2.2	-1.8	-1.3	-1.3	-1.4	0.02
2	GAEC-2 (0% set-aside)	1.3	-1.1	-0.7	-0.7	-0.8	0.01
3	Like scenario 2 but also full compliance to selected standards in all other analyzed sectors	1.3	-1.2	-0.8	-2.0	-0.8	0.02

The US, Argentina and Australia fill the gap caused by the reduction in net exports from the EU. The impact of the evaluated GAEC constraints on cereal prices is negligible. As such other 'shocks' to the sector (like for example the price increases caused by the new demands for biofuels) easily outpace and dominate the impact of the GAEC standard, irrespective of whether one uses conservative or progressive estimates of the standard impacts.

Conclusion

Cereals are a key crop for the EU arable sector as is reflected in its share in the total utilizable agricultural area of specialized arable farms, which for most Member States is over 50%. The EU is a net exporter of cereals (in particular France is important as a net exporting country). Its key competitors in the world market include the US and Argentina.

Environmental and biodiversity standards affect EU cereal production mainly via the good agricultural and environmental practices (GAEC) requirements. Concluding, it can be noted that the increase in production costs of cereals associated with the GAECs remain less than 1%. Several factors explain this result. First, the additional costs per hectare are generally low, with an exception being the costs associated with keeping idled land in good agricultural condition. Second, the best estimates of the current degree of compliance are rather high. Farmers have already included a number of GAEC requirements into their existing farming practices. Farmers benefit from its implementation since they reduce the risk of soil erosion and maintain soil condition.

Farmers may participate voluntarily in programmes reducing soil erosion and are prepared to accept some costs. This is also confirmed in the case of Canada. Rather than following a command-and-control approach a voluntary approach might be effective, in particular when government is prepared to provide education and training, technical assistance and cost-offsets. Since in Canada participation is voluntary farmers are not likely to be faced with net cost increases affecting their competitive position.

The use of set-aside land and/or the rate of set-aside will affect the estimated percentage cost increases (see differences between GAEC-1 and GAEC-2 scenarios): using more set-aside land for cultivating special crops (e.g. bioenergy production) or having lower set-aside rates (e.g. due to low buffer zone-requirements) lead to lower costs. With the set-aside rate currently set to zero, the calculated percentage cost increases more or less halved as compared to a 10% set-aside scenario (GAEC-1). The 3% buffer strip requirement in France, whereas previously accounted for within the 10% set-aside requirement, can be interpreted as an 3% minimum effective set-aside requirement (which holds even when the formal rate goes down to zero).

The impact of the GAECs on the EU's external competitiveness varied from a 1.8% reduction in exports in GAEC-1 (set-aside rate 10%) to a reduction of 1.1% (set-aside rate 0%). EU imports increase with approximately a similar percentage as exports decline. Total world trade is hardly affected by the impact of the GAEC standard. (However, trade is likely to be affected by changes in the set-aside policy (changes in land base) that are not separately evaluated).

References

AGPB (various years) Statistics. http://www.agpb.fr/fr/chiffre/recolte_france.asp

Agriculture and Agri-Food Canada (2006) Community Pasture Program - Business Plan 2006-2011. URL http://www4.agr.gc.ca/AAFC-AAC/display-afficher.do?id=1298655130912&lang=eng.

Bignal, E.M. and D.I. McCracken (1996) Low-intensity farming systems in the conservation of the countryside. *Journal of Applied Ecology*, 33, 413–424.

Boatman, N., C. Stoate, R. Gooch, C.R. Carvalho, R. Borralho, G. de Snoo and D. Eden (1999). The environmental impact of arable crop production in the EU: practical options for improvement. Report prepared for DG XI of the European Commission. Allerton Research and Educational Trust. Study contract B4-3040/98/000703/MAR/D1.

Brouwer, F. and D. E. Ervin (eds.) (2002). *Public Concerns, Environmental Standards and Agricultural Trade*. Oxon: CABI Publishing.

Drinkwater, L.E., P. Wagoner and M. Sarrantonio (1998) Legume-based cropping systems have reduced carbon and nitrogen losses. *1DWXUH* 396: 262-265.

Dufumier, M. (2009), 'Sécurité alimentaire et développement durable. Repenser l'agronomie et les échanges internationaux', *Futuribles*, n° 352, mai 2009, pp. 25-42.

European Commission (2006). Cereal Straw Resources For Bioenergy in The EU. Pamplona 18-19 October, 2006. EUR 22626 2007.

Eurostat (1997) European Commission and Environment Agency. Europe's Environment: Statistical Compendium for the Second Assessment.

Eurostat, Statistics, electronic database, URL: http://epp.eurostat.ec.europa.eu/portal/page/portal/agriculture/data/main_tables.

Evans, R. (1996) Soil erosion and its impacts in England and Wales. Friends of the Earth. London.

García-Torres, L., J. Benites and A. Martínez-Vilela. (2003). *Conservation agriculture, a worldwide challenge: environment, farmers' experiences, innovations, socio-economy, policy*. Dordrecht: Kluwer Academic Publishers.

IEEP (Institute for European Environmental Policy, United Kingdom) (2000). The environmental impacts of irrigation in the EU, a report to the Directorate General for Environment of the EC, Brussels.

Kleinhanss, W., F. Offermann, J.-P. Butault and Y. Surry (2011) Cost of production for major agricultural commodities in the EU member countries using farm accounting data. Braunchweig, von Thünen Institute, Report for the OECD Agrifood and market Directorate, (see FACEPA project: http://www2.ekon.slu.se/facepa/).

OMAFRA, (2006) Canada-Ontario Environmental Farm Plan. URL: http://www.omafra.gov.on.ca/english/environment/efp/efp.htm.

Skinner, J.A., K.A. Lewis, K.S. Bardon, P. Tucker, J.A. Catt and B.J. Chamber (1997) An overview of the environmental impact of agriculture in the UK *Journal of Environmental Management*. 50: 111-28.

Thorne, F. (2005). Analysis of the Competitiveness of Cereal Production in Selected EU Countries. In: EAAE Copenhagen 2005, Copenhagen, Denmark, 26-Aug-2005, 16 pages 12108 B2.

Chapter 10

Trade effects of environmental and animal welfare standards in the hog sectors of the EU, US and Canada

Kees de Roest, Jonathan Winsten, Predreg Rajsic and Koen Dillen

Introduction

The hog sector is one of the most intensive livestock production activities in developed country agricultures. Compared to the number of hogs they raise, these farms often have a limited land area. Typically hog farms have to cope with severe manure disposal problems, in particular when this activity is concentrated in certain parts of a country. In order to reduce water and air pollution caused by hog farms, the EU, the US and Canada have issued environmental regulations. The type and strength of constraints imposed on the hog farms in these three countries depend on the severity of the environmental problems they cause. It is therefore of interest to analyse the differences in farm environmental compliance costs attributable to these regulations and to investigate the effects these cost impacts may have on trade in swine meat between the EU, the US and Canada and on the world market in general.

This chapter focuses on regulations related to reducing water pollution from excess nitrogen. In the EU, these regulations are embodied in the Nitrates Directive. In the US, the farm cost impact of the Clean Water Act will be assessed and for Canada an analysis will be presented of the economic consequences for hog farms of the Nutrient Management Acts of Ontario and Manitoba. The compliance costs will be assessed by using typical hog farms.

A separate analysis will be dedicated to the animal welfare regulations in the EU, which for hog farmers may potentially have a cost impact. Again the typical farm approach has been followed for calculating the extra costs inherent to comply with the specific obligation to raise sows in groups.

The enforcement of environmental and animal welfare regulations may have implications for the competitiveness of the swine sector, especially when the constraints in one country are much stronger than in another. The calculation of the farm compliance costs will be therefore followed by an analysis dedicated to the possible trade effects of the environmental compliance for the hog sectors in the EU, the US and Canada, which will be carried out using the GTAP model as has been described in Chapter 1. The simulation will be presented both under the hypothesis of unilateral implementation in the EU as well as multilateral enforcement of the environmental regulations in all three commercial blocs.

Standards for nitrate pollution control and its cost implications

The main requirement established by the Nitrates Directive is to limit nitrogen application from manure to 170 kg N/ha per year (or 250 kg N/ha of manure on grasslands in Germany and the Netherlands), which may generate extra costs for manure disposal. In Nitrate Vulnerable Zones (NVZ) farms exceeding the maximum limit have to find extra land to spread excess manure, either by buying or renting land or paying a licence to spread manure on land of nearby farmers. Moreover, they are obliged to invest in manure storage facilities due to the prohibition on spreading manure in winter time.

An evaluation of the impact of compliance with the Nitrates Directive requires an analysis of the structure of swine farming systems. Such information is essential to identify the share of total number of enterprises as well as the share of the number of animals affected by the directive. These figure differ significantly among European countries. In order to assess how many pig farms potentially are affected by the Nitrates Directive the following stepwise procedure has been followed:

- Estimate the number of farms which have a stocking rate per hectare exceeding the limit of 170 kg N per hectare for the main producing countries in the EU.
- Estimate the number of pigs raised within Nitrate Vulnerable Zones.

Then, a literature review was used to establish the cost increase related to compliance with the Nitrates Directive for a typical hog farm in each of the selected countries. An analysis of the degree of compliance has allowed the estimation of the percentage cost increase at the sector level.

Affected pig production

In order to estimate the number of pig farms which are potentially affected by the Nitrates Directive the most important Member States in pigmeat production have been chosen: Germany, Denmark, Belgium, Spain, France, Italy, Netherlands and the UK. These eight countries represent 91% of the pig population in the EU-15 (see Table 10.1).

For the first step of the analysis, single country data were obtained from the Farm Structure Survey (FSS). These data were used to calculate the share of pig population raised in farms whose stocking rate is higher than 17 pigs per hectare, corresponding to the limit of 170 kg N/ha that is established for the NVZ. The assumption that the whole country is defined as NVZ is valid only for Germany, the Netherlands and Denmark while in the other five Member States only part of the territory has been declared as vulnerable. Therefore, in the second step the number of pigs raised in the NVZs of these countries was estimated by overlaying the NVZ map over the NUTS 3 regions using the REGIO databank of EUROSTAT. The final results of this analysis are reported in Table 10.1.

In the Netherlands and Belgium, the majority of hogs are raised in NVZs and on farms exceeding the 170 kg/ha per year N limit. Although Denmark is designated completely as vulnerable, only 11% of pigs are raised on farms with a stocking rate above the limits indicated, owing to the spatial distribution of pig farms and to the strict application of the

Nitrates Directive in that country. Denmark has required pig farms to own or rent the land on which the slurry is to be spread. Spain also has a low share of pigs affected, but this is due to the scattered pattern of NVZs.

Table 10.1 Share of pigs affected by the Nitrates Directive in selected countries (2005)

	BE	DK	DE	ES
Total pigs (head)	6,318,220	13,466,290	26,857,800	22,776,690
Share of pigs raised in NVZ (%)	100.0	100.0	100.0	24.3
Share of pigs affected (%)	60.3	10.9	17.0	7.6
Pigs affected (head)	3,809,620	1,470,300	4,558,380	1,727,397
	FR	**IT**	**NL**	**UK**
Total pigs (head)	14,792,810	8,757,640	11,311,560	4,860,400
Share of pigs raised in NVZ (%)	74.5	64.2	100.0	66.1
Share of pigs affected (%)	23.5	37.2	79.4	23.8
Pigs affected (head)	3,474,032	3,264,236	8,978,870	1,157,649

BE = Belgium, DK = Denmark, DE = Germany, ES = Spain, FR = France, IT = Italy, NL = The Netherlands, UK = United Kingdom.

Source: Elaborated by CRPA on Eurostat Farm Structure Survey and REGIO databank 2005

Impact of improved compliance with the Nitrates Directive on EU pig farms' costs

Compliance costs for hog farms have been reported in the literature for several countries which are subject to the Nitrates Directive. A literature review identified compliance costs for the Nitrates Directive for each of the eight countries (Christiansen, 2007; Gourmelen and Rieu, 2006; De Roest *et al.*, 2007; De Hoop *et al.*, 2004; Penlington, 2007). The percentage increase per kg pigmeat has been estimated aggregating these cost estimates to the pig production cost calculations carried out by Interpig, a network of European research institutes coordinated by the Meat and Livestock Commission and the British Pig Executive (see also Table 10.2).

The estimated cost increases represent future additional costs for those pig farms which do not yet comply with the Nitrates Directive. For farms already in compliance, these costs have already been incurred. The degree of compliance among the Member States differs significantly. In Italy, where only recently an acceleration of the implementation is taking place the degree of compliance can be estimated in only 20%. In contrast, in Denmark where a rather strict application has been imposed on livestock farmers for some time, the degree of compliance is very high. According to the assessments carried out in the country reports, other countries reach degrees of compliance ranging between these two extreme values.

In order to obtain a percentage cost increase at sector level for each country, two crucial variables have been considered:

- The degree of compliance with the Nitrates Directive.
- The share of affected pigs as a percentage of the total number of pigs in the country.

It turns out that Italy, the Netherlands and Belgium will still have to face a cost increase which varies from 1.1% to 2.5% (Table 10.3). Other countries show limited cost increases, either because of a limited number of affected pigs (Denmark), or because of a relatively small number of pigs present in NVZs (Spain) or due to a high degree of compliance (United Kingdom).

Table 10.2 Production costs of EU pig farms (€/kg) and cost increases due to compliance with the Nitrates Directive

	BE	DK	DE	ES	FR	IT	NL	UK
Feed	0.685	0.680	0.696	0.852	0.730	0.994	0.693	0.814
Breeding, vet/med and energy	0.094	0.105	0.159	0.122	0.109	0.115	0.143	0.109
Labour	0.125	0.165	0.200	0.124	0.177	0.137	0.174	0.199
Building, finance and miscellaneous	0.342	0.468	0.475	0.409	0.397	0.367	0.452	0.617
Total costs	1.245	1.418	1.530	1.508	1.413	1.613	1.461	1.740
Compliance costs with Nitrates Directive (€/kg)	0.076	0.046	0.059	0.076	0.087	0.134	0.044	0.090
% cost increase	3.50	3.21	3.86	8.87	6.16	8.29	2.98	5.18

Sources: total pig production costs: BPEX (2006)

Table 10.3 Cost increase for affected pig farms and EU pig sector due to the application of the Nitrates Directive, 2005

Share (%)	**BE**	**DK**	**DE**	**ES**
Share of pigs in NVZ	100.0	100.0	100.0	24.3
Share of pigs affected by cost increase	60.3	10.9	17.0	7.6
Degree of compliance	50	80	85	20
Relative cost increase per kg meat	3.5	3.2	3.9	8.9
Relative cost increase for sector	1.1	0.1	0.1	0.5
	FR	**IT**	**NL**	**UK**
Share of pigs in NVZ	74.5	64.2	100.0	66.1
Share of pigs affected by cost increase	23.5	37.3	79.4	23.8
Degree of compliance	70	20	50	90
Relative cost increase per kg meat	6.2	8.3	3.0	5.2
Relative cost increase for sector	0.4	2.5	1.2	0.1

Source: Elaborated by CRPA on BPEX, and EUROSTAT

Assuming no cost increase for the remaining Member States, the EU pig sector as a whole faces an additional compliance cost of 0.55% in order to reach full compliance with the Nitrates Directive. If we would assume that all farms having a pig density of more than 170 kg N/ha still have to adjust to the Nitrates Directive, i.e. the compliance rate is zero, then the total extra compliance cost would imply a 4.8% increase in pig production costs in the EU.

Compliance costs for US hog farms under the Clean Water Act

Under US Clean Water Act, which has been described in detail in Chapter 8, hog farmers are required to obtain a permit under the National Pollution Discharge Elimination System (NPDES). This involves the implementation of a comprehensive nutrient management plan (CNMP). The CNMP must be focused on the control of nitrogen (N) or phosphorus (P), based on state or local priorities. The implementation of CNMPs often results in operations seeking additional land on which to spread manure and/or the use of alternative nutrient control strategies for manure. It is important to note that the CAFO regulations currently affect a small percentage of livestock operations, but potentially a large part of the US pig herd. The US Department of Agriculture (USDA) inventory for 2006 shows that farms with more than 2,000 head are only 7.8% of all pig operations in the US, but account for 80% of the total pig population.

In the US, 90% of the pig herds are raised on farms having more than 1,000 pigs, and 54% are raised on operations with more than 5,000 head. The greatest concentration of pigs exists in Iowa, Minnesota and North Carolina.

The present analysis has estimated the costs of compliance with the Clean Water Act (CWA) for pig farms typical of Iowa. This state was chosen because it is the leading hog producing state in the US. The average hog operation size in this state is approximately 4,300 head, well above the threshold size to be considered a Concentrated Animal Feeding Operation (CAFO) according to the Clean Water Act.

The USDA has estimated the costs for implementing a successful CNMP. The CNMP costs associated with swine operations are an average cost of all swine farms in the US and are presented per animal unit (AU), which is equivalent to 454 kg of live weight.

The national estimates for the CNMP were adjusted from a per AU basis (US$44) to a per 45.4 kg of gain basis to be consistent with the available financial data for pig farms (see Table 10.4). Using this national average cost per AU of US$44, the average cost per 45.4 kg of gain is estimated to be US$1.65. This is calculated by multiplying the average weight gain per pig of 90.9 kg by 6.67 pigs by 2 turns per year, which results in a weight gain per pig of 1,210 kg of gain per AU. The final result is that the cost of compliance translates into a 3.54% increase in total production costs for farms affected by the CWA. As 54% of the US pig population is raised on farms with more than 5,000 pigs and the degree of non-compliance equals 56.4% the total sector cost increase for pigs in the USA to be attributed to the CWA is 1.08%. If we compare this to the additional cost increase of full compliance with the Nitrates Directive in the EU which has been calculated as 0.55%, we notice that the CWA obligations in the US have a stronger economic impact on the pig sector, in particular due to the lower rate of compliance with the CWA. If we compare the EU and the US using a zero compliance rate in both countries the cost increase in the EU would be 4.8% as against 3.54% in the US. The higher total compliance costs in the EU are due to the higher share of farms which has to invest in order to comply with the Nitrates Directive as compared to US pig farms affected by the CWA.

Nutrient management regulations in Canada: Cases of Ontario and Manitoba

While all Canadian provinces have legislation related to management of livestock production systems, it is beyond the scope of this chapter to examine each province's legislation in detail. According to the 2006 Census of Agriculture, Ontario and Manitoba produce about half of the Canadian pork output, each contributing roughly equal shares. Ontario farms are representative of the eastern Canada hog production systems, found also in Quebec and to a lesser extent in the Maritimes. These are farms with smaller land holdings and higher input intensity compared to the western Canada farms, found in Manitoba, Alberta and Saskatchewan. Consequently, estimates of environmental compliance costs in Ontario and Manitoba would likely be representative at the national level as well. The next two sections assess the nutrient management regulations in the two provinces and the effect on costs at the farm and industry level.

Table 10.4 Production costs for a typical Iowa swine operation, 2005

Direct operating costs	US$ per 45.4 kg of gain
a. Feed	13.94
b. Feeder pigs	21.32
c. Hired labour	0.52
d. Veterinary and medicine	0.58
e. Other operating costs	2.67
Total operating costs	39.03
Overhead costs	7.64
Total costs	46.67
Annual CNMP cost	1.65
Percentage of total costs (%)	3.54

Ontario Nutrient Management Act (2009) sets standards for different aspects of nutrient management: land application standards, siting and construction standards, requirements for farm preparations of farm level nutrient management plans and strategies. It uses a combined nitrogen-phosphorus limitation (total of 224 kg/ha per year for nitrogen and not more than 78 kg/ha per year in excess of phosphorus balance). The main targets of the regulation are the existing farms that generate more than 300 Livestock Units (LU), operations expanding to and beyond 300 LU and new operations that generate 150 LU or more.

As Salvano *et al.* (2006) elaborated, the new nutrient regulations in Manitoba, based on phosphorus soil test require farms with soils that contain more than 120 parts per million (ppm) of phosphorus to apply manure in the amount that does not exceed crop phosphorus removal capacity. Farms whose soils contain between 60 and 120 ppm of phosphorus can apply up to 2 times the crop phosphorus removal rate, while those with soil test indicating less than 60 ppm of phosphorus present in the soil can apply manure based on the crop nitrogen requirement. In addition, new regulations include stipulations regarding establishment of buffer strips, prohibition of winter application, incorporation of fall applied manure, and building new storage on land adjacent to watercourses and in Red River Valley, termed as Special Management Areas (SMAs).

Affected pig production

Western Canada: Manitoba

Manitoba's nutrient regulations are based on soil phosphorus content. Farms in which soil phosphorus content exceeds 120 ppm or where phosphorus is being accumulated are subject to the new phosphorus-based regulations. In order to get a measure of how much of the provincial swine production would be affected, we used the 2006 Census of Agriculture data to calculate the agricultural land nutrient balance for each township in Manitoba by subtracting nutrient removal by crops and pasture from nutrient input from manure. The underlying assumption is that the soil in townships with currently positive phosphorus balance is either accumulating phosphorus or is already saturated. Thus, farms in these townships would be subject to the new regulation.

The calculated nutrient removal varied across townships, depending on the ratio of cropland to pasture. The cropland nutrient removal coefficient was calculated using the Potash and Phosphate Institute (2001) estimates for North American wheat, barley, canola and peas. The typical crop mix for the province was 40% wheat, 20% barley, 30% canola and 10% peas. The pasture removal coefficient was 5 kg of phosphorus per hectare, based on Hattey *et al.* (2003). Phosphorus output is a product of the township-level number of pigs and the respective coefficients for phosphorus content of swine manure, found in Moore and Gamroth (1993). The phosphorus output in the townships with a positive phosphorus balance was much higher (2 to 4 times) than the provincial average. The calculated phosphorus balance was positive in 14 townships. There were 336 swine farms in the 14 townships, accounting for 28% of all swine farms in the province and 58% of Manitoba's hog inventory. Thus, more than half of the provincial production could be affected by the new phosphorus-based nutrient regulations.

Eastern Canada: Ontario

The main targets of the Ontario nutrient management regulations are the existing farms that generate more than 300 LU and any expanding operations. In order to produce an estimate of the share of the provincial swine production affected by the Ontario Nutrient Management Act, we used the 2006 Census of Agriculture data to calculate the average per farm number of swine livestock units for all townships in Ontario. The livestock units for different types of swine animals units were aggregated using coefficients found in the Minnesota Department of Agriculture Animal Unit Calculation Worksheet. Sows and gilts for breeding were assigned 0.3 LU per animal, boars 0.4 LU, nursing and weaner pigs 0.05 LU, and 0.3 LU per animal were assigned to grower and finishing pigs.

There were 950 farms located in townships for which the average hog farm size exceeded 300 LU. Even though these farms account for about 23% of all Ontario swine farms, they house almost half of Ontario's pig inventory (47%). It should be noted that this estimate does not include the expanding operations with less than 300 LU, which are also subject to the nutrient management regulations. This indicates that any cost effect resulting from the Nutrient Management Act would affect at least 50% of the province's pork supply.

Economic cost of nutrient regulations

Oginskyy (2005) estimated the opportunity cost of the 2002 Ontario Nutrient Management Act under different swine dietary options using a mathematical programming model of a typical swine farrow-to-finish operation in western Ontario - a mixed hog-crop production-type system, with hogs as the main market product and crops used as feed or for sale depending on market conditions. Herd and land size parameters were varied in sensitivity analysis. The estimated monetary opportunity cost of compliance, if expressed as a percentage of total production cost, ranged from 0 to 8.1% of total costs for year 2007, depending on the selected level of nutrient standard, pig diet and farm size. As expected, less stringent nutrient level standards involve lower compliance costs. Similarly, adding lysine and phytase to the pig diet reduces compliance costs.

Salvano *et al.* (2006) estimated the economic effect of the new phosphorus-based regulations on swine operations in two most swine-intensive townships in Manitoba: La Broquerie and Hanover. The baseline model assumed either having enough owned land to comply with the nitrogen-based manure application standard or being able to lease from neighbours. For the farms with sufficient land area, the additional costs were associated with spreading manure over a larger area. Those operations with insufficient land would have an option to transport manure to another farm or to reduce manure nutrient content by chemical treatment. The change in monetary costs between nitrogen-based and phosphorus-based scenarios were estimated. The estimated change in costs corresponded to a range from 0.04% to 4.94% of total production cost for 2007, depending on land area available, manure output and compliance strategy.

Table 10.5 summarizes the cost structure of Ontario and Manitoba swine production in Canadian Dollars and Euros, using 1.5 Canadian Dollars per Euro exchange rate. The Table also shows nutrient management compliance costs as a percentage of total costs.

Table 10.5 Production costs for Canadian hog farms and cost increases due to compliance with nutrient management regulations

Costs	**$/kg of live weight sold**			**€/kg of live weight sold**		
	Ontario	Manitoba	Average	Ontario	Manitoba	Average
Feed[1]	0.839	0.764	0.801	0.559	0.509	0.534
Breeding, vet/med and energy[2]	0.175	0.115	0.145	0.116	0.077	0.097
Labour	0.137	0.158	0.148	0.091	0.106	0.099
Building, finance and miscellaneous[3]	0.362	0.373	0.367	0.241	0.249	0.245
Total cost	1.512	1.410	1.461	1.008	0.940	0.974
Compliance cost	0.063	0.035	0.049	0.042	0.023	0.033
% of total cost	4.140	2.489	3.343	4.140	2.489	3.343

Notes: [1]Includes: breeding herd feed, nursery feed and grower-finisher feed. See also Manitoba Agriculture, Food and Rural Initiatives (2009); [2]Includes: net replacement cost for gilts, health, breeding (artificial insemination and supplies), utilities (hydro, gas);[3]Includes: marketing, miscellaneous, manure disposal, repairs and maintenance, operating loan interest, depreciation interest, taxes and insurance. See also Salvano et al. (2006).

The average compliance costs were estimated at $0.063 and $0.035 per kg of live weight sold in Ontario and Manitoba respectively. This corresponds to about 4.1% of total production costs in Ontario and 2.5% in Manitoba. Assuming each province is representative of nutrient management compliance costs for its respective region, Ontario for eastern and Manitoba for western Canada, the national farm-level compliance costs would be somewhere in the range of 2.5 to 4.1% of total production costs.

Animal welfare: evaluated standards and cost implications

Increasingly the protection of animal welfare has become a relevant issue and hog farms in the EU have to adjust their management plans and housing systems to the new requirements imposed by EU regulations. Actually, the welfare of hogs in the EU is ruled by the directives 91/630/EEC, 2001/88/EC and 2001/93/EC. The main aspects of these directives provide for a:

- Prohibition of tying sows and gilts.
- Prohibition of slatted floors for sows and gilts.
- Prohibition of isolation of sows during the period between 4 weeks after insemination and the week before farrowing with a minimum space allowance of 2.25 per m^2 for sows and 1.64 m^2 for gilts.
- Maximum stocking rates for different pig categories.
- Minimum standards for slatted floors.

Table 10.6 Estimating cost increases for the Canadian hog sector due to the application of nutrient management regulations

Cost figures	Ontario	Manitoba
Average cost per farm (% of total cost)	4.140	2.489
Share of provincial production affected	47	58
Degree of compliance (min)	0	0
Degree of compliance (max)	100	100
Industry level additional compliance cost (max)	1.932	1.432
Industry level additional compliance cost (min)	0	0
Mean industry level additional compliance cost (% of total cost)	0.966	0.716

Of all above mentioned measures, the most important pending cost increase likely arises from the requirement for group housing of sows, since the other measures already have a high rate of compliance in the EU (Enting *et al.*, 2006). Therefore, the following analysis focuses on the cost increase per kg of pig meat associated with the switch from individual crates for sows to group housing, taking also into account the minimum space per head required by the EU regulation.

Impact of improved compliance with animal welfare on EU pig farm costs

Two practical examples illustrate the consequences for a closed cycle pig farm, where breeding sows and gilts are raised together with fattening pigs. No piglets are purchased from other pig farms as they are produced within the closed cycle farm. The examples include:

- a comparison between the old conventional housing system with sows housed in individual crates with an investment in a new pig farm where pigs are housed in groups 4 weeks after insemination; and
- an adjustment of an existing pig farm to the new housing requirements for sows.

The new requirements would cause an increase in investment in the two pregnancy phases of the sows of 12% per sow place, but as these two phases represents only 10.8% of the total investment of a new closed cycle pig farm the total investment cost increases by only 1.3%. Using an amortization rate of 3% on buildings and 12% on equipment, the expected increase in total production costs (per kg pig meat) is limited to 0.1%.

Table 10.7 Investment costs of new building of closed cycle pig housing (old individual crate system against the new group house system)

Investment		Housing system with individual crates for sows		Group housing of sows after four weeks of insemination	
	Number of pig places	(€/pig place)	Investment (€)	(€/pig place)	Investments (€)
Insemination 1st phase	46	1,884	86,664	2,025	93,150
Pregnancy 2nd phase	61	1,027	62,647	1,257	76,677
Farrowing	30	3,684	110,520	3,684	110,520
Weaning	285	259	73,815	259	73,815
Gilts in first phase	24	600	14,400	600	14,400
Growing up to 50 kg	368	320	117,760	320	117,760
1st phase fattening	450	480	216,000	480	216,000
2nd phase fattening	720	700	504,000	700	504,000
Other investments			370,862		370,862
Total investment			1,556,668		1,577,184

Source: CRPA

An adjustment of an existing pig farm to the new requirements has a more significant impact on the economy of the farm, as the end solution is not as optimal compared to a new building. The production capacity has been presumed to be unchanged and the pig farmer is only adjusting the pregnancy parts of the farm building. The effect on production costs of this solution is presented in the Table 10.9.

Although the cost increase is higher for the adjustment of an existing farm, the cost impact also in this case remains rather limited and well below 1%. The animal welfare directives do not appear to have a significant impact on production costs. It should be stated however that the farms that have to adjust to group housing do face a financial problem related to the search for resources to finance the investment, which for small farms can be a significant burden. Considering the pig production cost calculations carried out by Interpig, and the different degrees of compliance among the main EU producer countries, the adjustment of existing individual crate pig farms to pig farms with group housed sows would generate a 0.11% increase for the EU pig sector (Table 10.10).

Table 10.8 Difference in production costs due to investment in a new pig farm which complies with the group housing requirements for sows (€/kg)

Costs	Before adjustment, use of individual crates	After adjustment, using group housing of sows
Feed	0.687	0.687
Labour	0.217	0.217
Other variable costs	0.254	0.254
Total monetary costs	1.158	1.158
Interest on capital	0.014	0.014
Interest and depreciation	0.162	0.164
Total costs of production	1.335	1.337

Source: CRPA

Table 10.9 Difference in production costs due to an adjustment of an existing pig farm to the new animal welfare requirements of sows (€/kg)

	Before adjustment, use of individual crates	After adjustment, using group housing of sows
Feed	0.687	0.687
Labour	0.217	0.217
Other variable costs	0.254	0.254
Total monetary costs	1.158	1.158
Interest on capital	0.014	0.014
Interest and depreciation	0.162	0.172
Total costs of production	1.335	1.344

Table 10.10 Production costs for pigmeat and percentage cost increase due to compliance with animal welfare directive

	BE	DK	DE	ES	FR	IT	NL	UK
Feed	0.685	0.680	0.696	0.852	0.730	0.994	0.693	0.814
Breeding, veterinary, medicines and energy	0.094	0.105	0.159	0.122	0.109	0.115	0.143	0.109
Labour	0.125	0.165	0.200	0.124	0.177	0.137	0.174	0.199
Building, finance and miscellaneous	0.342	0.468	0.475	0.409	0.397	0.367	0.452	0.617
Total costs (€/kg)	1.245	1.418	1.530	1.508	1.413	1.613	1.461	1.740
Cost increase (€/kg)	0.011	0.011	0.011	0.011	0.011	0.011	0.011	0.011
Cost increase (%)	0.90	0.79	0.74	0.75	0.80	0.70	0.77	0.65
Share of closed cycle farms in pig production (%)	50	50	50	50	50	50	50	50
Degree of compliance (%)	80	90	80	50	50	50	50	95
Relative cost increase sector (%)	0.09	0.04	0.07	0.19	0.20	0.17	0.19	0.02

BE = Belgium, DK = Denmark, DE = Germany, ES = Spain, FR = France, IT = Italy, NL = The Netherlands, UK = United Kingdom.

Source: Elaborated by CRPA

Standards and external competitiveness

Expected impact of the full compliance with animal welfare rules and the Nitrates Directive on pigmeat trade flows

Increasing costs induced by a full Nitrates Directive compliance in the EU, certainly may affect international trade flows in pig meat and the market shares held by the main exporter countries. These effects have been simulated with the GTAP model. The standard GTAP 6 database does not differentiate between pig and poultry meat. As this is essential in the competitiveness analysis, the differentiation is introduced through the Splitcom application. This application allows dividing existing GTAP sectors, i.e.: meat products, into new sectors based on secondary production and trade data. Based on FAOSTAT data we used Splitcom to introduce the commodity 'pig meat' in the GTAP model.

Full compliance with the Nitrates Directive, assuming no compliance costs in either the US or Canada, leads to a 3% decrease of both market share and exported volumes of the EU. The effect on the level of production could cause a general increase of the EU imports of pigmeat by 4.4% higher demand for imported pigmeat (Table 10.11).

Table 10.11 Effects of N-Directive on global bilateral pig meat export (variation on imports and exports, in %)

	Importers						
Exporter	**USA**	**EU15**	**Japan**	**ROECD**	**ROW**	**Total**	**% variation of exports to world market**
Canada	0.41	5.06	0.27	0.98	0.64	0.64	0.32
USA	0.00	4.14	0.27	0.99	0.52	0.69	0.69
Brazil	0.35	4.34	0.25	1.00	0.63	1.03	1.03
EU15	-3.10	0.00	-1.15	-3.83	-3.64	-3.03	-3.03
ROECD	0.39	4.65	0.25	0.00	0.62	2.25	2.25
ROW	0.40	4.25	0.26	0.97	0.57	1.90	1.76
Total	-0.40	4.37	-0.05	-0.71	-0.86	0.14	0.00
% variation of imports from world market	-0.54	4.22	-0.19	-0.85	-1.00	0.00	

In the Japanese market, which is the EU's largest export market, the loss of about 1.15% in the quantities traded by EU would stimulate an increase of the demand for pigmeat coming from Brazil, the US, Canada and other countries at a rate of 0.27% each. In the other OECD countries the decrease in EU exports has been estimated as 3.8%, which favours the competitive position of Brazil, Canada and the US by about 1% each. Brazil's export share gain would be higher than those of Canada and the US. The gains of the three top exporters range from a maximum of 1% in the case of Brazil to a minimum of 0.3% for Canada.

Animal welfare regulations should affect EU competitiveness to a lesser extent since its effect on pig farm production cost has been estimated to be limited to only a 0.1% increase. This would entail a 0.8% growth of EU imports and a decrease of exports of 0.7% (Table 10.12).

Summing up the effects (see Table 10.13) of both standards' requirements would result in an overall decrease of EU exports by 3.7%. The loss in quantity on the Japanese market would be equal to 1.5% while exports towards other OECD would decrease by 4.5%. As far as imports are concerned, a 5.24% growth in imports is expected in large part due to the higher imports from Brazil (+5.2%). Brazil would gain 1.2% of its global market share, while a 0.4 and 0.8% increase in exports on the global market is to be expected for Canada and the US.

Table 10.12 Effects of animal welfare regulations on global bilateral pig meat export (variation on imports and exports, in %)

Exporter	Imports USA	EU15	Japan	ROECD	ROW	Total	% variation of exports to world market
Canada	0.08	1.27	0.08	0.20	0.14	0.10	0.09
USA	0.00	0.75	0.09	0.19	0.12	0.15	0.15
Brazil	0.00	0.79	0.08	0.22	0.14	0.21	0.21
EU15	-0.60	0.00	-0.38	-0.71	-0.83	-0.69	-0.69
ROECD	0.08	0.90	0.08	0.00	0.14	0.45	0.45
ROW	0.08	0.81	0.08	0.18	0.14	0.38	0.37
Total	-0.08	0.83	-0.02	-0.13	-0.20	0.01	0.00
% variation of imports from world market	-0.09	0.82	-0.03	-0.14	-0.21	0.00	

Table 10.13 Combined effects of the Nitrates Directive and animal welfare regulations on global bilateral pig meat export (variation on imports and exports, in %)

Exporter	Imports USA	EU15	Japan	ROECD	ROW	Total	% variation of exports to world market
Canada	0.49	5.70	0.35	1.18	0.79	0.56	0.40
USA	0.00	4.94	0.35	1.18	0.64	0.84	0.84
Brazil	0.47	5.20	0.33	1.22	0.77	1.25	1.25
EU15	-3.66	0.00	-1.53	-4.51	-4.44	-3.70	-3.70
ROECD	0.47	5.59	0.33	0.00	0.76	2.71	2.71
ROW	0.49	5.10	0.34	1.17	0.66	2.30	2.13
Total	-0.47	5.24	-0.07	-0.83	-1.05	0.16	0.00
% variation of imports from world market	-0.63	5.07	-0.23	-0.99	-1.21	0.00	

Expected impact of the full compliance with CWA requirements in the US

Full implementation of the Clear Water Act requirements for the US pig sector would produce higher impacts than those expected for the EU Nitrates Directive. This is due firstly to the high share of the US pig herd raised in farms that are eligible to be classified as

CAFOs and secondly, because of the relatively recent implementation of the CWA, which explains a compliance rate close to zero.

The high costs of full compliance with CWA regulations by most intensive pig farms, would affect the competitiveness of the US on the pig world market entailing a 7.3% decrease in US pig meat exports. This gap left will be covered by the EU (+1.85%) and Brazil (+1.18%), which would both gain market share, in particular on the Japanese market. Higher gains in market share are foreseen in the case of Canada, whose exports would be stimulated by the increasing demand coming from USA (Table 10.14).

Table 10.14 Effects of US Clean Water Act on global bilateral pig exports (variation on imports and exports, in %)

	Imports						
Exporter	**USA**	**EU15**	**Japan**	**ROECD**	**ROW**	**Total**	**% Variation of exports to global market (%)**
Canada	4.46	0.00	2.11	2.03	1.14	3.30	3.97
USA	0.00	-9.15	-7.20	-7.35	-7.13	-7.27	-7.27
Brazil	4.43	0.05	2.12	2.10	1.16	1.18	1.18
EU15	4.47	0.00	2.12	2.06	1.25	1.85	1.85
ROECD	4.46	0.04	2.11	0.00	1.18	1.03	1.03
ROW	4.49	0.08	2.13	2.05	2.42	1.47	2.13
Total	4.47	-0.69	-0.17	-2.38	-1.78	-0.65	0.00
Relative variation of imports from world market (%)	5.15	-0.04	0.48	-1.74	-1.14	0.00	

Conclusions

As the pig sector is one of the most intensive livestock production activities in the EU it would be expected that it faces significant impacts from the Nitrates Directive. The extent to which the Nitrates Directive may create extra costs to the pig sector depends on the pig density per hectare in each Member State, on the percentage of pigs present in NVZs and on the degree of compliance of pig farmers to the Nitrates Directive. These three data differ very much from country to country and explain primarily the very different sector cost increases for the pig sectors of EU Member States. The overall EU additional cost increase to be attributed to the pig sector due to full compliance with the Nitrates Directive has been estimated at 0.55%. If we assume however a zero compliance rate the total compliance costs of the EU with the Nitrates Directive would be 4.8%.

Such a cost increase has a significant impact on the EU trade balance of pigmeat. Total EU exports will decline by 3% and imports will increase by 4.4%. This decline in EU competitiveness will favour the exports of Brazil (+1%), the US (+0.7%) and Canada (+0.3%) on the world market. EU imports will increase in particular from Brazil (+4.3%), Canada (+4.1%) and the US (4.1%).

From a comparison with the impact of the CWA in the US it turns out that this act raises the cost for the American pig sector by 1.08%, almost twice the comparable cost effect that we attribute to the Nitrates Directive in the EU. The reason for this substantial rise of costs has to be attributed to the large percentage of pigs affected by this measure and its rather recent application to US pig farms, which still implies a rather low degree of compliance. This rise in costs will cause a reduction in US exports of pigmeat of 7.3% and a decline of exports of 4.5%. Canada would gain the most from this situation increasing its exports by 4% on the world market, of which 4.5% more to the US and a 2.1% increase on the Japanese market.

Our calculation of competitiveness effects of animal welfare regulations for pig farmers in the EU shows that the cost increase is very limited. The reasons for this minor cost impact are an already high degree of compliance with the standards and the limited rise in costs for farmers which still have to adapt their farm to the new legislation. At farm level the cost increase is well below 1% and this generates a rise of costs at sector level of 0.11%. This cost impact evidently causes a growth of imports of only 0.8% and a decline of exports of 0.7%.

As has been expected the Nitrates Directive for the pig farms creates the most substantial burden of costs, in particular in EU Member States with a high pig density and a low degree of compliance.

References

BPEX (2006) *Pig cost of production in selected EU countries*, British Pig Executive, Interpig, Meat and Livestock Commission, Milton Keynes.

Christiansen, M.G. (2007) Danish Meat personal communication.

De Hoop, D.W., H.H. Luesink, H. Prins, C.H.G. Daatselaar, K.H.M. van Bommel and L.J. Mokveld (2004) Effecten in 2006 en 2009 van Mestakkoord en nieuw EU-landbouwbeleid. LEI Rapport 6.04.23, Den Haag.

De Roest K., C. Montanari and E. Corradini (2007) *Il costo della Direttiva Nitrati*, BIT SpA - Cassa Padana, Leno (Brescia).

Enting J., I. Vermeij, A.J.J. Bosma and A. van Buiten (2006) *Level playing field in de varkenshouderij, Implementatie, interpretatie en controle op naleving van EU-richtlijnen voor de varkenshouderij*, Animal Sciences Group, WUR, Wageningen.

Gourmelen, C. and M. Rieu (2006) Evaluation du coût de la gestion des effluents dans différents types d'exploitations porcines, *Journées Recherche Porcine*, 38, 263-70.

Hattey, J.A., G.W. Horn, C.R. Krehbiel, C.A. Stasia and E.G. Krenzer (2003), 'Phosphorus removal by different wheat-stocker cattle production systems. OSU Animal Science 2003 Research Report.

Manitoba Agriculture, Food and Rural Initiatives (2009), 'Guidelines for estimating swine, farrow-finish costs based on 300 sows and 6,204 pigs sold', Available : http://www.gov.mb.ca/agriculture/financial/farm/pdf/copswinefarrowfinishcosts300sows2007.pdf [Accessed: August 31, 2009].

Moore, J.A. and M.J. Gamroth (1993), 'Calculating the fertilizer value of manure from livestock operations', OSU Extension Service, Corvallis, OR.

Oginskyy, A. (2005), 'Modeling the impacts of nutrient pollution regulations on Swine farrow-to-finish farms', MSc Thesis, University of Guelph.

Ontario Nutrient Management Act (2009), *Revised Statutes of Ontario*. Toronto: Queen's Printer for Ontario. Chapter 4.

Penlington, N. (2007) Interpig – Costs of Nitrates Directive, Meat and Livestock Commission, Milton Keynes.

Potash and Phosphate Institute (2001), 'Plant nutrient use in North American Agriculture', PPI/PPIC/FAR Technical Bulletin 2002-1, Norcross, GA.

Salvano, E., A.N. Rousseau, G. Debailleuil and J.P. Villeneuve (2006), 'An environmental benefit-cost analysis case study of nutrient management in an agricultural watershed', *Canadian Water Resources Journal,* 31(2), 105-22.

Chapter 11

Environmental standards in the fruits and vegetables sector of Spain[1]

Consuelo Varela-Ortega and Paloma Esteve

Introduction

Spain is one of the main producers and exporters of fruits and vegetables in the EU. This chapter gives a general overview of Spain's fruit and vegetable sector and examines the impacts of environmental standards, including Good Agricultural and Environmental Conditions (GAECs) on the competitiveness of the fruits and vegetables sector of Spain. The application of the Single Farm Payment system as well as cross compliance regulations is uneven across industries in the fruit and vegetable sector. However, the last reform of the Common Market Organization for Fruits and Vegetables (CMO F&V) and the 2003 CAP reform (the CAP Health Check) are expected to lead to substantial changes in this sector.

The chapter is organized into five sections. Following the introduction, the second section describes production and trade for the fruit and vegetable sector. The third section describes the EU policy framework for fruits and vegetables. The effects of the 2003 reform of the Common Agricultural Policy (CAP) on land use and irrigated crops in Spain are used to illustrate the implications of this policy framework. The fourth section presents case studies of selected crops, regions and the effects of specific policy measures on competitiveness. The last section of the chapter contains a summary of the main conclusions.

The global dimension of fruits and vegetables

International trade in fresh fruits and vegetables is complex and dynamic. World production of fruit and vegetables in 2008 was over 1,560 million metric tons (FAOSTAT, 2011a) in which fruits represent around 37% and vegetables around 63%. In aggregated terms, Asia is the leading production region with a share of 66%, followed by Africa (9%), the EU (8%), North and Central America (7%) and South America (6%) (Table 11.1).

[1] The authors would like to acknowledge the careful reading, general comments and discussion about the CAP reform and its application in Spain by Dr. Ignacio Atance Muñiz (Spanish Ministry of Environment and Rural and Marine Affairs).

Table 11.1 Fruit and vegetable production by country (2008)

Countries	Production (1,000 metric tons)	Share of world production (%)
World	1,567,999	100
China	615,746	39.27
India	159,646	10.18
USA	64,917	4.1
Brazil	48,844	3.12
Turkey	40,037	2.55
Italy	30,995	1.98
Iran, Islamic Republic of	29,381	1.87
Egypt	28,914	1.84
Spain	28,818	1.84
Mexico	28,234	1.80
Indonesia	24,824	1.58
Philippines	21,459	1.37
Nigeria	20,330	1.30
Russian Federation	17,161	1.09
Japan	14,399	0.92
ROW*	394,292	25.15

** ROW = rest of the world.*
Includes all fresh (non-preserved) fruits and vegetables

Source: Own elaboration from FAOSTAT (2011a)

In the EU, fruits and vegetables account for about a sixth of the value of EU agricultural production, making the EU the third largest producer in the world. It is also the largest importer. Consumption of fresh fruit and vegetables in the EU is generally stable with 43 million metric tons and 46 million metric tons (respectively) consumed within the EU. Total vegetable production in the EU-27 was about 64 million metric tons in 2008. The leading vegetable producing Member States are Italy and Spain (with 13.7 and 12.67 million metric tons respectively), followed by Poland (5.4 million metric tons) and France (with 5 million metric tons). Fresh fruit production in the EU in 2008 was 61 million metric tons. Italy was the leading Member State (17.6 million metric tons), followed by Spain (16.2 million metric tons) and France (8.5 million metric tons).

Tomatoes are the leading vegetable product in the EU, in terms of volume, with a total 16.4 million metric tons produced in 2008. Of this volume, almost 6 million metric tons are produced in Italy, 4 million metric tons in Spain, 1.3 million metric tons in Greece and 1.1 million metric tons in Portugal. Other important vegetables crops in 2008 in the EU were cabbages (5.5 million tonnes), dry onions (5.5 million metric tons), carrots (5.3 million metric tons) and lettuce (3.3 million metric tons). Regarding fruits, with the exception of grapes, apples is the most important fruit product in the EU (in terms of production), with a total of over 12 million metric tons produced in 2008. Apple production takes place primarily in Poland (2.8 million metric tons), Italy (2.2 million metric tons), France (1.9 million metric tons) and Germany (1 million metric tons) (FAOSTAT, 2011a). Total citrus fruit production in the EU was 11.3 million metric tons in 2008. Oranges accounted for 6.6 million metric tons and small citrus fruits (tangerines, mandarins and clementines) for 3.2 million metric tons. Spain is the main producer of citrus fruit with 6.4 million metric tons in 2008 (FAOSTAT, 2011a), followed by Italy (3.5 million metric tons) and Greece (1 million metric tons). Production of peaches and nectarines was 4 million metric tons and production of pears was 2.4 million metric tons.

The Netherlands, the US and Spain are the three leading fruit and vegetable exporters in the world, by revenue, with a total value around US$15.7, 15.6 and 15.3 billion respectively in 2008 (see Figure 11.1), which accounts for 10% of global exports, followed by China with an export value of fruits and vegetables of US$12.6 billion. China is the largest fruit and vegetable producer in the world (Table 11.1). Developing countries such as China, India, Indonesia and Turkey are important potential competitors for the EU.

The EU is the world's largest importer of fruits and vegetables, when the region is considered in aggregate (FAOSTAT, 2011b). Together, the EU and the US have a share of 64% of world fruit and vegetable imports. The world's largest importing country in 2008 was Germany. The US, the UK, France and the Netherlands, in that order, were the next largest importing countries in the world. Latin American countries are the main suppliers of fruits and vegetables for Europe, supplying 27.4% of EU imports. The Mercosur countries (Argentina, Brazil, Paraguay and Uruguay) supply some 20% of EU imports. Important products are Chilean apples and Argentinean lemons. Mediterranean countries supply 14.7% of the total fruit and vegetable imports to the EU (Turkey, 30.7% and Morocco 27.6%). The main products imported from Morocco are citrus and tomatoes which constitute 70% of the total. Turkey supplies mainly citrus and dried fruits.

Figure 11.1 Export value of fruits and vegetables by country (2008)

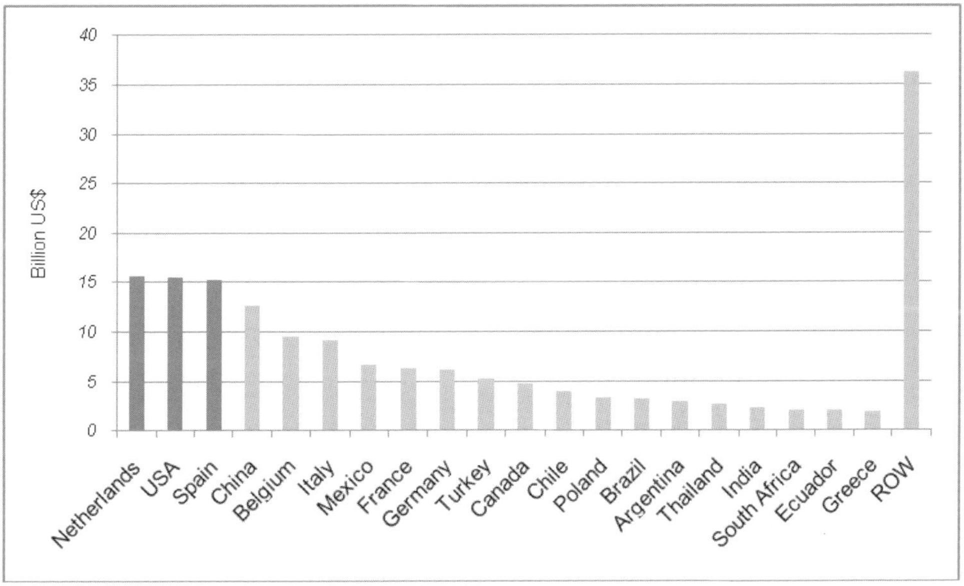

** ROW = rest of the world.*
Includes all fresh (non-preserved) fruits and vegetables

Source: FAOSTAT (2011b)

Fruits and vegetables constitute an important part of total agricultural production of Spain. In terms of value of production, fruits and vegetables account for 56% of the value of national agricultural production in 2008 (MARM, 2010). The largest products, in terms of volume, are citrus accounting for more than 6.3 million metric tons, including oranges at 3.4 million metric tons, and tomatoes, with more than four million metric tons. Valencia is the most important region for citrus production in Spain, accounting for 70% of total national production. Castilla-La Mancha is another important region for fruit and vegetable production. Water scarcity in the area is the limiting factor for expanding production of vegetables and therefore most farms have to rely on mixed cropping patterns combining vegetable production with less water intensive crops like winter cereals. Vineyard is the main crop in terms of area and, very often, it is combined with vegetables as well.

Reform of the fruit and vegetables regime in the EU

The CMO F&V differs from CAP regulations for other agricultural products. Established in 1962, the F&V CMO, institutes a minimum price for growers. A system of Community aid granted to certain citrus fruits was laid down in 1996. The CMO F&V was reformed in 2007 in the Council Regulation 1182/2007 (CEC 2007a, 2007b, 2008). This reform was intended to bring the fruit and vegetable sector closer to the rest of the reformed CAP framework. However, the new CMO continues to use Operation Programs (OPs) as the main instrument to support Producers' Organizations (POs). Through the OPs the POs can receive a grant of up to 4.6% of the total value of the marketed production. This grant is 50% co-financed by the POs (40% in case of organic production) and it should be invested in the related OPs of each PO. The new scheme includes specific measures for the prevention and management of production-related problems. The former crop harvest set-aside was abolished, as well as the obligation to dedicate a minimum 10% of the program's budget to environmental measures. The new regulation seeks to comply with the 2003 CAP reform in several respects. All fruit and vegetable areas will be eligible under the Single Farm Payment (SFP) scheme. Member States can consider exemptions to these entitlements and determine the period of reference and the payment amount for each vegetable or fruit product. A transitional stage was implemented to facilitate growers' adaptation to the new scheme. With this reform, all agricultural area that had fruit and vegetables during the reference period, including permanent crops and land under table potatoes, is eligible in all 27 Member States.

The introduction of fruit and vegetable areas in the Single Payment Scheme means that the cross compliance rules will apply to all fruit and vegetables producers receiving direct payments, which are decoupled. Support payments for fruits and vegetables have been increased for certain processed products (such as tomatoes, citrus, peaches, Williams and Rocha pears, raisins, dried figs and prunes). The aid scheme is based on direct aid to producers according to national thresholds. Penalties apply if processed volumes increase beyond fixed limits. The new CMO abolishes this type of support aid which is substituted by a consolidated national production 'envelope' under the SFP scheme. Under this modality, each member state can decide how to manage the SFP funds. These can be distributed among the former producers of processed fruits and vegetables for an equivalent payment of the amount they received under the previous regime or, alternatively, redistribute the new payment among all fruit and vegetable producers.

The general CAP reform approach allows Member States more flexibility in their application of CAP regulations. In Spain, the new fruit and vegetable regime has been applied under the National Strategy for sustainable OPs (MARM, 2008) and the Royal Decrees 262/2008 (BOE 2008a), 864/2008 (BOE, 2008b) and 1972/2008 (BOE 2008c). These national statutes define a transitory period so that fruit and vegetable land was not eligible for the SFP until December 31 2010. This limitation responds to the concerns among producers on the effects of the decoupled aid, mainly for flexible crops that can be consumed fresh or processed, like oranges and tomatoes. Moreover, the new possibility to produce fruits and vegetables on areas eligible for the SFP has raised serious concerns in relation to competitiveness. A three year transition period (2008, 2009 and 2010) was established for processed tomatoes with a 50% partial decoupling scheme. Following this transition period, a full decoupled scheme applies based on a reference period of years 2004/05, 2005/06 and 2006/07.

For citrus production, support payments were area based (per ha) fully coupled to production for a two-year transition period (2008 and 2009). Following this period, support payments were fully decoupled and transformed into a SFP to all farms in operation by September 30, 2006, regardless of the area dedicated to processed citrus crops. For all other fruits and vegetables, full decoupling of aid is not subject to a transition period, establishing a reference period of years 2004/05, 2005/06 and 2006/07, except for dried figs, peaches and Williams and Rocha pears for which the reference period is 2003/04, 2005/06 and 2006/07.

The main trade measure outlined in the reform proposal was the abolition of export refunds. The rest of the provisions related to trade with third countries remain unchanged until the outcome of the WTO negotiations, including entry prices, special safeguard clause and the general safeguard clause (USDA, 2007). Entry prices in the CMO F&V account for almost half of the total amber box subsidies declared by the EU in the WTO.

The impact of the CAP reform on land use

Historically, the CAP encouraged intensification of agricultural production across the EU. This has been more apparent in the Spanish central regions and in the Mediterranean Andalucía that grow significant volumes of CAP supported crops. Over the years, the CAP has evolved to respond to the international trade agreements and to include environmental requirements. Under the 2003 CAP reform, direct payments have been replaced by a Single Farm Payment, almost fully decoupled from production in Spain and tied to the 'cross compliance' of specific environmental regulations (among others). The last CAP reform of 2010, the so-called CAP 'Health check' (CEC 2009 a,b,c) includes stronger and simplified cross compliance provisions for the protection of ecosystems, including the requirement to respond to the new challenges of climate change and water management by developing adaptation strategies in the agricultural systems. For the first time, the new CAP specifically foresees cross compliance measures for protecting water courses and for achieving better water management practices in irrigated agriculture and it gives a further impetus to the integration of water policies and agricultural policies. The analysis that we have carried out in section 4 to assess the cost of compliance of irrigated farms reflects this emerging policy reality.

The CAP has sought to promote a multifunctional environmentally-oriented sustainable agriculture in the EU. Several authors have studied the potential of CAP policy instruments to attain the dual objective of agricultural production and water resources conservation in irrigated agriculture (Brouwer, 2004; Varela-Ortega *et al.*, 2002; Varela-Ortega, 2007; Varela-Ortega *et al.*, 2006).

We have compared the distribution of irrigated crops in a pre-reform year (2004) and post-reform year (2008) at the regional level. At the national level, the new decoupled SFP induces a crop shift away from intensively irrigated crops, such as maize and legumes, in favour of other low water-consumption crops, such as winter cereals. Maize no longer benefits from the higher yield-based subsidies of the previous CAP programmes. Land cropped for maize has diminished by 23% and there has been a 17% increase in winter cereals area. These are low water demanding crops that have an equivalent subsidy under the new CAP. There has also been a sharp increase in lands used for olive (also under CAP subsidies) and vine (not a CAP crop) (38% and 26% respectively) production. The expansion of these crops responds to the higher profits obtained in specialized farming well adapted to modern-technology drip irrigation and market opportunities.

The effects of CAP reform vary significantly across regions in Spain. CAP-dependent crops are grown predominantly in the central inland areas and in Andalucía. Andalucía, in the Mediterranean south, grows a variety of crops. The continental inland regions, with more limited cropping potential, grow predominantly continental CAP supported crops. In all regions, the maize cropping land has fallen in favour of winter cereals. However, in Andalucía, the larger than national average reduction in maize area (46%) has coincided with a sharp increase in intensive olive production (35%).

Costs of compliance with good agricultural and environmental conditions

The GTAP model used elsewhere in this book does not include the fruit and vegetable sector, so a case study representative farm level approach is used in this chapter to study the effects of compliance costs with the Good Agricultural and Environmental Conditions (GAECs) in the Spanish fruit and vegetable sector. For the fruit sector, we focus on citrus farms, because of the important role of citrus both in production and in export value. The main Spanish region in citrus production is Valencia, so that the case study area for the citrus farms is in that region. For the vegetable sector, the region selected for the case study is Castilla-La Mancha. This region is one of the most important nitrate vulnerable areas in Spain. In addition, aquifer overexploitation is an important issue in the area, so that the GAEC requirement regarding overexploited aquifers is also relevant here. The case study will focus on farm systems located on the Western La Mancha Aquifer, in the Upper Guadiana river basin, which combine problems of nitrate pollution and overexploitation, where the Water Abstraction Plan (WAP) establishes water quotas for the farms in the aquifer. The surface of the aquifer is 5,500 km^2 and irrigated agriculture in the aquifer represents a surface of 140,000 hectares. General characteristics of the 5 farm types included in the case studies are reported in Table 11.2. For the citrus fruit farms in the region of Valencia, compliance costs with pesticide regulations and the Nitrates Directive, and the impact of compliance on farm gross margins were evaluated. Data and information were obtained from interviews with experts and data from the Ministry of Agriculture and the Regional Government of Valencia as well as several studies that were available.

For the mixed vegetable farms of the region of Castilla-La Mancha, the analysis of the cost of compliance has been carried out using an integrated economic-agronomic simulation model developed for this area. The structure of this model is described in Figure 11.2. The products analysed include melons, peppers, potatoes and garlic, and the selected policy measures for this area are GAEC to control water use and the Nitrates Directive, considering that this area is a nitrate vulnerable zone. For the GAECs we have selected a specific measure that Spain has introduced - water use limitations in areas of overexploited aquifers as is the case in the region. Both measures are of major importance in Mediterranean countries and regions where groundwater is the major source of water for irrigated agriculture. In this case, the cost of compliance with the Nitrates Directive and with the GAECs water use control, is not a direct cost to farmers but rather a reduction in yields, due to a limitation in the use of nitrate fertilizers and water, and consequently a reduction in farm gross margins.

The basic analytical framework developed to analyse fruit and vegetables sector can be summarized as follows:

(i) Selection of the regions, sub-regions, agronomic areas and municipalities (counties) that represent the agricultural production systems of the products considered.
(ii) Selection of the most relevant policy measures.
(iii) Selection of representative farms within the study regions that will allow the analysis of cost structure at farm level.
(iv) Assessment of compliance costs in the pre and post establishment of the selected measures.
(v) Development of a model-based analysis integrating an economic (mathematical programming model) and agronomic model (CropSyst) for policy simulations and assessment of the cost of selected measures (Nitrates Directive and water use) (for the case study of Castilla-La Mancha).
(vi) Analysis of competitiveness and policy standards at farm level.

Figure 11.2 Steps to assess costs of compliance in the region of Castilla-La Mancha

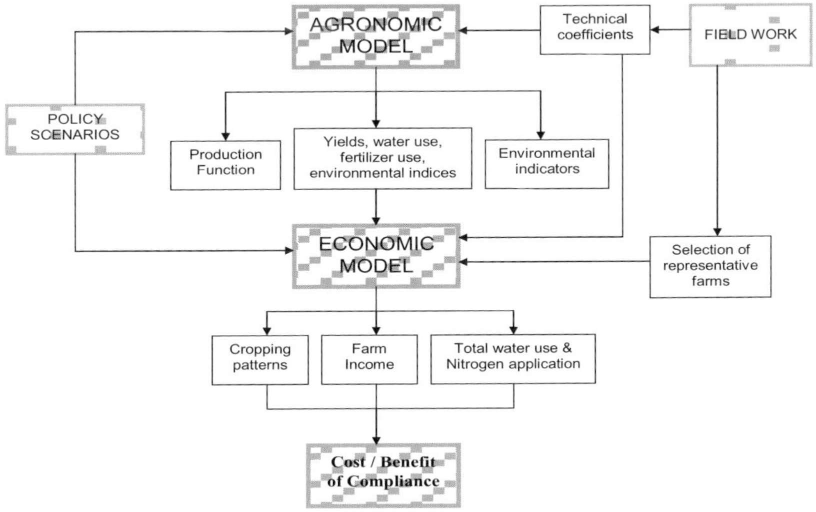

Table 11.2 Study regions and selection of farm types

Region	Farm	Area (ha)	Coverage in the study area (%)	Cropping pattern
Valencia	F&V_F1	5	71.7	100% irrigated citrus orchard
	F&V_F2	150	32.65	43% rain fed / 37% extensive irrigated crops / 20% horticulture
Castilla-La Mancha	F&V_F3	70	18.42	10% rain fed / 57% extensive irrigated crops / 33% horticulture
	F&V_F4	19	18.52	10% rain fed / 74% extensive irrigated crops / 16% horticulture
	F&V_F5	40	27.63	5% rain fed / 24% extensive irrigated crops / 31% horticulture / 40% vineyard

Cost of compliance in citrus farms in Valencia

Two aspects of compliance costs were analysed: (i) the cost reduction due to the required decrease of specific inputs use, and (ii) the impact on yields of input use reductions and of other specific conservation practices. Citrus orchards in the Valencia region apply nitrogen at an average rate of 280 kg N/ha/year. The application level allowed under the Nitrates Directive in citrus production in Valencia is 240 kg N per ha per year. A reduction of 40 kg N per ha might be the maximum amount of N that growers should decrease in order to comply with the Nitrates Directive. As such this generates an average reduction in fertilizer expenditures of €18.76 per hectare per year. Reductions in costs for pesticides uses for this region are not easy to estimate for different reasons. The most important changes on pesticides use during the last year have been related to insecticides, based on concerns about the effects of active ingredients on non-target organisms and human health. Thus, a variety of alternative products have appeared during the last years. For our case study on citrus, we have compared the certification method EUREPGAP with the conventional citrus orchard production in the region of Valencia.

The effects of compliance on citrus production cannot be defined as a cost of compliance since a reduction in inputs could not necessarily lead to a yield decrease in a clear manner. There is no evidence of yield decrease because of reductions in the average amounts of pesticides used or nitrogen application rates. Pesticide regulations and nitrogen application rate restrictions do not clearly modify revenue in citrus production first because the nitrogen dosages or pesticides application are currently many times above the optimal level and thus a reduction could even benefit the crop, and second, because there is a large variety of production protocols and techniques that allow producers to chose acceptable alternatives which minimize the impact on crop profitability.

According to information of local experts, the risk of control of the nitrates and erosion measures perceived by local producers in Valencia is not high, and even if the main citrus production areas belong to the nitrates vulnerable zone in the region of Valencia, requirements related to the GAEC and Nitrates Directive seem to be easy to comply with for most citrus farmers. Therefore, these measures might not imply a noticeable change in the costs at the farm level.

Cost of compliance of vegetable farms in Castilla-La Mancha

This portion of the analysis focuses on the impact of compliance with the Nitrates Directive and with water consumption measures for recovering overexploited aquifers. This specific legal requirement is part of the GAECs defined in the Spanish transposition of the EU cross compliance legislation (MAPA, 2004). There are six nitrate vulnerable zones in this region. The action programme (DOCM, 2001) establishes the maximum nitrogen application allowed for each crop in these vulnerable zones and specifies the agricultural practices required in order to minimize nitrogen leaching. The use of groundwater in the western La Mancha aquifer allowed the development of the irrigated agriculture during the 1960's and 1970's. The joint effect of the CAP and of Spanish rural development policies, which encouraged irrigation intensification, together with technological development, which reduced water pumping and abstraction costs, produced a huge increase of water abstraction from the aquifer, which led to a provisional declaration of overexploitation in 1987 and a permanent declaration in 1991. According to the article 54 of the Spanish Water Law of 1985, the river basin authority established a regional water plan with a compulsory water quota system (Water Abstraction Plan). The Water Abstraction Plan is updated every year and it is based on farm size, setting higher restrictions for larger farms. This water conservation policy produced important social and economic impacts, but has not succeeded to achieve the aquifer recovery. Agriculture is the main cause of overexploitation and of wetland degradation in the region and, therefore, its inclusion among the GAECs is highly relevant.

In accordance with the GAEC with water conservation requirements, farms with irrigated area in overexploited aquifer zones are required to have water abstraction permits and the installation of water flow meters is compulsory for every well on every farm.

The simulation modelling approach used for analysis for farms in this region consisted of an agronomic model CropSyst, calibrated for this area by Azaña (2007). This model calculated changes in yields for each crop type for different levels of nitrogen application. These results are then used as an input in the economic model that calculates changes in the farm's gross margin and cropping pattern. The economic model used is a farm-based non-linear single-period mathematical programming model of constrained optimization, developed by Varela-Ortega *et al.*, (2006). The model describes the behaviour of the representative farmers selected confronted by different policy scenarios. Following previous work in the area of study (Varela-Ortega *et al.*, 1998, 2002), the model incorporates new risk parameters and maximizes a utility function (U) subject to technical, economic and policy constraints (g). The utility function is defined by a gross margin (Z) and a risk vector (R) that takes into account climate as well as market price variability. The model can be summarized as follows:

Maximize U = $f(x)$, $f(x)$ = Z - R

Subject to the following constraints $g(x) \in S_1$,

$$x \in S_2$$

Where '*x*' is the vector of the decision-making variables or vector of the activities defined by a given crop-growing area and by an associated production technique, irrigation method and soil type (S). The model is solved using GAMS (General Algebraic Modeling System). The technical coefficients and parameters of the model were obtained from field work carried out during 2006 and 2007 (Varela-Ortega *et al.*, 2007) in the study area, consisting of surveys and interviews with farmers, irrigation community representatives, technical experts, river basin managers, and regional government officials. The model was calibrated and validated, using the risk aversion coefficient as calibration parameter and the comparative data on crop distribution, land and labour parameters in the study area (Varela-Ortega, 2007).

Model results showed that the impact of the Nitrates Directive varies considerably among different crops and therefore the impact of nitrogen reductions at the farm level changes depend on the farm's cropping pattern. The necessary nitrogen application rate reductions are considerably larger for some intensive crops such as peas, potatoes, vines or maize (65%, 62%, 56% and 42% respectively) and smaller for melons (28%) and wheat (31%). The results of the agronomic model show that the impact on crop yields is high for sugar beet (23%), sunflower (20%) and barley (19%). Among the horticultural crops, potatoes and garlic are the most sensitive crops (yield decreases about 10% and 6% respectively).

The results of the economic model show that, given the farmer's capacity to change cropping patterns as a means of adapting to regulations, those changes can minimize the impact of the Nitrates Directive. Compliance with this directive does not inflict any income loss to farms F2 and F4 (with flexible cropping patterns), results in a slight reduction in income for farm F3 (3%) and a considerable reduction for farm F5 (15%), due to a lower capacity to adjust its cropping pattern, since it has a high proportion of permanent crops.

The impact of the GAEC's water conservation measure is studied based on the application of the region's Water Abstraction Plan (WAP), a strict water quota system for recovering the overexploited Western La Mancha aquifer. Water abstractions are reduced from an average volume of 4,270 m^3/ha (historical water rights) to an average volume of 1,700 m^3 per ha ranging from a maximum of 2,640 m^3 per ha (for farms under 30 ha) to 1,000 m^3 per ha (for vineyards). This water policy produces a shift in cropping patterns from high water demanding crops (maize, sugar beet) to low water demanding crops (barley and sunflower). Horticulture crops are produced using modern irrigation techniques (sprinklers and drip irrigation) that have increased water use efficiency in the area.

Figure 11.3 shows that farms are more vulnerable to reductions in water availability than to nitrogen fertilizer restrictions. Compliance with the Water Abstraction Plan results in an average reduction of farm income of 20% as compared to a 3% reduction in the case of compliance with the Nitrates Directive. Compliance with the WAP is comparatively more costly to farms that crop water intensive crops, such as F3 and F5 (up to 40% and 30% of their initial income respectively).

Competitiveness at the farm level

The assessment of short-run competitiveness at the farm level is based on the relative changes in gross margin (GM). Aggregate results are calculated as a weighted average considering the relative weight of each farm type in the area. The aggregate results for the irrigation zone of the Western La Mancha aquifer (the main nitrates vulnerable area in the region of Castilla-La Mancha) show an average gross margin loss of 21€ per ha, which is equivalent to a reduction of about 3% (see Table 11.3). Average competitiveness index is 0.97 varying from 0.84 for the farm with a low adaptation capacity to 1.11 for the farm with a flexible cropping pattern and high adaptation capacity. This shows the potential benefits of compliance in some farms in this region in Spain.

Figure 11.3 Effect of the Nitrates Directive and of the Water Abstraction Plan (GAEC requirement) on farm income (income index)

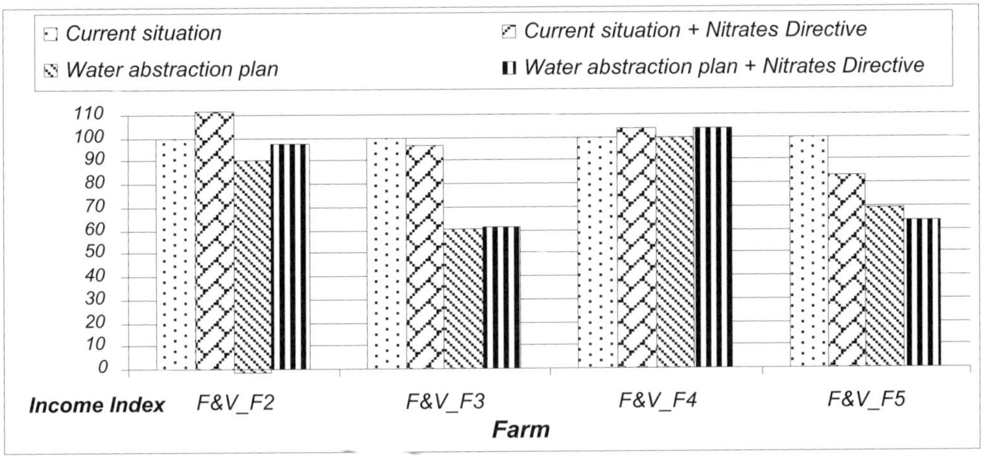

Table 11.3 Effects on competitiveness of compliance with the Nitrates Directive in the F&V sector

Regions/ crops	Farm type	Before implementation of the directive GM (€/ha)	During implementation of the directive GM (€/ha)	Absolute difference in gross margin variation (€/ha)	Relative difference in gross margin variation
Valencia (citrus)	F&V_F1	1862	1881	19	1.01
	F&V_F2	625	696	70	1.11
Castilla-La Mancha (mixed-vegetables)	F&V_F3	739	713	-27	0.96
	F&V_F4	754	781	28	1.04
	F&V_F5	1039	878	-161	0.84
	Aggregate average	804	783	-21	0.97

The same approach was used to analyse the effect of compliance with the water policy measures in the region of Castilla-La Mancha. Table 11.4 compares the pre and post situation of the establishment of water quotas (Water Abstraction Plan) for each farm type in the region. The results for the Western La Mancha aquifer show an average gross margin loss of 164 €/ha and a decrease of profitability of 20%, which represents a substantial decrease in competitiveness. The average competitiveness index is 0.80. The impacts on competitiveness of citrus from food safety and pesticide regulations were assessed by comparing the cultivation method of the selected farm type with the EUREPGAP cultivation method. The results, which represent a situation in which all SMR standards on pesticides regulations are covered, are reported in Table 11.5. Certified production schemes like EUREPGAP comply with standards on pesticides and food safety, and show a potential reduction in costs (150 €/ha). Using an Integrated Crop Management protocol, growers may experience a reduction in yields (reductions might be of about 30%), along with a reduction in costs and premium prices but gross margin per hectare increases.

Table 11.4 Effects on competitiveness of compliance with GAECs for water conservation

Regions/ Crops	Farm type	Before introduction of historical water rights GM (€/ha)	After introduction of water use quotas GM (€/ha)	Absolute difference in gross margin variation (€/ha)	Relative difference in gross margin variation
Castilla-La Mancha mixed-vegetables	F&V_F2	625	573	-58	0.91
	F&V_F3	739	447	-292	0.60
	F&V_F4	754	754	0	1.00
	F&V_F5	1039	719	-320	0.69
	Aggregate average	804	640	-164	0.80

Table 11.5 Effects on competitiveness of compliance with standards on pesticides and food safety

	Farm type	Pre-standard situation GM (€/ha)	Post-standard situation GM (€/ha)	Absolute difference in gross margin variation (€/ha)	Relative difference in gross margin variation
Valencia citrus	F&V_F1	1,211	1,362	150	1.12

Conclusions

The results of this chapter suggest that, in general terms, the cost of compliance with the environmental regulations for the fruits and vegetables sector in Spain would probably be slight (or even zero or beneficial in some specific cases). It has to be noted, however, that due to the limitations of the study, not all the fruits and vegetables production regions have been considered and, therefore, these general conclusions may differ across regions, farming systems and types of farms. In highly productive areas of single-crop production, such as the citrus groves of the region of Valencia, the assessment of the cost of compliance for pesticides and nitrates show that benefits of compliance might be possible. For these farms, many practices banned by European or Spanish policies have an alternative method that is economically attractive. Methods developed recently for pest management show clear improvements in pest control efficiency almost without any extra cost.

The impact of compliance with the Nitrates Directive on vegetable crops in the region of Castilla-La Mancha showed that the cost of compliance at the farm level varies with farm type and also depends on the opportunities that farmers have to change the mix of products that they grow. When a farmer's adaptation ability is low, the most vulnerable farmers would incur income losses of around 15%. When farmers adjust the cropping patterns to comply with the Nitrates Directive by changing to less nitrate-demanding crops and eliminating over-fertilization, the income loss is low and, in some cases, incomes may even increase.

Competitiveness of citrus production might be unaffected or even improved if changes in production practices result in the same yields with lower applications of inputs, so there could be a farm level benefit of compliance with standards. Growers tend to use nitrate fertilizers exceeding the maximum permitted dosage. Nevertheless, there is no evidence of yield decreases when fertilization is reduced to the directive's limits. In consequence, production costs decrease and gross margins in citrus production might increase slightly.

For mixed vegetable farms of Castilla-La Mancha, the effects of the application of the Nitrates Directive on the overall competitiveness of the region's farms also varies across farm types. Farms with a fixed cropping pattern have a lower capacity to adjust and therefore face higher compliance costs. Conversely, larger farms cultivating annual cash crops can adjust more readily to environmental regulation by changing their cropping mix. There is some evidence that compliance can have beneficial effects on certain types of farms in Spain. Water quotas established in the region have a significantly larger impact on farm profitability than the Nitrates Directive. However, there are synergies between water and agricultural policies, since water use limitations are promoting the substitution of water demanding crops with less water intensive crops which require also lower nitrogen applications. On the other hand, as the land cropped under fruits and vegetables will be eligible for the Single Farm Payment, it is expected that mixed-production farms that combine vegetable production will increase the land area dedicated to vegetables.

The fruit and vegetable sector in the EU faces important challenges in international trade from input use restrictions, such as the Nitrates Directive as well as pesticide registration and use regulations. On the other hand, certification schemes have the potential to improve the competitiveness of this sector through improved market access, and occasionally from decreases in costs and even premium prices. But competition from exports from less developed countries which may have production cost advantages, including different environmental compliance costs, will continue to be an important challenge for growers in Spain.

References

Azaña, L. (2007). Aplicación del modelo CropSyst a los sistemas agrícolas de la Cuenca Alta del Guadiana. Senior Thesis. Research directed by Carlos H. Hernández and Consuelo Varela Ortega. In Newater Project (n° 511179-2)-DG Research, European Commission. Universidad Politécnica de Madrid. Madrid (Spain).

BOE (2008a) num. 47 Real Decreto 262/2008, de 22 de febrero, sobre la integración de las frutas y hortalizas en el régimen de pago único y el establecimiento de los pagos transitorios para los sectores de cítricos y tomates enviados a transformación.

BOE (2008b) num. 126 Real Decreto 864/2008, de 23 de mayo, sobre fondos y programas operativos de las organizaciones de productores de frutas y hortalizas.

BOE (2008c) num. 289 Real Decreto 1972/2008, de 28 de noviembre, sobre reconocimiento de organizaciones de productores de frutas y hortalizas.

Brouwer, F. (ed.) (2004). Sustaining agriculture and the rural environment; governance, policy and multifunctionality. Cheltenham, Edward Elgar.

CEC (2007a). Council Regulation (EC) No 1182/2007 of 26 September 2007 laying down specific rules as regards the fruit and vegetable sector, amending Directives 2001/112/EC and 2001/113/EC and Regulations (EEC) No 827/68, (EC) No 2200/96, (EC) No 2201/96, (EC)

No 2826/2000, (EC) No 1782/2003 and (EC) No 318/2006 and repealing Regulation (EC) No 2202/96.

CEC (2007b). Commission Regulation (EC) No 1580/2007 of 21 December 2007 laying down implementing rules of Council Regulations (EC) No 2200/96, (EC) No 2201/96 and (EC) No 1182/2007 in the fruit and vegetable sector.

CEC (2008). Commission Regulation (EC) No 292/2008 of 1 April 2008 amending Regulation (EC) No 1580/2007 laying down implementing rules of Council Regulations (EC) No 2200/96, (EC) No 2201/96 and (EC) No 1182/2007 in the fruit and vegetable sector.

CEC (2009a). Council Regulation (EC) No 72/2009 of 19 January 2009 on modifications to the Common Agricultural Policy by amending Regulations (EC) No 247/2006, (EC) No 320/2006, (EC) No 1405/2006, (EC) No 1234/2007, (EC) No 3/2008 and (EC) No 479/2008 and repealing Regulations (EEC) No 1883/78, (EEC) No 1254/89, (EEC) No 2247/89, (EEC) No 2055/93, (EC) No 1868/94, (EC) No 2596/97, (EC) No 1182/2005 and (EC) No 315/2007.

CEC (2009b). Council Regulation (EC) No 73/2009 of 19 January 2009 establishing common rules for direct support schemes for farmers under the common agricultural policy and establishing certain support schemes for farmers, amending Regulations (EC) No 1290/2005, (EC) No 247/2006, (EC) No 378/2007 and repealing Regulation (EC) No 1782/2003.

CEC (2009c). Council Regulation (EC) No 74/2009 of 19 January 2009 amending Regulation (EC) No 1698/2005 on support for rural development by the European Agricultural Fund for Rural Development (EAFRD).

DOCM (Diario Oficial de Castilla-La Mancha) (2001) num 73. Orden de 15 de Junio de 2001 de la Consejería de Agricultura y Medio Ambiente, por la que se aprueba el Programa de Actuación aplicable a las zonas vulnerables a la contaminación por nitratos de origen agrario en la Comunidad Autónoma de Castilla-La Mancha. Available at: http://docm.jccm.es/portaldocm/verDisposicionAntigua.do?ruta=2001/06/26&idDisposicion=123062555482130416

FAO (2007). FAO Statistical Yearbook 2005-2006. FAO Statistics Division.

FAOSTAT (2011a). Crop production. Updated: 17 May 2011. Available at : http://faostat.fao.org/

FAOSTAT (2011b). Trade. Updated: 22 November 2010. Available at : http://faostat.fao.org/

MAPA (Ministerio de Agricultura, Pesca y Alimentación) (2004) REAL DECRETO 2352/2004, de 23 de diciembre, sobre la aplicación de la condicionalidad en relación con las ayudas directas en el marco de la política agrícola común. Ministerio de Agricultura, Pesca y Alimentación, Madrid, Spain.

MARM (Ministerio de Medio Ambiente y Medio Rural y Marino) (2008). Estrategia nacional de los programas operativos sostenibles para el sector de frutas y hortalizas. Available at: http://www.mapa.es/es/agricultura/pags/estrategia_frutas/introduccion.htm

MARM (Ministerio de Medio Ambiente y Medio Rural y Marino) (2010). Anuario de Estadística Agroalimentaria 2007. Ministerio de Medio Ambiente y Medio Rural y Marino. Subdirección General de Estadísticas Agroalimentarias. Servicio de Directorios Estadísticos Agroalimentarios. Madrid, Spain.

USDA (2007). EU-27 Trade Policy Monitoring Responses to European Commission Fruit and Vegetable Reform Proposal. US Department of Agriculture. Foreign Agricultural Service. GAIN (Global Agriculture Information Network) Report Number: E47023. Available at: http://www.fas.usda.gov/gainfiles/200708/146292116.pdf

Varela-Ortega, C. (2007). Policy-driven determinants of irrigation water use and environmental protection: a case study in Spain. In F. Molle and J. Berkoff (eds). *Irrigation Water Pricing Policy in Context: Exploring the Gap Between Theory and Practice*. Comprehensive Assessment of Water Management. CABI Publishing, Wallingford, UK, pp. 328-346.

Varela-Ortega, C., J.M. Sumpsi, A. Garrido, M. Blanco and E. Iglesias (1998). Water pricing policies, public decision making and farmers response: implications for water policy. *Agricultural Economics*, (special issue of the XXIII Conference of the IAAE, Sacramento, USA, August 10-16 1997). Vol. 19-2: 193-202.

Varela-Ortega, C., J.M. Sumpsi and M. Blanco (2002). Water availability in the Mediterranean region. (Chapter 6) In: Brouwer, F. and Van der Straaten, J. (eds): *Nature and Agriculture Policy in the EU*. International Library of Ecological Economics. Edward Elgar Publishing Ltd. Cheltenham, UK, pp. 117-140.

Varela-Ortega, C., I. Blanco and G. Carmona (2006). Agro-economic model for analyzing policy scenarios and cost-effectiveness of policy measures, linking water and agricultural policy. Development of a prototype model. General Report - Newater WP 1.7 Methods for transition to Adaptive Management. Newater Project (nº 511179-2)-DG Research, European Commission. Universidad Politécnica de Madrid, Madrid (Spain).

Varela-Ortega, C., I. Blanco, G. Carmona and P. Esteve (2007) Field work analysis report in the Upper Guadiana Basin (Spain). Newater WP 1.7 Methods for transition to Adaptive Management. Newater Project (nº 511179-2)-DG Research, European Commission. Universidad Politécnica de Madrid, Madrid (Spain).

Chapter 12

Evaluating WTO institutions for resolving trade disputes involving non-tariff measures: four cases involving Brazil

Sílvia Helena Galvão de Miranda

Introduction

The World Trade Organization (WTO) system is not completely suitable if developing country expectations in terms of solving international trade controversies, particularly those related to sanitary and technical issues, are to be met. The situation is even worse if the environmental requirements currently affecting trade flows are considered. The Technical Barriers to Trade (TBT) Agreement and the Sanitary and Phytosanitary (SPS) Agreement are currently being used to address international sanitary, technical and even environmental trade issues and will probably continue to be used in this respect. It is expected that the WTO and the Multilateral Environmental Agreements (MEAs) will eventually make an improved regulatory system available to deal with trade and environmental disputes.

This chapter is intended to provide insight into the role of the WTO in the solution of SPS and TBT trade disputes from the perspective of developing or middle-income countries, in contrast to the developed country perspective offered by the other chapters of this book.

Sanitary and technical barriers[1] to commerce have become one of the most challenging issues in the field of international economics, not only because they can restrict trade, but

[1] There are several arguments about the understanding of the word 'barrier' as it applies to trade. Some people prefer the word 'obstacle.' This discussion is still alive, particularly by the Technical Barriers to Trade Committee at the WTO. One of the arguments is that when one refers to a sanitary or technical barrier, it is assumed that this measure makes trade impossible. On the other hand, the word obstacle means that the measure makes trade difficult while not necessarily halting trade flow. For the private sector, as highlighted in Miranda and Barros (2005), a barrier is any kind of sanitary or technical requisite that jeopardizes business (increases costs of compliance and/or reduces trade flows or prices). From this perspective the barrier does not have to be a regulation, it can also be a private standard. However governments do not consider private standards to be barriers because they do not have the force of law. A second feature essential to understanding the discussion in this chapter relates to the recognition of a sanitary or technical requirement as a barrier or obstacle, and how this is connected with the legitimacy issue. A barrier is legitimate if it is non-discriminatory, is supported by scientific evidence and enhances food safety, the environment or animal, human and plant health. If the purpose of the regulation is provided for by an agreement's provisions and procedures, based on science and legitimate goals, even though this measure impacts trade, it will not be considered an illegitimate barrier. On the other hand, if a regulation's requirement is based on protectionism and distorts trade, then it is considered an illegitimate barrier and the domestic regulation will probably face criticisms and disputes between the enacting country and its trade partners. The determination of a regulation's legitimacy is complex, and each regulation or standard should be considered on a case-by-case basis. This chapter will not address the legitimacy issue in depth. A more detailed discussion is provided by Miljkovic (2005) and Miranda and Barros (2005); the main legitimacy criteria come from GATT, SPS and TBT Agreements (www.wto.org).

because economic agents, particularly consumers, are increasingly sensitive to the concerns they address. Several of these measures raise costs and reduce competitiveness while others exclude goods from the market. This rather negative view of the effect of sanitary and technical measures often prevails when the topic is discussed in developing countries; however, authors such as Roberts and Unnevehr (2005) have recently emphasized the possible beneficial effects on welfare generated by the international regulation of sanitary and technical standards.

Surveys taken in several countries have shown that a significant number of recently established sanitary and technical requirements are private voluntary standards rather than government regulations and that there is an increasing demand for even stricter sanitary and technical regulation. New, stricter standards seem to benefit all consumers; however, these stricter standards create more difficulties in the developing world than in the developed world.

The developing world is impacted by financial limitations, technological and workplace restrictions, cultural and religious factors, and infrastructure deficiencies less evident in developed countries. These handicaps impede implementation of international sanitary and technical regulations. Even within each developing economy, compliance is unequal. In Brazil, for example, many fruit producers are adopting the Fruit Integrated Production[2] (PIF) label, a national quality certification stamp; but in some regions and sectors the process of adjustment has lagged.

Part of the controversy surrounding international regulation of sanitary and technical practices stems from difficulty encountered when attempting to identify this type of regulation's economic and social costs and benefits. Measurement of the effects of sanitary and technical regulation is exceedingly complex as it involves public health, food safety, and protection of environment from new pests and diseases rather than the logistics of international trade. Adding to the controversy is the problem encountered trying to disentangle legitimate regulation from protectionism, a caveat that justifies the following section's brief discussion of the legitimacy issue.

Table 12.1 gives some examples that illustrate the importance of trade disputes arising from the application of sanitary and technical measures. The examples come from a review of relevant literature, press releases, and WTO Committee releases.

Conceptual and legal aspects

The legal framework for analysing sanitary and technical trade measures is basically contained in the SPS and TBT Agreements, both within the scope of the WTO. The two agreements establish provisions that guide negotiation and regulation of sanitary and technical measures. These provisions are reviewed in works by Roberts *et al.* (1999) and Burnquist *et al.* (2005).

The basic difference between the SPS and the TBT agreements is in their scope. The SPS Agreement covers all measures intended to protect

- human or animal health from food-borne risks;
- human health from animal or plant-carried diseases; and

[2] In the original Portuguese, Produção Integrada de Frutas.

- animals and plants from pests or diseases, whether or not these are technical requirements.

Table 12.1 Examples of disputes related to sanitary and technical issues

Subject of disputes	Country imposing the regulation or measure	Country raising the complaint
Risk assessment and risk management procedures imposed to wheat imports in 1997[1]	Brazil	US
Ban on beef exports because of hormones [2]	EU	US, Canada
Varietal tests and quarantine requirements on fruits (apples, cherries, walnuts and nectarines) [3]	Japan	US and Brazil
Requirements on levels of aflatoxin in Brazil nuts since 2003[1,4]	EU	Bolivia, Brazil and other 11 countries
Food and Mouth Disease (FMD) in bovine - recognition of regionalization[5]	US	Brazil

Sources: [1]WTO (2007), [2]Victor (2000), [3]WTO (1998), [4]Silveira et al. (2007) and [5]Miranda (2001)

The TBT Agreement covers all technical requirements, consumer information (e.g. labelling), voluntary standards and the procedures to ensure that these requirements and standards are met (conformity assessment procedures), except when these are measures specifically covered by the SPS Agreement.

The SPS Agreement commits signatories to implement sanitary and phytosanitary measures based on internationally established guidelines and risk assessment procedures. The international guidelines are established by three scientific organizations: the Codex Alimentarius (Codex) for food safety measures, the World Organization for Animal Health (OIE), and the International Plant Protection Convention (IPPC). According to Roberts (1998), these international organizations are taken as providers of benchmark or 'safe harbour' standards.

Implementation of multiple, divergent SPS measures to regulate both processes and limits has caused the costs of trade to increase substantially and led to attempts at harmonization. After analysis of Maximum Residue Limits regulations, Telteboim *et al.* (2007) found that despite the WTO efforts to harmonize technical and sanitary standards, individual countries continue to set and apply their own limits. These country limits, if they significantly diverge from an accepted norm, cause additional costs and other difficulties for exporters of Brazilian fruits.

However, a positive effect of the SPS Agreement's adoption has been the application of the 'regionalization' principle (article 6 - Adaption to Regional Conditions, Including Pest- or Disease-Free Areas and Areas of Low Pest or Disease Prevalence). Applying this principle, if a country can demonstrate that all or at least part of its territory is free from a hazard, it can circumvent sanitary and technical restrictions blocking all their exports of the affected product. For instance, Brazil established the Cattle Circuits process by which regions free of foot and mouth disease (FMD) were differentiated from infected regions. Miranda (2001) and Miranda and Barros (2009) reported that this tool allowed an increase of Brazilian beef exports to the EU (EU) from regions free of FMD with vaccination.

The TBT Agreement states that countries should prevent the adoption of technical regulations if they create unnecessary obstacles to trade. Conflicts generated by technical obstacles are related to labelling requirements, conformity assessment procedures, traceability, content of products, testing and sampling requirements, and other issues that are sometimes referred to as sanitary issues. According to the Agreement, countries are not supposed to adopt technical regulations that are more trade-restrictive than needed to fulfil a legitimate goal, which can be to meet national security requirements, to prevent deceptive practices, to protect human health or safety, to preserve animal and plant health, or to protect the environment.

The Agreement states that it 'deals only with technical regulations, standards and conformity assessment procedures related to products or processes and production methods. Standards as defined by ISO/IEC Guide 2 may be mandatory or voluntary. For the purpose of this Agreement standards are defined as voluntary and technical regulations as mandatory documents. Standards prepared by the international standardization community are based on consensus. This Agreement covers also documents that are not based on consensus' (TBT Agreement/WTO).

The WTO's main tool to guarantee transparency is Notification, due whenever a government proposes a new regulation or modifies an already existing one if it differs from an international standard or if it potentially might affect trade. Once notified, the SPS or TBT Secretariat is responsible for forwarding the notification to other WTO members to assure them an opportunity to comment on that regulation.

A remarkable difference between the SPS and TBT approaches, pointed out by Popper *et al.* (2004), is that unlike the SPS Agreement, the TBT Agreement treats 'sound science' as an element for consideration but not a fundamental basis for establishing policy. 'The absence of more rigorous requirements for risk assessment under the TBT Agreement leaves considerable room for manoeuvre when determining the acceptability of regulations, standards, and conformity-assessment procedures, including necessity.'

Even though international agreements set out to prevent countries from using technical and sanitary regulations to restrict trade or protect domestic suppliers from foreign competition, this goal is not always achieved. Many authors have suggested that the problems of protectionism and unjustified trade restriction could be ameliorated if more light were shed on unilaterally adopted sanitary and technical regulations.

Roberts *et al.* (1999) proposed a system that classifies technical measures by policy instrument and scope first and then by regulatory goal, which indicated how and why domestic demand and supply would change as a result of the success or failure of a specific technical measure. This system would allow the determination of whether a regulation is welfare-reducing or welfare-enhancing and if it arises from a risk-reducing or non-risk reducing policy. Josling *et al.* (2004) updated that methodology so that it first identified the regulatory agents, then the dimensions of the measure, and then the tools to be employed by the regulatory measures. In a study of Brazilian beef trade focused on regulations related to foot and mouth disease and BSE, Machado (2007) adapted these methodologies to address sanitary measures.

Popper *et al.* (2004) also focused on the classification of technical measures due to their concern about these measures' consistency with TBT provisions. The authors prepared a report to the US National Institute of Standards & Technology (NIST) proposing criteria for specifically identifying TBT-like practices. They determined that in practice it is often difficult to discern WTO inconsistency when addressing TBT measures and concluded that the determination of legitimacy is possible only on a case-by-case basis.

More recently, Miranda and Barros (2005) developed a procedure to qualitatively and quantitatively evaluate the level of legitimacy of a sanitary or technical requirement. They proposed a checklist encompassing three evaluative steps:

- preparation of an economic impact analysis;
- carrying out an analysis of the requirement's legality in respect to compliance with the provisions of international agreements; and
- analysing the sanitary or technical regulation's scientific justification.

In final form, their checklist, re-titled the Relevance Index, contains two components: an economic impact analysis and a legitimacy analysis, which is a combination of the earlier checklist's legality and scientific analyses (Miranda and Barros, 2005).

Advances in the methodology and data collection used for this legitimacy analysis will probably turn the legislative adoption of a sanitary or technical measure into a more comprehensive process that includes a discussion on the measure's potential benefits and costs.

The application of SPS and TBT agreements in developing countries

This section examines implementation of the SPS and TBT Agreements in developing countries, specifically focusing on the caveats those countries face meeting SPS and TBT requirements and highlighting the effects of these measures' imposition by countries or firms (private standards) on trade and welfare.

Burnquist *et al.* (2005) analysed notifications of sanitary regulations to the SPS Agreement Committee. Their analysis included 34 countries in the Western Hemisphere and covered the period from January 1995 to December 2001. About 47% of the 34 nations did not notify the SPS Committee of any SPS regulation changes. The authors noted that sanitary regulation might have been underreported, possibly due an individual country's determination that the regulation or its change would not impact trade. Moreover, most Western Hemisphere countries would have faced difficulties meeting Enquiry-Point requirements to take part in the WTO's transparency process.

Burnquist *et al.* (2005) noted a pattern in the notification process when they correlated notifications by Western Hemisphere country with that country's World Bank income classification. They found that the hemisphere's higher-income countries (US and Canada) submitted SPS notifications much more frequently than lower income countries.

Henson *et al.* (1999) conducted ten case studies of developing countries to identify the relative importance of several factors in respect to their ability to meet SPS requirements when exporting agricultural and food products to the European Community. The studies were based on responses to questionnaires. The most important complaints coming from the respondent countries were 'insufficient access to technical and scientific expertise' and 'incompatibility of SPS requirements with domestic production/marketing methods.' Less frequently cited problems were 'poor awareness of SPS requirements within the agricultural and food industries' and 'poor access to information on SPS requirements'.

Several authors have studied the effect of the SPS Agreement's implementation on trade. Stanton (1999) noted that the purpose of the Agreement is to promote international trade by limiting the use of SPS regulation as disguised barriers to trade. This purpose is explicit in the TBT Agreement, albeit not so clearly in the SPS Agreement.

In order to examine how the SPS Agreement affects a country's sanitary policies, one must have an understanding of how international standards are established and identify the exceptions that allow a country to deviate from those standards. An exception to harmonization occurs whenever nations adopt SPS measures less stringent than current international measures. Victor (2000) holds that the weakening of existing criteria could be thought to be much less of a threat to trade than strong measures but that the problem of conflict between the SPS Agreement's stipulations and the weakening of SPS criteria has probably not yet arisen because SPS criteria is most likely to be lowered by developing countries; and these countries have not yet fully implemented the Agreement.

Numerous researchers have studied the effect of TBT requirements on importation. Ferraz Filho (1997) took a survey of Brazilian exporting companies in the 1990s. The author found that slaughterhouses and electrical equipment industries were most directly affected by technical requirements and considered these requirements a very important obstacle to expansion in the international marketplace.

The World Bank conducted a TBT survey that included 689 firms, 20 industrial sectors, and 17 developing countries (Wilson and Otsuki, 2004). About 70% of these firms reported that they faced technical regulations in their export markets, with EU and US regulations generally considered to be the most important. The study showed that to meet EU and US standards, the firms invested in additional plant and equipment, redesigned their products, and hired additional labour for production, testing, and certification. It is clear that technical requirements increase production costs and certainly complicate bringing a product to market.

In Brazil, Faria and Burnquist (2006) evaluated the effect of technical requirements on Brazilian exports based on a survey applied to 77 exporting firms. The study variables represent 6 types of technical measures that affect exported products and are distinguished by the measure's sphere of concern: packaging, the production process, labeling, conformity, quality, and environmental effect. About 55% of the total companies sampled were domestic, most of them supplying both developing and developed countries. Adherence to quality requirements had the greatest effect on Brazilian exports, followed by conformity testing and processing requisites. Labeling was found to be of lesser importance (Faria and Burnquist, 2006). Similar results were verified by a World Bank study of the effects of technical measures on the exports of several developing countries at the end of the 1990s.

The many factors affecting developing countries as they try to comply with sanitary and technical measures has led to renewed effort in the difficult task of correctly evaluating the current and potential economic and social impacts of this type of regulation.

SPS and TBT disputes - Brazilian background

This section presents four trade cases involving Brazil and its trading partners. The cases highlight differences between developing and developed countries in respect to sanitary and technical requirements.

Risk assessment and risk management procedures imposed by Brazil on wheat imports from the US (1997)

In 1997, the US questioned new Brazilian requirements that essentially stopped the importation of US wheat into the country. The new standards were established to prevent introduction of the fungus *Tilletia controversa* (TCK bunt or Dwarf bunt) and as part of a harmonization effort within the MERCOSUR regional trade bloc.

The US complaint became a Specific Trade Concern (STC) under the SPS Technical Committee's scope. The US contended that a 1996 bilateral agreement between the US and Brazil contained an understanding that the fungus could not be established in Brazil and that this understanding had never been scientifically disproved. Brazil countered that they had only agreed that the fungus would probably not enter but that new Brazilian risk assessment legislation necessitated the new requirement and that the bilateral agreement did not preclude Brazil from applying internal risk assessment legislation (WTO, 2007).

This STC was finally resolved in July 2001 when the US reported that after several technical consultations, Brazil had issued new import instructions allowing the importation of certain classes of US wheat. At that time, the dispute was considered settled.

In March 2001, the Brazilian government did withdraw its ban on the importation of Hard Red Winter, Hard Red Spring and Soft Red Winter wheat grown in the US. Due to phytosanitary concerns, the ban remained in force for *durum* wheat and all wheat produced along the US West Coast (FAO, 2004), which remains a contentious issue. In August 2001, Brazil decided to ban the importation of wheat with more than 2 percent dockage. This wheat had previously been allowed to enter the country as 'out of standard' (substandard).

Wheat was again the focus of new debate in early 2002 when new Brazilian regulations on imported cereals and grains were announced. The new regulations required that all countries exporting to Brazil, except MERCOSUR[3] member countries, submit a Pest Risk Assessment within 180 days from the import date (FAO, 2004). Legislation requiring this type of analysis is a milestone in the organization of phytosanitary requirements and had been recommended in the SPS Agreement and by the IPPC.

Brazil is one of the world's largest importers of wheat. Storz *et al.* (2005) note that despite Argentina's advantages, there are significant opportunities for US wheat in the Brazilian market, particularly from May through September. The authors point out that in 1996, the US exported roughly 760,000 metric tons of wheat to Brazil with a value of US$174 million. In 2002, US wheat exports to Brazil had fallen to US$96 million; and by 2007 the volume exported to Brazil had declined to only 354,065 metric tons.

There have been no other disputes between Brazil and the US involving sanitary and technical measures to the importation of wheat brought to the attention of the WTO. However, by granting preferential tariffs to MERCOSUR members, Brazil has opened the door to future trade disputes with non-MERCOSUR affiliated countries.

---

[3] The MERCOSUR agreement had been in force since 1995. MERCOSUR members have become the most important suppliers of wheat to Brazil, and wheat is the main Brazilian agricultural import.

From the economic perspective, the wheat case is a prime example of the impact of phytosanitary measures when they become a trade measure. While the volume of wheat imported into Brazil is huge, the US, a major international wheat exporter, has a very small share of its market. From the legal perspective, it appears that Brazil's phytosanitary requirements met the SPS agreement's transparency provisions, in that the country consulted its trading partners before implementing the new regulations. By requiring that exporters provide a pest risk analysis, Brazil has met the agreement's 'scientific basis' requirement for the implementation of new phytosanitary regulations.

The wheat dispute offers a better understanding of the importance of phytosanitary requirements in international trade and indicates that the measures they erect can be negotiated. While phytosanitary measures may not be the most important obstacles to international trade, each new regulation of this type must be subjected to detailed study to determine its relevance and legitimacy.

Requirements on Brazilian mango exports to Japan

According to Brasil (1999), Japanese legislation requires that a phytosanitary certificate accompany imported plants and plant products. The certificate should be issued by a competent authority from the exporting country.

Essentially, Japan divides agricultural imports into three categories:

- forbidden products;
- products that require quarantine procedures; and
- products without quarantine requirements.

Because they are grown in regions that contain the Mediterranean fruit-fly, a pest not found in Japan, mangoes fall into the first category. The fruit-fly also causes a restriction on the importation of other Brazilian fruits.

A Brazilian report on export measures notes the Japanese pest protection policy requires that a different test be performed on each agricultural product type, even if the products are just different varieties of the same species. The testing is a very expensive, time-consuming procedure (Buainain and Batalha, 2007). Japanese authorities have ignored evidence showing that the same test can be easily extended to different varieties of the same species.

The US also argued against this regulation under the scope of the SPS Agreement through the WTO in April 1997. Brazil supported the US in this dispute, also raising concerns about the Japanese varietals test and accompanying quarantine requirements. The WTO panel affirmed the US position in 1998. The decision recommended that Japanese authorities modify their varietal test requirements for several fruits, as they were inconsistent with SPS Agreement provisions. The decision was corroborated by the WTO's Appellate Body in February 1999.

A report issued by Brasil (1999) noted a lack of transparency in the Japanese fumigation requirements. The provision requires the fumigation of fresh horticultural and fruit products whenever live insects are encountered, even if that insect is present in Japan. This system increases all costs and negatively affects fresh fruit and vegetable quality by delaying delivery. The strict exigencies Japan has been imposing on imported Brazilian mangoes arise, in most part, from Japan's effort to prevent infestation by the Mediterranean fruit-fly. In 2003, Brazil and India complained against the Japanese fumigation requirements during a SPS technical committee meeting.

According to the Brazilian complaint, for 18 years the country had been trying to get the Japanese to ease entry restrictions on mango importation, which entailed an exporter provided steam treatment in addition to other technical measures taken by Brazil to remove live Mediterranean flies from the fruit. Brazil alleged that treatments already provided by the growers gave results that were equivalent to those from steam treatment. Brazil also claimed that Japan had hindered negotiations with their demands for more information while ignoring results from prior scientific studies. Brazil considered that Japan's requirements were inconsistent with the SPS Agreement's equivalence, regionalization and technical cooperation provisions (WTO, 2008).

Before taking this issue to the SPS Committee, Japan and Brazil negotiated bilateral understandings and a commitment to technical cooperation. In 1986, Brazil requested technical assistance, and Japan agreed to the request. Brazil discontinued the assistance programme in 1990 as part of a push to develop its own hot-water based treatment. The new treatment was preliminarily approved by Japanese authorities in 1998. Final data was submitted by both Japan and Brazil in 2001; however, Japanese authorities requested additional information before they could conclude their technical studies of the equivalence of the two pest eradication methods.

Specific Trade Concerns documents from October 2003 describe a complaint by Brazil alleging that Japanese new restrictions on mango imports were unjustified, arguing that the fruit in question had been produced in an area 2,000 km from where fruit flies had been found. Japan replied that authorities were reviewing fruit-fly trapping information that had recently been received from Brazil. In March 2004, Brazil notified the WTO that the Japanese authorities had had a favorable reaction to technical data provided by Brazil in 2003, and a solution was expected. The solution was to include a protocol addressing the packing, storage and transportation of mangoes destined for Japan.

A Japanese technical evaluation of the Mediterranean fruit-fly problem was completed in 2004, and a pest risk assessment process was developed. At that time, a bilateral meeting was held to coordinate plant quarantine efforts and to formalize the requirements for hot water dipping (WTO, 2008). In March 2005, Brazil informed the SPS Committee that in September 2004, Japan had modified its phytosanitary regulations and established specific norms for the import of mangoes from Brazil. In January 2005, the first shipment of Brazilian Tommy Atkins mangoes was exported to Japan.

Currently, Brazil is negotiating with the Japanese government about the possibility of exporting other mango varieties to Japan, as the Tommy Atkins variety was not well accepted by Japanese consumers. The current negotiations may be as difficult to resolve as the earlier ones, which took almost 10 years to settle.

One point to take away from this discussion is that it takes a considerable amount of time to resolve disputes over phytosanitary regulation. Risk analysis, document submission and consideration, and decision making on the international level is not something that is rushed, consequently the economic impact of this type of dispute on the exporting country can be severe. WTO transparency and criteria assessment procedures regarding equivalence process are still under-regulated by the WTO, which should be of concern since 'equivalence' was proposed as a method of simplifying and expediting harmonization.

European requirements for Brazil nut aflatoxin levels

The dispute between the EU and Bolivia, Brazil and other countries over the aflatoxins content of foodstuffs is a case of a phytosanitary obstacle to trade based upon questionable technical requirements. In 1987, the Codex Alimentarius concluded that there was not enough information to establish an aflatoxin tolerance level. Although the Committee did set standards for peanuts in 1999, no level was set for other foodstuffs. The Codex Alimentarius is still working on the issue, and there is still no international standard for aflatoxin levels (WTO, 2009).

In March 1998, Bolivia, Brazil and 10 other countries raised a dispute before the SPS technical committee complaining that European regulations setting a new maximum level for aflatoxin contamination in foodstuffs would severely restrict trade without resulting in a significant reduction in consumer health risks. Additionally, the same countries contended that the new requirements were established without a proper risk assessment study and that the sampling procedure that exporters were required to employ represented an unjustifiably large expense. The complainants urged the EU to review its requirements.

The EU replied that its requirements reflected the level of protection required by its member countries. The EU also asserted that the complainant's less expensive sampling procedures were flawed in that the sample size was too small. The WTO's solution to this dispute was reported in March 2004. At that time, the WTO determined that some European sampling procedures and maximum aflatoxin levels should be revised.

After evaluating all the initial comments received through the WTO's consultancy process, the European Community reported that it had forwarded a revised proposal to its member country representatives in June 1998. This first revised proposal reduced some maximum aflatoxin levels and extended the timetable for introduction of the new requirements until January 1999. However, in September 1998, before the revised requirements were introduced, Bolivia informed the SPS Committee that the EU's revised proposal would severely, negatively affect its Brazil nut exports. The Bolivian complaint led to technical consultation between Bolivia and the EU. After consultation, Bolivia presented a plan to improve the quality of its Brazil nuts and requested application of the WTO's developing country 'special and differential treatment' provision. As a result, an assistance programme between Bolivia and EU was set up. In 2002, an accreditation and certification procedure was successfully negotiated.

In June 2002, it was determined that the largest Bolivian exporters were able to meet the EC requirements. In June 2003, Bolivia informed the WTO that a bilateral meeting had had a favourable outcome and that Bolivia should soon receive permission for the export of whole Brazil nuts to the EU. In March 2004, bilateral consultations between Bolivian and EU representatives led to the scheduling of an assessment visit by EU authorities to finalize the procedure for EU certification of Bolivian Brazil nuts for export to the EU. Over recent years, the volume and value of Bolivian Brazil nut exports to the EU have increased; although, the considerable expense involved in successfully completing the EU certification process forced smaller Bolivian Brazil nut exporters out of the EU market.

Despite understandings reached between Bolivia and the EU, there remained several other countries' unresolved complaints against the EU's aflatoxin regulations. One of the main points underlying the complaints was that the EU had not presented a risk assessment to support the need for its restrictive aflatoxin requirements.

As Bolivian exports of whole Brazil nuts and almonds to the EU grew, the export of Brazilian whole Brazil nuts to the EU dropped drastically. Silveira *et al.* (2007) showed that the European's maximum levels for aflatoxin contamination negatively affects Brazilian Brazil nut exports, with a 1% reduction of the European's maximum limit reducing Brazil's Brazil nut exports to the EU by 1.22%, and that difficulties encountered attempting to test for these miniscule levels excessively and unwarrantedly increase producer costs. Their study concluded that regulation of maximum aflatoxin levels needs to be harmonized globally and that the Brazilian government should make an effort to improve production, harvest, storage and transport conditions in their country to reduce mycotoxin contamination.

Several Brazilian Brazil nut shipments to the EU and US were barred at the border because of aflatoxin levels above regulatory limits. In July 2003, the EU suspended the importation of Brazilian Brazil nuts and requested that government authorities impose special conditions on import of Brazil nuts originating in or consigned from Brazil, for instance: official Brazilian aflatoxin certificates must accompany consignments; consignments may only enter the EU at specified points; and consignments be subjected to sampling and analysis and can be detained for a maximum of 15 working days from the point of entry into the Community before release onto the market (EC decision #493).

Results from scientific studies evaluating damage to the human organism from different aflatoxin levels have not been conclusive, which has led to great diversity among individual country aflatoxin regulatory requirements. By 2003, 61 countries had established their own specific limits on the amount of aflatoxin B1 that can be contained in human foodstuffs and 76 had specific limits to total aflatoxins: B1+B2+G1+G2 (FAO, 2004). Some countries have succeeded in harmonizing their regulation within trade blocs (Australia/New Zealand, EU, MERCOSUR) (Silveira *et al.*, 2007). Over time, aflatoxin limits have become lower as the sensitivity of analyses improves and knowledge about the effects of aflatoxins advances while divergence between producers and consumers has grown over what aflatoxin level effectively guarantees food safety.

In 1987, the European Commission's Codex Alimentarius committee concluded that there was not enough information to establish an aflatoxin tolerance level and recommended that the level of aflatoxin in food should be reduced 'as low as reasonably achievable' – ALARA. In 1996 and 1997 JECFA[4] conducted quantitative risk assessments to revaluate this issue and concluded that in regards to human health risks there was not a significant difference between assuming a maximum level of 10 μg/kg or 20 μg/kg of aflatoxin B1 (Boutrif, 1998). In July 1999, the Codex Alimentarius revised its proposal for the maximum tolerable total aflatoxins limit in peanuts and adopted 15 μg/kg as the international reference. No other limits were adopted at that time.

There is not a specific Brazilian limit defined for all aflatoxins specific to Brazil nuts. In 1977, the Brazilian Health Ministry passed Resolution no. 34/76 for Food, which established a maximum B1 + G1 aflatoxins limit of 30ppb (μg/kg) in foods for human consumption (Silveira *et al.*, 2007).

The European Brazil nut aflatoxin contamination limit of no more than 4 ppb of aflatoxin (2 ppb for aflatoxin type B1) is a very low limit compared to the Codex and to US

[4] Expert Committee on Food Additives (JECFA), a committee that involves the Food and Agriculture Organization (FAO) and the World Health Organization (WHO) and is the consultative group responsible for evaluation of toxic substance safety limits. JECFA provides scientific recommendations that are the basis for the WTO's Codex Alimentarius's development of food safety standards.

referents (US maximum aflatoxin content <10 ppb). Silveira *et al.* (2007) reviewed some of the criticisms of the imposition of such low aflatoxins limits: there has been no determination of what actually is a safe daily aflatoxin consumption limit; there has been too little discussion in a competent international forum of what this limit should be; daily Brazil nut consumption is unknown, other than it is much lower than the consumption of other foodstuffs containing aflatoxin; and the European aflatoxin limit is only very rarely, if ever, achieved through supplier efforts. A number of technicians believe that the acceptable level is more a political than a scientific determination, as there is no scientific evidence to support the European establishment's determination of the safe aflatoxin level.

At this time, the European aflatoxin limit on Brazil nuts appears to be unjustifiable on either a scientific or a consumer health basis. Silveira *et al.* (2007) concluded that the EU's Brazil nut aflatoxin specifications override market forces and do not allow the market to balance supply and demand via prices.

To open the European market, Brazil's government implemented initiatives to control aflatoxin contamination in the peanut and Brazil nut chains. The initiatives included the Good Manufacturing Practices resolution and the Ministry of Agriculture's Emergency Work Plan for Brazil Nuts and involved the productive sector, their representatives, and the country's exporting association. In May 2004, a new federal regulation providing rules for the sanitary certification of Brazil nuts was issued. The regulation was intended to improve Brazil nut quality in the domestic market and to meet some requirements of European Commission decision no. 493.

Brazil exports Brazil nuts to the major world markets, but the volume of these exports has been decreasing since 2003, particularly to the EU. In 2002, European imports of Brazilian Brazil nuts were 3,400 metric tons. Because of the restrictiveness of EU's technical requirements, these imports fell to only 16 metric tons in 2005. Brazil did succeed in re-directing some Brazil nut exports to other countries, like Bolivia, Canada, Norway and the US. In 2002, Brazil exported no Brazil nuts to Bolivia; in 2004 Brazil exported 6,300 metric tons of Brazil nuts to Bolivia.

Based on interviews and a review of relevant literature, Silveira *et al.* (2007) concluded that apparently Brazil did not choose to comply with European Commission's Brazil nut regulations. More to the point, Brazilian programs implemented to improve production chain procedures did not achieve their intended goals in terms of product quality. Brazil has since revised its nut export strategy, shifting the focus of exportation to shelled Brazil nuts rather than whole Brazil nuts, because shelled nuts present a lower level of aflatoxin, and focusing Brazilian nut exportation to markets that have enacted less stringent aflatoxin regulations.

Foot and mouth disease in bovines: the regionalization provision and its acceptance

Foot and mouth disease, and the trade restrictions it has brought about, has severely affected international bovine meat trade in recent years. Indeed, this is the iconic example of a sanitary dispute involving international markets and Brazil.

Within the Western Hemisphere, FMD-free countries (Canada, US and Mexico) restrict meat imports from countries that have not been able to completely eradicate FMD, such as Brazil. Due to fear of FMD contamination, the US only imports processed Brazilian beef.

Other Latin American countries have been impacted by FMD, but the impacts have been limited. Argentina reached FMD free status in May 1997 (OIE) by completing a vaccination programme. Although the country experienced a FMD outbreak in 2001, it still was able to export 20 thousand metric tons of fresh beef to the US that year, meeting its export quota.

The US also provided a 20 thousand metric ton quota to Uruguayan beef exporters (Miranda, 2001). Of course, every time there is a sanitary outbreak in MERCOSUR, Uruguayan and Argentinean beef exportation to the US is temporarily interrupted. After the outbreak is controlled, negotiations begin to re-start the process and exportation eventually resumes.

Lima *et al.* (2005) found that bans on Brazilian beef due to its FMD disease status cost the country US$7.5 billion in 2004. The non-tariff measures were for the most part erected by Pacific Rim countries: US, Canada, South Korea, Japan, Australia and New Zealand; all countries that, in general terms, do not accept 'regionalization' in respect to FMD.

The US, one of the world's major meat traders, has been considered FMD-free since it last had a FMD outbreak in 1929. The country has maintained a strict regulatory position in the commerce of beef and pork. In 2002, a US Department of Agriculture (USDA) risk analysis of beef importation from Uruguay found that if FMD was reintroduced into the US, the country would have suffered export losses of over US$3 billion in 2001 and total losses of between US$37 and US$44 billion (USDA, 2002). At the March 2004 Continental Conference for FMD Eradication in the western hemisphere held in Houston, North American officials estimated a total cost of US$52 billion if FMD was reintroduced, inclusive of the effects on grains markets (Lima *et al.*, 2005).

Kassum and Morgan (2002) estimated that world trade in animals and meat products was deeply affected by FMD and BSE in 2000 and 2001, when most countries closed their markets to at least 25% of all beef exports and almost 40% of all whole pork exports. Bans included stricter border control measures that affected even countries officially free of animal diseases. According to the authors, SPS/WTO FMD notifications led to 113 bans and serious restrictions on imports between 1995 and 2001. However, the authors are positive that the SPS Agreement, with its notification requirement, achieved its goal by providing the transparency needed to prevent some outbreaks and limit others.

The OIE has multiple objectives. It is charged with:

- improving animal health worldwide based upon the development of sanitary standards;
- collecting, analysing and disseminating scientific information on animal medicine;
- guaranteeing the transparency of country status in relation to animal diseases; and
- providing technical assistance and promoting international cooperation to control animal diseases.

The WTO and the OIE are linked: the SPS Agreement is enforced by the WTO, and provides the means to guarantee transparency in regards to sanitary measures implemented to protect animals from diseases and zoonoses, as far as these diseases affect human health; and the OIE has the technical knowledge needed to prevent and control these diseases.

In the case of FMD, OIE is responsible for the official recognition of disease-free areas, which are then differentiated into disease-free areas and disease-free with vaccination areas. This differentiation is essential to trade, since some nations (Japan and South Korea) have an aversion to imported beef from areas that have not eradicated the FMD yet.

Brazil is made up of more than one region, and there can be a difference in status between regions in regards to animal disease. One single case of FMD impedes the recognition of that area as FMD-free for at least two years. The same requirement is used to classify a region or country as FMD-free with vaccination, i.e. no cases during the last two years and no virus in the last 12 months.

FMD was first detected in Brazil in 1895 and was considered endemic until 1980. Since the 1970s, the Brazilian focus has been on quality vaccines and infected animal transit control to lead the advance against this disease. In 1992, Brazil's National Program to Eradicate FMD (PNEFA) was launched, and the country was divided into different sanitary regions (regionalization) based on FMD prevalence and the disease's effect on trade flow. In 1996, new criteria were added into the regionalization mix: risk of FMD transmission and possibility of FMD control. Over the years, PNEFA has succeeded in reducing Brazilian FMD outbreaks and improved the country's sanitary status.

The principle of regionalization, a non-binding WTO guideline, holds that restrictions on products from areas with disease do not apply necessarily to whole countries. This principle has not been recognized by all countries, with some still applying their FMD measures on a by-country basis alone. The US does not apply the regionalization principle in relation to Brazil.

The first Brazilian request to export beef to the US was submitted to the US Animal and Plant Health Inspection Service (APHIS) in 1999, just after the Brazilian states of Santa Catarina and Rio Grande do Sul were certificated FMD-free with vaccination by the OIE. In August 2000, there was a case of FMD in Rio Grande do Sul; and the certification process was suspended at Brazil's request. In 2002, the certification process was restarted. In 2003, a sanitary mission from the US, Canada and Mexico visited Brazil. It was concluded that the importation of fresh Brazilian beef into the US will not be permitted until APHIS conducts a risk assessment of Brazil's sanitary status, taking regionalization into account, and provides a health certificate for importation. Currently, exporters cannot ship unprocessed fresh and frozen Brazilian beef to the US.

Uruguay had to go through the same APHIS risk assessment process after FMD outbreaks in 2000 and 2001, but exportation has resumed (Lima *et al.*, 2005). From this perspective, the release of Brazilian beef exports to the US is taking an inordinately long time, perhaps due to unclear application of the regionalization principle.

The EU accepts regionalization and has recognized the OIE designated sanitary status of Brazil's regions in relation to FMD. Russia was the second largest importer of Brazilian beef by volume and the largest for pork. Unfortunately, Russian actions in the international meat market are frequently inexplicable. Although Russia claims to recognize the regionalization principle, it banned all meat imports from Brazil after FMD outbreaks in the Brazilian states of Pará and Amazonas (Brazil's North Region) in 2004.

Lima *et al.* (2005) reported that the EU has complained about the lack of transparency and coherence in the North American policy on regionalization, particularly during the sanitary crisis in 2001 when some FMD cases were reported in several EU Member States. The EU has also argued against the US policy on regionalization during and after an outbreak of swine fever.

One great value of bilateral sanitary agreements is that they can be used to establish and recognize process 'equivalence' as provided for in the SPS Agreement's Article 4. When countries reach a consensus that different sanitary or phytosanitary measures offer the same level of health protection, product flow hindered by sanitary or technical measures improves and trade tends to liberalize and expand. The process of negotiating a bilateral trade agreement provides the transparency and flexibility needed to reach this consensus. An equivalence accord as part of a bilateral sanitary agreement should lead to more efficient regulation of specific sanitary concerns. Although equivalence accords are useful, there is room to advance the debate about how to better implement their provisions.

Bovine meat regulations account for most SPS notifications. Machado (2007) discovered that FMD concerns caused the second most notifications, just after BSE notifications. From 1997 through 2005, FMD was the motive for 22 Specific Trade Concerns (STCs) registered with SPS/WTO, most of them occurring just after a sanitary crisis in the exporting country. Analysing the frequency and nature of STC registrations, Machado found that the majority of concerns were raised against the world's main beef importers, confirming the belief that major exporters strongly react to importer regulation. The author (2007) emphasizes that importers have had a more important role than exporters in determining sanitary control regulations and that importer regulations have a high probability of becoming world trade references.

Final remarks: challenges to developing countries and the role of WTO institutions

A review of pertinent literature shows that there is consensus in the conviction that non-tariff measures, which comprise sanitary and technical requirements, have a disproportionate effect on international trade. The United Nations Conference on Trade and Development database (TRAINS) shows that, on average, 5,620 tariff lines per country were subjected to some type of non-tariff measure in 2004, with technical requirements accounting for 58.5% of this total. It is important to reiterate that regulation in itself is not a barrier to trade, but some regulations have become obstacles to trade.

The harmonization of divergent sanitary and technical measures is a time consuming, often technologically complicated, human capital intensive, and legally convoluted process that may raise additional trade concerns and excessively burden undercapitalized businesses and countries. Examples discussed in this chapter have illustrated some of the inefficiencies and intricacies of this process.

Adding to this tangle, each country has a propensity to establish its own sanitary regulations according to their technological infrastructure, human-capital availability, sanitary status, and political priorities. Above all, economic interests and preferential trade agreements (formally or not formally agreed upon) may drastically affect each country's policy decisions.

Sanitary and technical requirements aligned with international scientific guidelines and non-discriminatory by region, country, or supplier are generally not assumed to be trade measures. Use of the SPS's notification requirement, a relatively unambiguous tool to meet the WTO's transparency test and facilitate analysis of any regulation's scientific basis or possible bias, helps insure that trade measures are not intrusive actions of a political or otherwise unwarranted nature.

Developing countries and their exporting industries are right to have concerns about their ability to comply with sanitary and technical requirements imposed in the international market. Meeting regulatory limits established by governments and voluntary standards demanded by firms and consumers, even if they are related to quality, is often difficult in developed countries and can be an unattainable goal in the developing world, aptly illustrated by the difficulties Brazil encountered in trying to comply with EU limits on its Brazil nut exports.

Henson *et al.* (1999) suggest that the effects of SPS measures are more expressive for developing countries than for developed countries due to the relative importance of developing country agricultural and food exportation sectors and the same countries' technological deficiencies. Successful implementation of SPS and TBT Agreement requirements, especially conformity and risk assessment requirements, also may require a disproportionately large infrastructure investment by developing country governments.

A book edited by Milan *et al.* (2007) shows the results of a research project to identify the levels of quality in Brazilian agribusiness chains and the challenges that must be overcome to improve these levels to those demanded by the international market. One of the greatest but least recognized obstacles confronting any attempt to improve Brazilian agricultural product quality is the significant asymmetry between regional laboratories and available human capital. To stay competitive, Brazilian industry and government must invest in laboratory facilities and trained personnel. This investment would significantly improve the likelihood that Brazilian agricultural products will comply with international quality, environmental protection, and sanitary standards.

It is remarkable that Brazilian exporting companies (many of them multinational) are pressured more by the certification processes (voluntary rules) than by government technical regulations. In light of this, quality standards may be more likely met if a voluntary rather than a mandatory process establishes them.

Over the mid-term, a better result for developing countries in the world trade arena would be achieved by increasing their participation in relevant international forums discussing sanitary and technical regulation and standardization; however, more active participation requires more investment. The current low level of developing country participation in this type of discussion gives undue weight to developed country scientific and political interests when sanitary and technical guidelines are debated. It is worth noting that political interests, more than science, guide international negotiations.

Another way to improved rates of sanitary and technical regulation compliance is to include a discussion of these regulations during regional integration and bilateral partnership negotiations. These more focused gatherings have been shown to provide the best agenda for dealing with sanitary and technical regulation at this time. During negotiations leading to a bilateral agreement, there is room for compromise over specific issues, thereby improving the chance of reaching an understanding. These negotiations also create an opening to apply for technical assistance. Whatever the forum, the harmonization of sanitary and technical requirements among countries that are at extremely different levels of development is very difficult.

References

Boutrif, E. (1998), 'Prevention of aflatoxin in pistachios', *Food, Nutrition and Agriculture,* 21: 32-8.

Brasil Ministério do Desenvolvimento, Indústria e Comércio exterior. MIDC. Secex/Funcex (1999), *Barreiras externas às exportações brasileiras – 1999.* 271p.

Buainain, A.M. and M.O. Batalha (Coord.) (2007), 'Cadeia Produtiva de Frutas'. *Série Agronegócios*, v.7. Brasília: IICA; MAPA/SPA. 102p.

Burnquist, H.L., G.S.A. Barros, S.H.G. Miranda and J.H. Cunha Filho (2005), 'Sanitary and Phytosanitary Requirements in Agricultural Trade', in Marcos Jank (ed), *Agricultural Trade Liberalization: Policies and implications for Latin America*, Washington: IDB, p.165-96.

FAO (2004), '*Worldwide regulations for mycotoxins in food and feed in 2003',* FAO Food and Nutrition Paper nº 81, Food and Agriculture Organization. Available at: www.fao.org/documents/show_cdr.asp?url_file=/docrep/007/y5499e/y5499e01.htm.

Faria, R.N. and H.L. Burnquist (2006), 'Análise do grau de restritividade de exigências técnicas às exportações brasileiras', *Revista de Economia e Agronegócio*, 4 (4): 411-30.

Ferraz Filho, G. (1997), 'Barreiras técnicas ao comércio internacional', *Revista Brasileira de Comércio Exterior,* 52: 47-56.

Henson, S., R. Loader, A. Swinbank, M. Bredahl and N. Lux (1999), 'Impact of sanitary and phytosanitary measures on developing countries', The University of Reading, Department of Agricultural and Food Economics. 106 p.

Josling, T., D. Orden and D. Roberts (2004), *Food Regulation and Trade: Toward a Safe and Open Global System,* Washington: Institute for International Economics.

Kassum, J. and N. Morgan (2002), 'The SPS Agreement: Livestock and Meat Trade'. 16p. (Draft).

Lima, R.C.A., S.H.G. Miranda and F. Galli. (2005), 'Febre Aftosa: Impacto Sobre as Exportações Brasileiras de Carnes e o Contexto Mundial das Barreiras Sanitárias', Estudo do CEPEA/ICONE. Available at: www.cepea.esalq.usp.br/pdf/CEPEA-ICONE_Aftosa%20(final).pdf.

Machado, R.Q. (2007), 'Análise do perfil das restrições comerciais à carne bovina nos acordos SPS e TBT'. Master's Dissertation, Escola Superior de Agricultura 'Luiz de Queiroz'- USP (In Portuguese).

Milan, M., S.H.G. Miranda, S. Dezen, E. Costa and L.A. Pinazza (Coord.) (2007). *Sistema de Qualidade nas Cadeias Agroindustriais – Qualiagro*. São Paulo: s/n. 208p.

Miljkovic, D. (2005), 'Sanitary and phytosanitary measures in international trade: policy considerations vs economic reasoning' (Viewpoint), *International Journal of Consumer Studies,* 29 (3): 283-89.

Miranda, S.H.G. (2001), 'Quantificação dos efeitos das barreiras não-tarifárias sobre as exportações brasileiras de carne bovina'. PhD Thesis, Escola Superior de Agricultura 'Luiz de Queiroz'- USP (In Portuguese).

Miranda, S.H.G. and G.S.C. Barros (2005), 'A proposal to evaluate the legitimacy of sanitary and TBT measures in Brazil', Paper prepared for the IATRC Winter Meeting, December 4-6, San Diego, CA.

Miranda, S.H.G. and G.S.C. Barros (2009), 'The application of intervention models to non-tariff trade barriers: a case study of Brazilian beef exports', *Journal of International Agricultural Trade and Development,* 5: 255-72.

Popper, S.W., V. Greenfield, K. Crane and R. Malik (2004), *Measuring economic effects of technical barriers to trade on US exporters*, Washington: NIST-US Department of Commerce, Technology Administration. Planing Report 04-3.

Roberts, D. (1998), 'Implementation of the WTO Agreement on the Application of Sanitary and Phytosanitary Measures. Agriculture in the WTO', WRS – 98-44, Economic Research Service, US Department of Agriculture.

Roberts, D. and L. Unnevehr (2005), 'Resolving trade disputes arising from trends in food safety regulation: the role of the multilateral governance framework', *World Trade Review*, 4 (3): 469-97.

Roberts, D., T.E. Josling and D. Orden (1999), 'A framework for analyzing technical trade barriers in agricultural markets', Market and Trade Economics Division, Economic Research Service, US Department of Agriculture. Technical Bulletin, n. 1876.

Silveira, L.T., F.R. Rodrigues and H.L. Burnquist (2007), 'Impactos econômicos dos limites máximos permitidos de aflatoxina sobre as exportações brasileiras de castanha-do –Pará', Paper presented for the XLV Congresso da SOBER. July 22-25, Londrina, PR, Brazil.

Stanton, G. (1999), 'Review of the operation of the SPS Agreement', Prepared for World Bank's Integrated Program of Research and Capacity Building to Enhance Participation of Developing Countries in the WTO 2000 Negotiations, Conference of Agriculture and The New Trade Agenda in the WTO 2000 Negotiations. October, Genève.

Storz, C.D., T.G. Taylor and G.F. Fairchild (2005), 'A primer on exporting to Brazil', University of Florida (UF). IFAS Extension. Available at: http://edis.ifas.ufl.edu/FE505.

Telteboim, M.C., S.H.G. Miranda, L. Oliveira and V.A. Ozaki (2007), 'Limites máximos de resíduos e suas implicações no comércio internacional de frutas', *Revista de Política Agrícola*, 16 (1): 102-12.

USDA. US Department of Agriculture (2002), Risk assessment - Importation of fresh (chilled or frozen) beef from Uruguay, November 2002.

Victor, D.G (2000), 'The Sanitary and Phytosanitary Agreement of the World Trade Organization: An assessment after five years', *New York University Journal of International Law and Politics*, 32 (4).

Wilson, J.S. and T. Otsuki (2004), 'Standards and technical regulations and firms in developing countries: new evidence from a World Bank technical barriers to trade survey' (Preliminary Draft). The World Bank.

WTO (1998), *Japan measures affecting agricultural products. Report of the panel*, WT/DS76/R/ 27.

WTO (2007), *Specific Trade Concerns,* G/SPS/GEN/204/Rev.7/Add.3.

WTO (2008), *Specific Trade Concerns*, G/SPS/GEN/204/Rev.8/Add.3.

WTO (2009), *Specific Trade Concerns,* G/SPS/GEN/204/Rev.9/Add.3

Chapter 13

Meeting standards in EU agriculture: the changing policy context

Vicki Swales

Introduction

The food we eat and, more importantly, how it is produced, is under greater public scrutiny than ever before. Increasingly, consumers demand not only that food is safe to eat but that its production has not harmed the environment and, where livestock products are concerned, that animals have been treated humanely. As a result, agricultural production is subject to an increasingly wide array of standards, introduced to ensure that public and consumer expectations are met.

The application of standards is a means to achieve a level of safety, quality or consistency in the products and processes that affect our lives. Standards are also vital tools of industry and commerce and often provide the basis for buyer-seller transactions. The level at which a standard is set depends on the desired outcomes. Minimum or baseline standards usually reflect 'socially accepted norms' and frequently seek to prevent negative outcomes that would generally be considered as detrimental to society, such as unsafe food or the pollution of water. Higher standards, by their nature, go beyond accepted norms and are generally adopted to secure positive outcomes that confer additional benefits; such standards are often used in order to achieve market advantage.

Standards relating to agricultural production are found most frequently in the fields of environment, public health, animal and plant health and animal welfare. Such standards may derive from legislation, be defined by market requirements or represent industry best practice. This suggests that different approaches can be taken to achieving standards in agricultural production, from public policy intervention on the one hand through to market led initiatives on the other. The approaches adopted by a particular country or region to setting, promoting and achieving compliance with standards are likely to be determined by a range of factors including: the levels of awareness and understanding of problems arising from agricultural production and the nature of the problems themselves; the availability and costs of technical solutions to those problems; administrative and delivery structures; and, the prevailing politics and culture.

The role of public policy in setting, promoting and achieving compliance with minimum or baseline standards is of particular interest in the context of EU agriculture and the environment. The introduction of cross compliance in the EU – discussed in Chapter 2 – established a link between some CAP payments and compliance with certain legislative

 The Economics of Regulation in Agriculture: Compliance with Public and Private Standards (eds F.M. Brouwer, G. Fox and R.A. Jongeneel)

requirements in order to promote more sustainable agriculture. The 2008 CAP Health Check resulted in a number of reforms to the CAP, including revisions to cross compliance, and further CAP reform is anticipated from 2013 in response to the EU Budget Review. Whilst any future CAP reforms are being driven primarily by budget considerations rather than environmental ones, it is also clear that the EU (as well as its competitor countries) needs to address some key environmental challenges including issues such as climate change, water quality and quantity and biodiversity loss. A growing world population means that these environmental challenges will need to be met whilst increasing food production and responding to consumer demands. Given these political, environmental and societal pressures, the future prospects for a policy such as cross compliance and the issue of how best to meet standards in EU agriculture more generally, warrant further consideration. The experiences and approaches of some of the EU's agricultural competitors, namely the US, Canada and New Zealand, to meeting standards, as discussed in Chapters 3, 4 and 5 respectively, are relevant here.

With these issues in mind, this chapter explores the role of, and need for, policy interventions to promote standards in relation to agriculture and environmental protection in the EU. It considers different policy approaches to meeting baseline standards, contrasting the EU approach with approaches in the US, Canada and New Zealand, and identifies what lessons can be learned. Finally, recognising that EU agriculture policy is likely to face reforms in future, this chapter considers the possible direction of such reforms and the likely implications for meeting standards in the agriculture sector.

Why the need for standards?

The modern day farm level processes that produce the food we eat give rise to a wide range of externalities. Whilst many of these externalities are positive or beneficial to society, such as the maintenance of attractive landscapes or the regulation of soil and water resources, many are negative and potentially harmful to society, such as pesticide residues in food or the pollution of drinking water. By setting standards, public institutions seek to prevent the negative externalities of food production and promote the positive ones.

Whilst food safety, animal and plant health and animal welfare are all issues which demand responses from policy makers, particular attention has been given in recent years to the need for environmental protection measures in the agricultural sphere. Some of the main environmental impacts that public policy is currently trying to address through standards in the agriculture sector are discussed below. The focus is on the EU with three of its main agricultural competitors, namely the US, Canada and New Zealand, examined as comparisons (see Chapters 2, 3, 4 and 5 for further discussion of the development of environmental legislation in the EU, US, Canada and New Zealand respectively).

Agricultural externalities in the EU and competitor countries

The relationship between farming and the environment is complex. Unlike some other economic activities, farming forms part of an ecosystem rather than being external to it. Agricultural commodities are produced by employing land, labour and capital, introducing inputs such as seeds, pesticides and fertilizers and by manipulating the environment through a range of different practices such as land drainage, tilling of soil, diverting natural water sources and grazing livestock. As a result, these practices can impact – both positively and negatively - on soil, air and water resources and on the physical landscape and the biodiversity found there. In many countries, agricultural activity has also been an important factor in the creation of valued landscapes and many species and habitats have become adapted to, and dependent on, certain forms of agricultural management. Increasingly, agriculture's role in relation to climate change is also acknowledged as being both a source of greenhouse gases, particularly methane from livestock production and through the use of fossil fuels (to run machinery and for the production of fertilizers), and a potential sink for gases such as carbon dioxide. Increasing amounts of agricultural waste are also of concern.

The OECD (2001) characterized the main environmental impacts of agriculture in relation to their contribution to:

- Soil quality (erosion, nutrient supply, moisture balance, salinity).
- Land quantity (area of ecological management of agricultural land).
- Water quality (nutrient, pesticide and sediment runoff and leaching, salinity).
- Water quantity (irrigation consumption, use efficiency, water retention capacity, flood prevention).
- Air quality (emissions of dust, odours, ammonia and greenhouse gas, absorption of carbon dioxide).
- Biodiversity (farm and indigenous animal and plant diversity).
- Wildlife and semi-natural habitats (diversity of animal and plant habitats associated with farming).
- Rural landscape (environmental features of areas shaped by farming, including those associated with historic buildings and landmarks).

The environmental impacts of agriculture in the EU are increasingly understood and documented. For much of its agrarian history, Europe has been dominated by relatively extensive, low input systems of farming which gave rise to a wide diversity of species and habitats and managed critical natural resources sustainably (Beaufoy *et al.*, 1994). Farmland which continues to be managed by such extensive farming systems today has been defined as High Nature Value (HNV) but is threatened by the opposing trends of intensification and specialization on the one hand and land abandonment on the other (EEA, 2004; IEEP, 2007a). Work is in progress to measure the full extent of HNV farmland but estimates suggest 30% or more of EU farmland may be HNV. Elsewhere, over the past fifty years or more, agriculture has become increasingly intensive and specialized with an emphasis on high input-high output systems which have resulted in a range of negative environmental impacts (EEA, 2005; Dwyer, 2007) summarized in Table 13.1.

Policy responses have resulted in some improvements in the quality of Europe's environment over the past thirty years but many serious environmental problems remain. Measures are needed to address past environmental damage and to protect, maintain and enhance existing environmental resources through agricultural management. Environmental standards have a critical role to play in this regard.

Are Europe's agricultural competitors experiencing the same or similar environmental problems as a result of the agricultural systems and practices that predominate there? Analysis by the Organisation for Economic Co-operation and Development (OECD) of the relationship between agriculture and the environment (OECD, 2004; OECD, 2008) illustrates that whilst there are many similarities there are also some notable differences which may have a bearing on the policy responses adopted and standards applied.

Table 13.1 Summary of the main impacts of agriculture on the environment in the EU

Environmental attribute	Main environmental impacts resulting from agriculture
Soil	Soil erosion by water and wind mainly in southern and eastern Member States. Soil compaction, contamination and acidification have lessened in New Member States in recent years but may increase with anticipated intensification.
Air	Air pollution from ammonia (leading to acidification especially in northern Member States) and odour from intensive livestock production. Declining emissions of methane and nitrous oxide since 1990 in EU 15 but picture more variable in new Member States.
Water	Water pollution by nutrients and sediment (affecting both ground and surface waters). Unsustainable water abstraction especially in southern States with 50% of population living in water stressed areas and expected to increase as a result of climate change.
Biodiversity	Habitat loss, degradation and fragmentation which has resulted in declines in the number and distribution of some species, most notably farmland birds and butterflies. New Member States retain large areas of undisturbed semi-natural habitats and HNV farmland but these are threatened by intensification and abandonment.
Landscape	Declines in landscape character and quality due to homogenization of land use activities and loss/degradation of landscape features.
Climate change	Production of greenhouse gases such as methane from livestock and carbon dioxide released from soils under certain management practices. Carbon storage and biofuels can help to mitigate climate change.

Source: Summarized from EEA (2006) and Dwyer (2007)

In both EU countries and their competitors (US, Canada and New Zealand) the general trend since 1990 has been one of increasing production on a reduced or near stable area of land achieved by intensifying production. However, the overall intensity of production is lower in the US, Canada and New Zealand than in most other OECD countries (but increasing in some cases) whilst the intensity of production in most of the EU 15 is above the OECD average (OECD 2004 and 2008)[1].

Soil erosion caused by both wind and water is increasingly problematic in both the US and Canadian wheat belt whilst water-related erosion problems are found in New Zealand, especially on steep land that has been cleared of natural vegetation.

Excess nutrients appear to be an increasing problem in the EU competitor countries with increases in both fertilizer use and livestock numbers resulting in overall nutrient surpluses. Water pollution is a particular problem in some regions of the US and at local levels in Canada and New Zealand. Greenpeace (2008) publicized record algal blooms caused by fertilizer run-off in 2007 in Quebec and on Lake Winnipeg, Canada whilst communities in New Brunswick and Quebec have been campaigning against Intensive Livestock Operations (mainly hogs) due to problems of odour and water pollution. Similar concerns have arisen from Concentrated Animal Feeding Operations (CAFOs) in parts of the US. Such intensive livestock systems are also significant contributors to air pollution

[1] Some care must be taken when comparing average intensities. Production can be concentrated in certain regions and be relatively intensive even in countries where the national average intensity of production is low e.g. France, US, Canada.

problems giving rise to odours and release of ammonia, leading to acidification, as well as releasing greenhouse gases such as methane. Other water issues include excessive groundwater extraction levels for irrigation particularly in some parts of the US.

Strong similarities can therefore be identified between the EU and its competitors in relation to agriculture's impact on natural resources of soil, air and water. But when it comes to biodiversity and landscapes, it appears there are notable differences between the two. While there appears to be an intrinsic relationship between farming, biodiversity and landscapes in Europe, with many species and habitats having co-evolved with traditional agricultural practices over time, in New Zealand and North America much of the biodiversity value is associated with natural areas such as grasslands, wetlands, native forests and bush. In other words, there appears to be a much stronger distinction between land occupied by farming and 'wild' or unmanaged land of high nature value. There is however evidence of these natural habitats being threatened and declining as a result of agricultural practices such as the conversion of natural grasslands and wetlands to cropland (OECD, 2004). Similarly, whilst traditional farming practices in Europe have helped to mould and shape the landscape over time giving rise to highly valued cultural landscapes, the same cannot largely be said of the US, Canada and New Zealand.

These brief insights suggest that policy makers in EU competitor countries are facing some similar challenges to those in the EU when it comes to dealing with the impacts of agriculture on the environment but also some distinctly different issues. So how are policy makers in the EU and competitor countries addressing these problems? Are similar or different policy mechanisms being used to meet both baseline and higher standards in agriculture?

Approaches to meeting standards

The EU's response to the environmental problems resulting from increasingly intensive and specialized production has been to introduce, over time, a series of measures designed to prevent environmental damage and degradation on the one hand and incentivize positive environmental action on the other, as discussed in Chapter 2. Environmental protection has largely been the remit of regulation whilst maintaining existing environmental assets and encouraging their enhancement has fallen mainly to incentives such as those offered under agri-environment agreements. Cross compliance, as introduced by the 2003 CAP reforms, is distinctive in creating a link between compliance with (mainly pre-existing) regulatory standards and some CAP payments (all direct payments under Pillar I of the CAP and payments under eight measures within Pillar II[2]). Non-compliance with standards can result in the withdrawal of a proportion of or all a recipient's CAP payments. The objective of establishing such a link between standards and payments is to promote sustainable agriculture and help to prevent land abandonment. Cross compliance also establishes the baseline standards that must be met before farmers can receive other incentive based payments, such as agri-environment payments. It therefore marks an important boundary or reference level between baseline environmental protection standards and higher standards designed to achieve more demanding environmental outcomes.

---

[2] The CAP is comprised of two distinct Pillars. Pillar I consists of market measures and direct aids to farmers, including the Single Farm Payment, and accounts for the largest share of the CAP budget. Pillar II consists of a series of measures under the umbrella of the European Agricultural Fund for Rural Development (EAFRD) including the agri-environment measure and the Less Favoured Area measure.

The EU's choice of policy measures reflects two important principles that have become enshrined in policy development – the 'polluter pays principle' and 'the provider gets principle'. Together, these principles help to explain the EU's approach to meeting environmental standards in the agriculture sector. The former principle underpins much EU environmental law and makes the case that a polluting party must pay for the damage done to the natural environment or bear the costs associated with avoiding such damage. The 'provider gets principle' recognises that agriculture produces a range of environmental goods and services that benefit society as a whole but which are insufficiently rewarded by the market and hence underprovided. A desire to correct such market failure in relation to the provision of environmental goods and services, many of which can be defined as 'public goods'[3], increasingly underpins the rationale for public intervention in the agriculture sector. The 'provider gets principle' is essentially the justification for payment of a wide range of land management incentives today. Whilst both principles are not without their problems, for example it is often difficult in practice to ensure that the polluter actually pays rather than passing on costs to consumers or to determine the value of public goods and therefore what the provider should receive for providing them, they are increasingly used to determine the nature and manner of public sector interventions.

The application of environmental legislation in the EU agriculture sector has evolved over time. Directives concerning the conservation of wild birds[4] and the protection of groundwater from pollution[5] date back to 1979 and 1980 respectively. In 1991, the Nitrates Directive[6] introduced restrictions regarding the application of manure in order to protect water from pollution while more recently the Water Framework Directive[7] introduced a range of controls in order to ensure that all water bodies achieve good ecological status by 2015. A framework Directive for Soil[8] is also now under consideration but political agreement has not yet been reached. Implementation of existing Directives by Member States has had a somewhat chequered history requiring the European Commission, in numerous cases, to open infringement procedures. Nonetheless, this growing body of legislation suggests a willingness on the part of the EU institutions to regulate agricultural activity in order to address key environmental problems. The more recent introduction of cross compliance which, according to Regulation 1782/2003, serves to 'incorporate in the common market organizations basic standards for the environment, food safety, animal health and welfare and good agricultural and environmental condition' is further evidence of the desire by EU policy makers to achieve environmental protection and an example of how compliance with standards can be reinforced through financial mechanisms. Indeed, the

---

[3] Public goods exhibit 'non-rivalry' and 'non-exclusiveness', meaning that one person's consumption of the public good does not diminish its availability for another person (e.g. me gazing at a landscape does not stop anyone else from doing so) and that it is not possible to restrict consumption to only those paying for it (e.g. I don't have to pay to gaze at a landscape). This latter condition effectively means that there is no market mechanism to reward producers of public goods, and so public goods are under-supplied.

[4] Council Directive 79/409/EEC of 2 April 1979 on the conservation of wild birds (OJ L103, 25.4.1979).

[5] Council Directive 80/68/EEC of 17 December 1979 on the protection of groundwater against pollution caused by certain dangerous substances (OJ L 20, 26.1.1980).

[6] Council Directive 91/676/EEC of 12 December 1991 concerning the protection of waters against pollution caused by nitrates from agricultural sources (OJ L 375, 31.12.1991).

[7] Directive 2000/60/EC of 23 October 2000 establishing a framework for Community action in the field of water. OJ L327/1.

[8] COM (2006) 232 final. Proposal for a Directive establishing a framework for the protection of soil.

Commission's mid term review of the Sixth Community Action Programme[9] pointed to the effectiveness of cross compliance in integrating environmental concerns into the CAP and stated that the Commission would investigate whether this policy tool could also be applied in other policy areas such as fisheries and transport.

But as pressure to regulate the agriculture industry has increased so too have calls for substantive reforms to the CAP and to increase support for public goods payments (BirdLife International, 1997, 2007; WWF, 2008). These calls, which coincided with other internal and external pressures for CAP reform, have led to progressive reforms to commodity regimes (reductions in price support and the gradual decoupling of subsidies from production) and the development of agri-environment policy. Agri-environment payments constitute payments to farmers who undertake agri-environment commitments (which go beyond mandatory standards) on a voluntary basis. Payments are intended to cover the additional costs and any income foregone resulting from the commitment made. The first EU wide agri-environment measures were introduced in 1986 as support for environmentally sensitive areas and progressed to Regulation 2078/92 'the agri-environment regulation' in 1992 as one of four 'accompanying measures' to the MacSharry CAP reform. Regulation 2078/92 later became part of a wider Rural Development Regulation in 1999 and today, agri-environment payments are one component of the European Agricultural Fund for Rural Development (EAFRD). This Fund also provides support to improve the competitiveness of agriculture, the quality of life in rural areas and the diversification of the rural economy. Agri-environment payments are not the only environmental payments in EAFRD; other measures such as natural handicap payments in mountain areas and Natura 2000 payments seek to encourage sustainable land management and achieve environmental outcomes. The priority given to such measures and the funds allocated to them varies considerably across the EU Member States but all countries must implement agri-environment measures and the uptake of these measures by farmers has increased substantially since their first introduction (EEA, 2005).

In the EU, we can see that environmental legislation, cross compliance and agri-environment measures are the main policy instruments used to respond to environmental issues in the agriculture sector. Together, these measures are used to achieve a range of environmental standards starting with environmental protection and rising to environmental enhancement. Environmental regulation operates irrespective of any payments made to farmers but cross compliance and agri-environment measures are central components of the CAP and linked directly to certain payments to farmers. How does this compare with approaches to meeting standards in EU competitor countries? Do other countries adopt similar or different approaches to achieving environmental standards?

Meeting standards in EU competitor countries

The OECD (2004) notes that all OECD countries, including the US, Canada and New Zealand, impose regulatory requirements to address the negative impacts of agriculture on the environment, ranging from outright prohibitions to standards and resource use requirements. The imposition of such requirements appears to have been extended or developed since the early 1990s. In the same way that EU Member States are required to

[9] COM (2007) Communication from the Commission to the European Parliament, the Council, the European Economic and Social Committee and the Committee of the Regions on the mid-term review of the Sixth Community Environment Action Programme.

transpose and implement EU wide legislation at national and regional level, OECD states that, 'An increasing number of regulatory requirements derive from state, provincial, regional or local measures under the framework of national umbrella legislation, in order to accommodate the local nature of many environmental concerns.' The Nutrient Management Act in Ontario and the Water Protection Plan in Manitoba, which set targets for N and P levels and regulate some activities such as the timing of manure spreading to land, are two examples of regulatory approaches applied at Provincial level in Canada to control water pollution from intensive livestock operations (see Chapter 4 for further discussion of regulatory approaches in Canada). In the US, the Clean Water Act requires CAFOs to have pollution permits and implement nutrient management plans (see Chapter 3 for further discussion). In New Zealand, water protection is driven by the Resource Management Act; farmers themselves remain responsible for resource management and authorities charge farmers in order to recover the costs associated with programs and applications (see Chapter 5 for further discussion).

Beyond regulation, a range of different approaches to meeting environmental standards are found in use in EU competitor countries (OECD, 2008). In Canada, the 2003-2008 Agriculture Policy Framework (APF) provides support to the farm sector under a range of different programs. The APF consisted of five pillars, namely: business risk management, food safety and quality, science and innovation, environment, and the renewal of the agricultural sector. Under each of these, various financial instruments were made available to support the agriculture sector. These included: income support under the Canadian Agricultural Income Stabilization (CAIS) scheme; production insurance; the Canadian Food Safety and Quality Program (CFSQP), under which it helped stakeholders to develop and implement food safety, quality and traceability systems throughout the food chain; a science and innovation programme including funding the sunrise industries engaged in the production of new commodities and bio-products such as bio-fuels and bio-materials; and, a number of environmental programs. Most APF agri-environmental programs are cost-shared between the Federal, Provincial and Territorial Governments. Programs under the environment chapter of the APF provide producers with assistance to improve their environmental management of soil, water, land and biodiversity by cost-sharing activities such as technical assistance, extension, research and demonstration activities.

In the US, agri-environment programs – also based on cost-sharing – form a growing component of agriculture policy. Measures such as the Conservation Reserve Program (CRP) and Wetlands Reserve Program (WRP) help farmers to remove highly erodable land (HEL) from production and protect and restore wetlands respectively. To achieve this, the programs fund the purchase of long-term agreements or conservation easements. A conservation easement is a legal agreement by which landowners convey certain property rights to a qualified entity for safekeeping, usually for a set period of time or sometimes in perpetuity. The easement essentially prohibits certain activities such as development, cutting down forests or converting wetlands to cropland. In recent years, greater emphasis has been given to supporting conservation practices. The Conservation Security Program (CSP) introduced in 2002 – renamed as the Conservation Stewardship Program by the 2008 Farm Bill – pays farmers who have met a high standard for environmental performance to adopt or maintain practices to further enhance environmental performance, such as improving soil and water quality or wildlife habitat. Meanwhile, the Grassland Reserve Program (GRP) introduced in 2003 aims to preserve and improve native grass species. Technical advice and research to support sustainable farming is also funded under the Farm Bill. Cross compliance is also in evidence in the US; the basic principle of making a link 'across programs' originated there during the late 1970s as a way of enforcing set-aside requirements (Baldock and Mitchell, 1995). Today, in order to receive payments under commodity programs such as price support loans and income support payments farmers must implement soil conservation systems ('sodbuster') on HEL and refrain from draining wetlands ('swampbuster'). In the same way that EU farmers can have payments under the Single Payment Scheme withheld for non-compliance, violations of either sod- or swampbuster result in the loss of all USDA programme payments for that year and for the years that follow.

New Zealand differs most among the countries under consideration here possibly as a result of having moved furthest towards a fully liberalized agriculture sector. Producer support represented just 2% of agriculture receipts as calculated by the OECD Producer Support Estimate (PSE) compared to an OECD average of 30% of receipts. Of the support given to agriculture, expenditure on agri-environment measures accounted for 15% in 2004 and 90% of this was spent on research and education (OECD, 2008). Consultation, outreach and education appear to be having an impact although significant environmental problems remain – especially soil erosion and water pollution – and new approaches are now being tested. Nutrient flows from agriculture into Lake Taupo were such that in 2004 national and local governments agreed to fund a total package of nearly NZ$82 (US$54) million to limit nutrient flows through restrictions on land use and allowing nitrogen trading to occur. Similar to easements, conservation covenants are used in New Zealand to protect a variety of open spaces, including forest remnants, wetlands, lakes, peat lakes, coastline, tussock grasslands, from damaging agricultural practices. Some farmers have entered into open space covenant agreements through the QE II National Trust, a statutory body with a budget of NZ$4 million. The Trust provides limited support to protect certain areas of farmers' land while they retain ownership. Community based approaches – known as Landcare or conservation clubs – are also found in New Zealand as they are increasingly in Canada and the US. Such approaches which encourage collective action to solve environmental problems, '…tend to take advantage of farmers' self-interest in environmental conservation and make use of local expertise in solving environmental problems' (OECD, 2004).

Lessons learned from EU competitor countries

This brief analysis suggests a number of similarities in approaches to achieving environmental standards in the EU and some of its competitor countries. Environmental regulation is in evidence in all countries although there may well be differences in the intensity of regulation reflecting, amongst other factors, differences in the intensity of agriculture. Forms of cross compliance can be found in both the US and EU with the threat of payment withdrawal being used as leverage to secure compliance with standards. There also appears to have been an expansion over the last decade or more of agri-environment type measures in both EU and non-EU countries although the exact nature of these varies from place to place. There are, however, some notable differences between the EU and the countries examined. In the US, the buying out of property rights through the extensive use of conservation easements is notable. OECD (2001) suggests that conservation easements may be attractive policy instruments because, '…purchasing a partial interest in a parcel of land is less costly than acquiring full ownership and they involve voluntary transaction. Moreover, compared with conservation incentives (taxes or annual payments) or tradeable development rights, easements do not entail large administrative burdens.' Cost-sharing programs and technical assistance are much more prevalent in the likes of the US, Canada and New Zealand whilst in the latter, the low level of agricultural support generally appears to have resulted in much greater emphasis on research, education and technology transfer. However, growing concerns regarding water pollution in New Zealand is leading to innovative approaches such as the introduction of tradeable nitrogen permits. Community based approaches are also increasingly predominant in the non-EU countries examined.

Examining policy approaches used in other countries can often be instructive; new ideas can be gained and lessons can be learned from experiences elsewhere – although the scope for transferring approaches needs to be considered carefully given the context-specific nature of environmental problems and solutions (Bemelmans-Videc *et al.*, 2003; Dolowitz and Marsh, 2000; Evans, 2006). In principle, some of the approaches used to help meet environmental standards in other countries could be relevant to the EU agriculture sector. The greater emphasis in other countries on research, education and technology transfer as well as cost-sharing programs and the use of easements to buy out property rights is of particular interest. Such approaches may warrant further consideration in an EU context, especially if reforms to the EU Budget and the CAP constrain the use of existing policy mechanisms. It is to these issues that we turn next.

Future prospects for meeting standards in EU agriculture

The CAP faces adjustments and reforms urged on by a combination of both internal pressures (such as consumer demands and budget constraints) and external ones (such as World Trade Organization (WTO) negotiations and world food demand). Such reforms may have wider implications for how environmental standards are achieved in the agricultural sector in future. With this is mind, this section reviews the main drivers of, and opportunities for, CAP reform, the possible policy outcomes that might emerge and their implications for meeting environmental standards. The extent to which the EU could adopt approaches to meeting standards used elsewhere is also considered briefly.

'New challenges' for EU agriculture

The CAP Health Check was only the most recent in a long line of reforms to EU agriculture policy. Launched by the European Commission on 20 November 2007[10], and concluded with an agreement on 20 November 2008[11], part of the justification for the reform proposals was explained as the need to meet 'new challenges' including: climate change; bioenergy; water management and biodiversity. Agriculture was noted as having the potential to help mitigate climate change but also being 'highly exposed' to it. Equally, it was suggested that without sustainable water management, '…the pressure on the quantity and quality of water for agriculture will increase considerably.' Agriculture's key role in the protection of biodiversity was also highlighted as was the EU's likely failure to meet a commitment to halt declines in biodiversity by 2010. The strengthening of rural development policy (EAFRD) and cross compliance were both agreed as responses to these challenges, through increased funding for EAFRD and changes to Good Agricultural and Environmental

---

[10] COM (2007) 722 Communication from the Commission to the Council and Parliament 'Preparing for the "Health Check" of the CAP reform'

[11] See http://ec.europa.eu/agriculture/healthcheck/index_en.htm for further details of the reform agreement

Condition (GAEC) within cross compliance. In other words, both regulatory and incentive based approaches were proposed, until 2013 at least, as part of a strategic approach to meeting standards in agriculture and promoting sustainable agriculture. Whilst this reform is likely to make a valuable contribution to addressing the environmental challenges identified by the Commission much remains to be done. Issues such as climate change and better water management are likely to continue as key drivers of future policy developments.

Environmental challenges are not the only ones likely to drive EU agriculture policy and production standards in future. The prospects of a growing world population and increasing demand for food have led to growing concerns about food security and a so-called 'food challenge'. FAO (2002) estimates that global food production needs to rise by 50% by 2030, and to double by 2050, to meet the demand of a rising world population and supply the eating habits of the growing middle classes in emerging economies. ELO (2008) suggests that improving productivity to meet these demands and, at the same time, improving the condition of the natural environment, is the 'Great 21st Century Land Challenge' and continues:

> 'During the next half Century the world faces fundamentally two challenges – to produce enough food for the growing numbers of, on average, better-off and longer living people, and at the same time to preserve natural resources and the environment. The two challenges are closely interrelated, and each is heightened by climate change. The EU is of such a scale that we must play a responsible role in rising to these two challenges. Furthermore, food and environment are core competences of the EU.'

The EU budget review

While the Health Check dealt with some rather specific issues which required adjustment to the CAP in the short term, a much broader debate about the future of the CAP as a whole is already in train. A key process driving this debate is the EU Budget Review which was initiated with the launch of a document entitled, 'Reforming the budget, changing Europe'[12] and the opening of a public consultation which ran until April 2008.

The Budget Review owes its origins to heated discussions at the EU Summit in December 2005 when the last EU financial framework for 2007-2013 was agreed. Tony Blair, the then UK Prime Minister, raised fundamental questions about EU spending priorities and attacked the CAP in particular, much to the displeasure of France, historically a staunch defender of the policy. As a result, in May 2006, an Inter-institutional Agreement on budgetary discipline and sound financial management between the European Parliament, the Council and the Commission invited the Commission:

> '...to undertake a full, wide ranging review covering all aspects of EU spending, including the CAP, and of resources, including the UK rebate, to report in 2008/9'.

The Budget Review asks some fundamental questions about European spending and, in particular, the effectiveness, efficiency and transparency of budget delivery. The outcomes of this Review are likely to determine the next EU financial framework for 2014 onwards.

[12] Communication from the Commission, reforming the Budget, Changing Europe: A Public Consultation Paper in View of the 2008/2009 Budget Review, SEC (2007) 1188, 12.9.2007

Discussions about the CAP, within the context of the budget, may well be far reaching with questions asked about the overall purpose of the CAP and the benefits achieved by currently spending 47% of the total EU budget on agriculture. This expenditure finds itself increasingly in competition with other demands on the EU budget. Some commentators (IEEP, 2007b) consider it is almost a given that the CAP budget will be cut in real terms in 2013. If this is the case, the CAP of the future may look rather different to the CAP of today. One of the strongest arguments for the continuation of CAP funding and for public payments to farmers is that it is needed to ensure the provision of public goods such as the maintenance of wildlife species and habitats and the management of landscapes. Responding to the 'new challenges', such as climate change, is also seen as a legitimate reason for providing certain kinds of support to the agriculture sector. Equally, there may be arguments for support to achieve social objectives within rural areas, although not necessarily targeted at the agricultural community as has largely been the case in the past. As the objectives of the CAP are debated so too will be the means of achieving such objectives. The extent to which objectives (and standards) can be achieved through the current framework of CAP Pillar I versus Pillar II payments, or whether alternative approaches are needed, are likely to be the focus of much discussion. Possible prospects for the CAP beyond 2013 are discussed below.

The WTO negotiations

The future direction of the CAP, as well as being determined by internal processes such as the Budget Review, may also be influenced in the longer term by external processes such as the WTO negotiations.

A new round of trade negotiations began in November 2001 – the so-called Doha Development Agenda. The Doha Declaration defines the basis for these negotiations and includes commitments to:

- Substantial reductions in trade-distorting domestic support.
- Substantial improvements in market access.
- Reductions of, with a view to phasing out, all forms of export subsidies.

In August 2004, a Framework Agreement was reached that committed WTO members to eliminate '...by a credible end date, all forms of export subsidisation and for the progressive reductions of tariffs and domestic support'. Negotiations to define the necessary details of such an agreement have been protracted. Negotiations were suspended in July 2006 and later resumed, only to end again without agreement in July 2008. In July 2009, WTO agriculture negotiators began technical work to identify what data will be needed and how to present it when members eventually draft their commitments to cut agricultural tariffs, open up quotas, reduce domestic support and scrap export subsidies. For a large majority of the WTO's 151 members, major reductions to trade-distorting farm support in rich countries remains central to delivering on the development dimension of the round.

The outcome of these negotiations could yet have implications for future EU agriculture policy and may have some bearing on any CAP reform decisions beyond 2013. Potter (2007) has suggested that, in order to satisfy Green Box requirements (for non-trade distorting subsidies), CAP Pillar I payments may need to become more like basic agri-environment type payments justified on the grounds of delivering public goods rather than income support. Potter has also suggested that with an agreement in place, and assuming the EU secured recognition for Geographical Indications (GIs)[13], Europe's farmers, or at least some of them, are likely to focus on producing high-value products rather than undifferentiated commodities. In this context, production standards would be of particular importance, albeit many of them driven by market requirements rather than legislation. Under such a scenario, Potter suggests cross compliance could be expected to migrate to a limited system underpinning other types of payment.

The CAP beyond 2013: future prospects for meeting standards in agriculture

New food and environmental challenges, pressure on the EU budget and a possible new global trade deal all suggest that further reform of the CAP is inevitable. Whilst the nature of any reform is, at this stage, difficult to judge, numerous ideas and proposals have been put forward by farming and environmental stakeholders, some EU Member States, think tanks and academics (see for example BirdLife International, 2007; COPA-COGECA, 2008; Ministry of Agriculture, Nature and Food Quality, 2008; Notre Europe, 2008; Swinnen, 2009). Ideas range from maintaining the existing CAP structure but 'greening' Pillar I income support by adding new environmental conditions to phasing out such support over time and substantially increasing support under Pillar II, with numerous other variations. The European Commission is expected to issue a communication in mid 2010 setting out some possible options for CAP reform beyond 2013.

The future prospects for meeting environmental standards in agriculture are therefore rather uncertain at this stage. It is expected however that the CAP budget will be reduced and this will have a significant influence on future funding streams under Pillars I and II and what can be achieved through those Pillars in terms of environmental standards. These issues can be helpfully explored by considering possible alternative directions for a future common EU agriculture policy[14]. What follows are three possible alternatives for CAP reform based on a combination of changes to the budget and to Pillars I and II. For each of these, the implications for meeting standards and addressing environmental challenges are considered. The suggestions put forward are described as 'reform options' but are intended to be illustrative only and to help think through CAP reform implications.

CAP reform option 1

- The Pillar I budget is reduced moderately e.g. by no more than 20-30%.

---

[13] Geographical indications are place names (in some countries also words associated with a place) used to identify the origin and quality, reputation or other characteristics of products (for example, 'Champagne', 'Tequila' or 'Roquefort'). Only products produced in the defined areas and with the relevant characteristics can be labelled accordingly and the measure effectively protects products from unfair competition. There is extensive debate within the WTO regarding GIs and the level of protection afforded to such products with the EU arguing that registered GIs must be protected in other WTO member countries.

[14] The ideas discussed here draw on previous work on policy options for cross compliance. See: Farmer, M. and Swales, V. (2007), 'Future Policy Options for Cross Compliance'. Background paper 26 April 2007 Cross Compliance Policy Seminar Brussels. D23 of the Cross Compliance Network. IEEP, London; Farmer, M. and Swales, V (2007) Facilitating the CAP Reform: compliance and competitiveness of European agriculture. Project No SSPE-CT-2005-006489. D15: Guidelines for policy: A Policy Perspective on Cross Compliance and Competitiveness.

- All direct payments are fully decoupled from production and paid as a flat-rate area payment.
- As a result of these changes, the value of the direct payment is less than at present.
- Pillar II continues largely as at present in terms of content and budget allocation.

Under this option, income support payments in the form of the SPS are reduced but remain significant. With no expansion of Pillar II and hence constraints on how far 'good' environmental behaviour can be incentivised, EU policy makers may look to strengthened regulation and enhanced conditionality of Pillar I in order to address key environmental problems. This is essentially a 'greening' of Pillar I. Cross compliance therefore remains a key policy tool for achieving environmental standards under this option.

Possible options for enhancing cross compliance could include the addition of new items of EU environmental legislation, such as the Water Framework Directive (WFD) to the SMRs under cross compliance. GAEC, in particular, could be broadened with new standards added, such as those that are normally associated with agri-environment schemes e.g. requirements for farmers to set-aside a proportion of their land as 'ecological compensation areas'. In such areas conservation land uses, including buffer strips alongside features, would be required. This could be a simple way to manage environmental pressures such as soil erosion, diffuse agricultural pollution and biodiversity conservation. This could provide a replacement for compulsory set-aside which was abolished following the CAP Health Check. Issues such as climate change and better water management could also be further incorporated in GAEC.

CAP reform option 2

- The Pillar I budget at EU level is reduced substantially e.g. by about 60%. Direct payments that remain are targeted to farms in marginal areas where land abandonment would result in detrimental impacts to the environment.
- The Pillar II budget is increased substantially from savings made by reducing the Pillar I budget (but Pillar II does not receive a full transfer of funds allowing some reduction in the overall CAP budget).
- Rural development measures are expanded to combat climate change, achieve better water management and reverse declines in biodiversity.

Under this option, Pillar I is reduced substantially. Only farming systems located in High Nature Value or more marginal farming areas, where the consequences of land abandonment would result in significant loss of public goods, receive any basic income support. Farmers in receipt of targeted direct payments must comply with the requirements of EU legislation (as must all farmers) and adhere to some basic environmental standards appropriate to the type of farming systems being supported (e.g. in relation to appropriate stocking densities). Baseline standards must also be met by farmers who wish to claim Pillar II payments. Whilst the rationale for CAP support may be easier to justify under this option, some potential problems are likely to arise. Cross compliance no longer becomes an effective mechanism since relatively few farmers qualify for the direct payment and Pillar II measures are voluntary. The leverage cross compliance offered to encourage compliance with standards over a large area of farmland therefore declines requiring policy makers to use alternative mechanisms. The enforcement of legislation by relevant inspection bodies and the application of fines become increasingly important under this Scenario. As in EU competitor countries such as the US, Canada and New Zealand, greater emphasis on research, extension and advice could be used to help to support compliance with legislation, for example helping to identify best and least cost management practices. There could also be a case for the introduction of cost-sharing programs where the burden of legislation was felt to be too great in some situations. Such programs could potentially be funded from the Pillar I budget. Options such as tradeable permits for nitrogen use could also be explored.

CAP reform option 3

- The CAP budget is reduced by 30-50%; the CAP now consists of only one Pillar and is re-named the EU's Common Rural and Environment Policy (CREP) or similar.
- The CREP does not provide income support for farmers, beyond providing a minimum safety net mechanism, supplemented by private risk insurance.
- CREP payments are designed to deliver specific public benefits and support the sustainable development of rural areas.

Under this option, the rationale for public payments to the agriculture sector is one of delivering public goods. Payments can only be made for activities that go beyond mandatory requirements set by legislation. The emphasis of payments is likely to be on maintaining existing environmental assets and achieving environmental enhancement beyond the baseline. Particular emphasis would be given to addressing key environmental challenges such as climate change, bioenergy, better water management and biodiversity. As for Option 2, the enforcement of baseline legislation would rely on systematic inspection and the application of fines. With no Pillar I or cross compliance mechanism to draw on, standards are achieved through application of the polluter pays principle on the one hand and the provider gets principle on the other. However, additional support to improve compliance with legislation and promote best practice, e.g. through advice and training, could be provided through CREP. Environmental taxation, e.g. a nitrogen or pesticide tax, could also be considered as a means of achieving environmental protection. Community based approaches to land management could also be explored, possibly supported and facilitated through CREP funding.

What these three options help to demonstrate is that different budgetary outcomes and changes to the architecture of the CAP Pillars have implications for the choice of public policy instruments employed to help meet environmental and other standards in the EU agriculture sector. Environmental regulation is likely to remain important in all cases, and may need to be strengthened, but the scope for use of cross compliance and incentive payments varies depending on the outcomes of any reforms regarding Pillars I and II. The application of other policy instruments such as information provision (including research, extension and advice), capacity building (including training and facilitating community based land management), alternative market incentives (such as tax breaks, taxes, user charges and conservation easements) and market based instruments (such as tradeable quotas and permits) may become increasingly necessary. Instruments and policy approaches used in EU competitor countries may have some relevance in the EU and warrant further consideration.

Conclusions

The EU is not alone in needing to employ policy measures to address environmental problems in the agriculture sector. Competitor countries such as the US, Canada and New Zealand are also experiencing the negative impacts of intensive agriculture on soil, air, water and biodiversity resources and face the same needs to address issues such as climate change. Similarities in policy approaches are apparent in all countries, for example, the use of regulation to prevent damaging activities and to set minimum standards for water or air quality. But beyond regulation, the approaches employed appear to differ, owing something to the underlying policy framework and the nature of support given to the agriculture sector. In all cases, policy makers appear to rely not just on one measure but on a suite of measures to achieve standards in agriculture, perhaps a reflection of the complex interactions between agriculture and the environment, and of the challenge of achieving sustainable food production.

Within the EU, the CAP plays a dominant role in helping to achieve environmental standards (both baseline and higher standards) within the agriculture sector through instruments such as cross compliance and Pillar II measures. These are underpinned by a substantive body of environmental legislation. But the CAP faces possible reform beyond 2013 as a result of budgetary pressures and the need to respond to new food and environmental challenges. The CAP of the future could look substantially different from the CAP of today.

The three CAP reform options presented here show the relationship between CAP payments and the delivery of standards is rather delicate. Changes to the CAP may determine the extent to which policy mechanisms such as cross compliance remain relevant or appropriate to achieving such standards in future. As long as income support payments remain, the principle of making a link between such payments and compliance with certain standards remains valid. Farmers have an incentive to comply with standards or face withdrawal or loss of payments. If such payments were significantly reduced or abolished, the link would be weakened and other means of achieving compliance with desirable standards, such as regulation, information provision (including research, extension and advice) and market based instruments, are likely to become more important. The level of funding remaining for Pillar II type measures will also be critical in determining the extent to which farmers can be incentivised to meet certain standards, especially those which go beyond baseline regulatory requirements. Experiences in EU competitor countries, and the policy approaches employed to meet standards, are highly relevant in the EU context and deserve further consideration. However, the scope for transferring approaches needs to be given careful consideration due to the context-specific nature of environmental problems and solutions.

As pressure on our natural resources increases – from factors such as population growth, changing consumer demands and climate change – our response to protecting those resources and ensuring their sustainable use must grow ever more sophisticated and effective. Setting, promoting and achieving compliance with standards is an essential part of the future challenge for global agriculture.

References

Baldock, D. and K. Mitchell (1995), *Cross Compliance within the Common Agricultural Policy: a review of options for landscape and nature conservation,* London: Institute for European Environmental Policy.

Beaufoy, G., D. Baldock and C. Clark (1994), *The Nature of Farming: Low intensity farming systems in nine European Countries.* London: IEEP, WWF and JNCC.

Bemelmans-Videc, M-L., R.C. Rist and E. Vedung (Eds.) (1993) *Carrots, Sticks and Sermons: Policy Instruments and their Evaluation.* Comparative Policy Analysis Series, New Jersey: Transaction Publishers.

BirdLife International (1997) *A Future for Europe's Rural Environment: Reforming the Common Agricultural Policy.* Brussels: BirdLife International.

BirdLife International (2007) *New Challenges, New CAP: BirdLife International's vision for the future of the EU Common Agricultural Policy*. Brussels: Bird Life International.

COPA-COGECA (2008) *Visions for the Future of Agricultural Policy in Europe*. Brussels: COPA-COGECA.

Dolowitz, D. and D. Marsh (2000), 'Learning from Abroad: The Role of Policy Transfer in Contemporary Policy Making', *Governance,* 13 (1): 5–24.

Dwyer, J. (2007), 'The state of the rural environment in Europe: what challenges and opportunities for future policy?', Paper prepared for the LUPG/BfN conference 'Future Policies for Rural Europe 2013 and Beyond' held on 19 and 20 September 2007, Brussels.

EEA (2004), *High Nature Value Farmland: Characteristics, trends and policy challenges*, EEA Report No 1. Copenhagen: European Environment Agency.

EEA (2005), *Agriculture and Environment in EU 15: The IRENA Indicator Report*, Copenhagen: European Environment Agency.

EEA (2006), *Integration of Environment into EU Agriculture Policy - The IRENA Indicator-based Assessment Report*, EEA Report No 2. Copenhagen: European Environment Agency.

ELO (2008), *The 21st Century Land Use Challenge*, Brussels: European Landowners Organisation.

Evans, M. (2006), 'At the Interface between Theory and Practice - Policy Transfer and Lesson Drawing', *Public Administration*, 84 (2): 479–89.

FAO (2002), *World Agriculture: Towards 2015/2030*, Rome: Food and Agriculture Organization of the United Nations.

Greenpeace (2008), *'Dead Zones: How agricultural fertilizers kill our rivers, lakes and oceans'*, Greenpeace.

IEEP (2007a), *Final Report for the Study on HNV Indicators*. Contract Notice 2006- G4-04. Study for the European Commission.

IEEP (2007b), 'Towards the CAP Health Check and the European Budget Review: The proposals, options for reform, and issues arising', Paper prepared to inform discussions at a 'CAP Retreat' organised by the German Marshall Fund of the US and the William and Flora Hewlett Foundation on October 2007.

Ministry of Agriculture, Nature and Food Quality (2008) *European Agricultural Policy 2020: The Dutch Outlook.* The Hague, Ministry of Agriculture, Nature and Food Quality.

Notre Europe (2008), *CAP Reform Beyond 2013: An Idea for a Longer View*. Paris.

OECD (2001), *Improving the Environmental Performance of Agriculture: Policy Options and Market Approaches,* Paris: Organisation for Economic Co-operation and Development.

OECD (2004), *Agriculture and the Environment: Lessons learned from a Decade of OECD Work*, Paris: Organisation for Economic Co-operation and Development.

OECD (2008), *Environmental Performance of Agriculture in OECD Countries since 1990*, Paris: Organisation for Economic Co-operation and Development.

Potter, C. (2007), 'Agricultural Trade Liberalisation and the Rural Environment', Paper prepared for the LUPG Conference 'Future Policies for Rural Europe: 2013 and beyond' held in Brussels, 19-20 September 2007. See: http://www.lupg.org.uk/Default.aspx?page=109.

Swinnen, J. (2009), 'The Future of Direct Payments: Better targeting, phasing-out, new objectives or time for a 'Green Deal' for EU agriculture?', Paper prepared for the BEPA Workshop, February 26, 2009. University of Leuven and Centre for European Policy Studies, Brussels.
WWF (2008), No More EU-trophication: How CAP Reform can Save the Baltic Sea, WWF, Sweden.

Chapter 14

The pollution haven hypothesis and the location of livestock production: two North American case studies

Alfons Weersink

Introduction

The previous chapters have examined how policies targeted toward addressing environmental concerns within agriculture have impacted various regions and commodity sectors. This chapter builds upon those comparisons by assessing the influence of environmental regulation on the location decisions for livestock production within a local region (south-western Ontario, Canada) and at a national scale (US). The North American livestock sector has experienced a geographic concentration of production in fewer regions and a shift in production to areas with little prior livestock experience. One of the reasons may be the increasingly important role of the processing sector and the integration of this sector back into production. Processing plants operating under economies of size are becoming larger and fewer, and scattered around the country with clusters of livestock farms around them (Apland and Anderson 1996; Abdalla *et al.*, 1995). Such clusters tend to move to localities with better natural endowments, labour market conditions, and business environment due to agglomeration economies or tax policies (Roe *et al.*, 2002).

Changes in the spatial distribution of livestock production may also be directly affected by differences in the stringency of environmental regulations across administrative regions. The spatial differences in environmental pressures might create 'pollution havens' where lenient regulations in some regions may attract livestock producers to build their facilities in such localities. It is possible that by introducing or maintaining lax environmental regulations relative to competing regions and allowing tardy enforcement of those regulations, one region can lure 'dirty' industry investments, which are important in employment creation and regional economic development (Kunce and Shogren, 2002; Levinson, 2000; Jafee *et al.*, 1995). If governments really do engage in a race to the bottom, certain regions would have an inefficiently high number of livestock operations. Because the assimilative capacity of the environment is deliberately undervalued in a region where a race to the bottom has occurred, the heavier concentration of livestock farms is the cost to the local community.

Both US and Canada allow national and sub-national governments to enact relatively independent policies in the environmental sector. A disparity in regulatory stringency among US states arose in the 1980s when the federal government delegated the function of devising regulatory regimes to state authorities (Kraft and Vig, 1994; Levinson, 2000).

Martin and Zering (1997) argue that the differences in regulations across states in the US were responsible for the shift in large-scale intensive pork production to southern states such as North Carolina and Arkansas. While there are differences in regulatory stringency across provinces in Canada as across states in the US, there also existed differences across lower tier governments within a province. For example, local municipalities in Ontario until recently could directly influence the form and location of livestock barns through controls on the land use and the issuance of building permits. This decentralized regulatory approach resulted in a range of environmental laws with some municipalities going to the extent of imposing caps on livestock numbers while neighbouring ones may have had few restrictions (FitzGibbon *et al.*, 2002). The difference in environmental pressure across jurisdictions presents a natural experiment to test the pollution haven hypothesis in agriculture.

This chapter reviews the results of two studies examining the factors affecting farm location with specific emphasis on the existence of pollution havens in the livestock sector. One study by Weersink and Eveland (2006) examines the impact of differing municipal by-laws at the local level in south-western Ontario. The other study by Herath *et al.* (2004) determines the effect of environmental regulations on changes in the spatial distribution of production levels of hogs, dairy and fed-cattle by state in the US. The chapter begins by discussing the conceptual framework on factors affecting farm location that underlies both studies. The results at the local level are then presented and followed by the national analysis. Implications for the influence of environmental pressures of farm location are then discussed.

Conceptual framework

The local and national studies both employ a similar framework to determine what influences farm location. In both studies, livestock intensity is assumed to be a function of four major categories of explanatory variables:

- the degree of environmental regulations facing farmers;
- relative output and input prices;
- the attractiveness of a region for a firm considering locating in the area or business climate; and
- agglomeration economies and regional support for farming.

The following sub-section defines how livestock intensity is measured in each of the two studies. The specific factors within each of the major categories of explanatory variables are then defined along with their hypothesized effects on farm location.

Dependent variable

Spatial production changes can be captured by the number of new farms or the intensity of production within a region. Bartik (1988) argued that aggregate measures of regional economic activity, such as inventory levels, reflect a number of different types of economic decisions by agents. Production levels can change due to the expansion or contraction of existing facilities, the introduction of new facilities, or the closing of old ones. Since new firms considering locating in a region tend to face harsher environmental constraints than existing firms due to grandfathering arrangements, the opening up of new facilities will be lower in a region with more stringent environmental regulations (Bartik, 1988). Measuring farm location through new barn building or inventory change is one aspect of overall farm structure that may include factors such as farm size distribution and marketing arrangements.

Local analysis

The local study follows the recommendation by Bartik and uses primary data on building permits granted to individual farmers to build an addition to an existing barn and build a new barn. Such permits are required for all new buildings, primarily to ensure construction codes are met for safety but increasingly as a means of enforcing environmental restrictions. Although over 200 municipalities within 8 counties of the traditional livestock production region of south-western Ontario were contacted, usable results were obtained for 42 municipalities. Of those municipalities that did provide data, 1,424 building permits were issued between 1996 and 2001. Over 40% of these permits were for swine facilities with approximately 26% for dairy barns and 19% for poultry operations. The majority of these permits (996) were for new barns while only 30% represented additions. Over 75% of the permits issued for the swine (428) and poultry (221) sectors were for new barns. The majority of permits for both new facilities and additions (760) were for barns less than 100 livestock units (LU) but most of the permits issued for the swine sector were for facilities larger than 100 LU. Further descriptive statistics on the barn permits is in Eveland *et al.* (2005).

The focus of the analysis is on the number of permits that were granted per 10,000 ha of farmland by each municipality. The average value is 11.5 permits per 10,000 ha with the high density building areas tending to be located in the traditional livestock producing municipalities of Perth and Huron counties. In addition to the density of total barn building permits, two other dependent variables were considered in the local study. One measured the density of permits for new facilities and the other counted the density of permits (new and additions) for swine barns. The former is used to assess whether environmental regulations are tougher for new operations which also tend to be more mobile and thus more sensitive to such regulations. The density of permits for swine operations is used to determine whether differing municipal by-laws are particularly effective in altering location decisions of hog farms for which public opposition is most vocal.

National analysis

While the number of new livestock operations may be the best measure of regional production changes due to environmental laws, it is not available for an extended period of time at a larger, aggregate level. Thus, livestock inventory is used as the measure of spatial production in the national US study. Data on hog, dairy and fed-cattle production levels from 1975 to 2000 were collected for each of the 48 contiguous states. In order to control for cyclical fluctuations of livestock inventories, annual inventory shares for each of the three livestock sectors from 1976 to 2000 were collected resulting in 25 observations over time for each state for each livestock sub-sector.

The hog and dairy sectors in the US are increasingly siting production towards western states and away from traditional production regions in the east. In contrast, production levels in the fed cattle sector have increased over the last generation only in the three main producing states (Texas, Kansas and North Dakota). The spatial changes in US livestock inventory over time are described further in Herath *et al.* (2004).

Explanatory variables

Environmental stringency

The weakness of most previous studies examining the effect of regulatory pressures on farm location is the measurement of environmental stringency. Both the local and national studies have constructed unique variables on the environmental pressure facing livestock farmers.

Local analysis

Environmental stringency at the municipal level in Weersink and Eveland (2006) is proxied by five possible explanatory variables surrounding the possible requirements that must be in place for any proposed livestock barn to get a building permit:

- required manure storage days;
- minimum distance from waterway;
- minimum lot size;
- nutrient management plan (NMP) requirements based on density; and
- trigger for expansion.

An increase in the number of manure storage days required by new livestock facilities allows the producer to apply waste less frequently, but it also means a greater initial investment for new livestock facilities, which could deter farmers from starting their business in certain areas. The minimum distance from the waterway and minimum lot size for new barns imply the farmer must set aside a greater amount of land in construction. Increasing the distance between livestock operations and neighbours may increase financial costs and thus decreases the likelihood of barn construction occurring but it may also reduce the likelihood of conflicts that represent future monetary and non-monetary costs to the farmer. The effect of the stocking density restriction from the NMP depends on land availability. If the constraint is binding, the cost of the NMP is the amount of land required multiplied by the land rental rate.

The two remaining potential environmental stringency variables in the local study relate to the triggers at which producers must complete a nutrient management plan (NMP) as part of receiving a building permit. A NMP indicates the sources, uses, and application testing/methods of organic and inorganic nutrients on the farm, and represents a significant cost to farmers, particularly smaller farms. The trigger is defined here in terms of the density of the proposed facility expressed in animal nutrient units. The lower the trigger value for a municipality, the greater the chance the farmer will have to file a NMP for a given size of operation and thus the less likely they are to build within the municipality. While it varied across municipalities, a common trigger that is currently used for the standard provincially is 300 animal nutrient units (NU) with 6 finishing pigs representing one NU. A final dummy variable was defined depending if the municipality had a trigger for the requirement of a NMP.

National analysis

The national study by Herath *et al.* (2005) constructed a single regulatory stringency measure for each state over time rather than use a one-period cross sectional stringency measure as has been done in other studies testing the pollution haven hypothesis in agriculture. State values of 14, one-period environmental stringency measures developed for various years by alternative studies were gathered. Examples of such indices include the Green Index by Hall and Kerr (1991) and per capita environmental quality control expenditure by state from the US Dept of Commerce. A measure for the year 2000 for all states was calculated using the approach Metcalfe (2000) employed in estimating regulatory pressures facing farmers for a subset of states in 1994 and 1998. The presence (or absence) of seven regulations for each of the 48 states were summed to form the 2000 stringency index. In order to capture the temporal changes of regulatory stringency across states, the individual state values for all the above indices were normalized by dividing through with the arithmetic mean of the respective index with missing years estimated through interpolation of nearby years. The normalized index value represents the position of the state relative to the arithmetic mean of each index and allows the different indices to be combined into a single regulatory variable with a consistent scale measure.

Relative prices

Increasing output prices or decreasing input prices can both increase relative profitability, which is assumed to be one of the main drivers of the production expansion and spatial distribution of livestock operations. For example, the livestock to feed price ratio has been shown to positively affect spatial distribution of livestock production in the US. Previous studies have also found firms are more likely to locate in regions with lower energy costs (McConnell and Schwab, 1990) and lower labour costs (Levinson, 2000). Relative output to feed price indices at the state level are used for the US study but are not appropriate for the local Ontario study since producers in this relatively small geographic area face the same output prices for their livestock as most are either marketed under a supply management scheme (dairy and poultry) or a pooling system (swine).

The one price used in both studies is the value of farmland. Studies assessing spatial distribution of national livestock production have found evidence of inventories shifting to locations with cheaper land prices and thus cheaper construction costs (Mo and Abdalla, 1998). This effect within a local area may be countered by the lower feed costs associated with higher-yielding, and consequently, higher-priced farmland. Thus, the a priori effect of farmland value on the density of barn permits granted is ambiguous.

General business climate

Business climate variables capture the receptiveness of the community towards a new firm locating in the region and the attractiveness of the region for the prospective firm. The socio-economic variables that could influence the establishment of a livestock operation include: population, unemployment rate, education and wealth.

Increasing the number of people and businesses can increase the amount of available labour, increase the demand for associated products, and reduce costs by increasing the extent of public infrastructure (Eberts and McMillen, 1999). However, the increasing role of the 'not in my backyard' (NIMBY) attitude is expected to dominate any positive economies of urbanization so that population density is hypothesized to have a negative effect on livestock production density.

The level of unemployment can have an influence on farm location through labour supply and the receptiveness of the community towards new livestock operations. A region with a high unemployment rate is more likely to have excess labour available to work in agriculture. In addition, areas with higher unemployment may seek livestock operations as a means to generate economic opportunities.

The skill set of the workforce as proxied by the education level has been found to be a major attraction for new non-agricultural firms locating in a given region (Bartik, 1988; McConnell and Schwab, 1990). While increasing the education level may increase the productivity of potential farm labourers, education is expected to have a negative influence on the density of barn permits issued. A more educated population is more likely to be concerned about environmental quality and thus more likely to be aware about potential negative externalities associated with livestock production.

Another variable related to the NIMBY hypothesis is wealth. Wealth is measured in terms of average dwelling cost and average per capita income. The environmental Kuznets curve suggests the value placed on environmental quality increases with income. Thus, wealthier regions will not want potentially polluting industries in their backyard and have access to the resources to fight proposed and present livestock operations. Furthermore, higher income regions can rely on other sources of economic growth besides livestock production. Thus, wealth and income are assumed to be negatively related to the density of building permits.

Livestock infrastructure and support

Agglomeration economies refer to the positive effects that a farm can experience as a result of having a high concentration of other farms in the region (Eberts and McMillen, 1999). For example, Roe *et al.* (2002) found agglomeration economies and market access to be the significant factors influencing the number of hogs raised at the county level in the US Midwest. Agglomeration economies were also found by Isik (2004) to be important determinants of changes in production levels in the US dairy sector. The effects include lower input prices from suppliers and lower transaction costs in finding those suppliers who are more likely to be concentrated in livestock intensive regions. In addition, there is greater access to information on production and management practices in regions with a greater concentration of farmers.

In the local Canadian study, infrastructure is measured by the density of hog, dairy, and beef production in the municipality in 1996 as measured by the Canadian agricultural census and the growth in these livestock sectors between the 1991 and 1996 census years. The direct and indirect support for agriculture was also measured by three variables attempting to capture the importance of farming to the municipality. One measured the percentage of the population living in rural areas. A second is the share of all land in the region that is allocated to agriculture and the third variable is the share of total labour force employed in agriculture. It is hypothesized that increases in the 'ruralness' of the region will increase the number of barn building permits issued in the municipality.

State farmland area is an important determinant for both general receptivity of farming operations and assimilative capacity for land-based manure disposal. States with greater farming areas are assumed to be more receptive to livestock operations. Gollehon *et al.* (2001) found that about 72% of large livestock operations had inadequate land capacity to utilize all the manure based organic nitrogen produced from their operations and required alternative disposal arrangements. Thus, the costs of manure disposal are likely to be lower in states with more available farmland.

Production shares are likely to increase in regions where the distance to market is smaller, since transportation and transaction costs will be lower. For example, access to slaughtering facilities was found to be positively related to the intensity of hog production within 15 states by Roe *et al.* (2002). Market access is measured in the national US study by the number of hogs and beef slaughtered within the state. Market access for dairy processing capacity is captured by whole milk equivalent in thousand pounds used in manufactured dairy products at the state level.

Results

Local analysis

The regression results explaining the density of total barn building permits, new barn permits, and permits for swine facilities in south-western Ontario are reported in Table 14.1. The adjusted R-square values are reasonable given the cross-sectional nature of the data but relatively few coefficients are statistically significant across the reported regression coefficients due to the panel nature of the data and the correlation among independent variables.

The first, third and fifth columns of Table 14.1 list the estimated coefficients when all explanatory variables are included with the exception of the density and growth rates for the dairy and beef sectors which are highly correlated to the respective values for the hog sector. The high degree of correlation between the variables representing the importance and growth of different farm types suggests that the municipalities under consideration do not specialize in just one type of livestock. In addition, the percentage of the population that is rural and the share of land allocated to agriculture are highly correlated with the share of the workforce in agriculture and so only the latter is used as a measure of ruralness. While the variables tend to have signs consistent with the initial hypotheses, the only statistically significant coefficient when all explanatory variables are included tends to be the density of hog production in the period prior to the collection of building permit data. In other words, barn building occurred primarily in the traditional production regions.

The subsequent columns in Table 14.1 report the results of dropping additional variables on the basis of their degree of correlation with other variables and their insignificance in the previous regression equations. The second column for each of the three dependent variables contains only the explanatory measures that consistently remain statistically significant. For example, there was a high degree of correlation (if one type of regulation is imposed on livestock farmers, other requirements are also likely to be imposed). The variables that remain statistically significant in the case of the density of permits for all livestock facilities are the size at which a nutrient management plan is required, the rate of growth in population density, and the growth rate in hog density. Increases in all three variables were found to increase the density of barn building permits issued by a municipality.

Table 14.1 Regression results explaining barn building permit density in Southwestern Ontario

Variable	Total permits		New facilities		Swine	
Intercept	-5.236 (0.846)	-1.911 (0.658)	35.098 (0.154)	-1.094 (0.646)	6.951 (0.568)	5.308 (0.017)
Environmental stringency						
Manure storage	0.018 (0.627)		0.028 (0.233)	0.030 (0.001)	0.011 (0.340)	0.018 (0.000)
Minimum distance from waterway	-0.062 (0.671)		-0.086 (0.370)		0.011 (0.807)	
Minimum lot size	0.010 (0.921)		-0.003 (0.962)		-0.034 (0.262)	-0.060 (0.009)
NMP Trigger (NMD)	1.225 (0.471)	1.515 (0.007)	-0.286 (0.798)		0.338 (0.549)	
Trigger for expansion	-1.899 (0.739)		1.403 (0.702)		-1.697 (0.361)	
Relative prices						
Farmland value	0.002 (0.647)		0.004 (0.140)		0.00192 (0.172)	0.0015 (0.060)
General business climate						
Population density	0.240 (0.269)		0.444 (0.025)	0.315 (0.005)	0.033 (0.723)	
Population growth rate 91-96	0.936 (0.039)	0.722 (0.033)	0.194 (0.508)		0.212 (0.159)	
Unemployment rate	0.082 (0.953)		-0.360 (0.700)		0.398 (0.400)	
Share of population with university	0.018 (0.914)		-0.021 (0.916)		0.128 (0.211)	
Average income	-0.008 (0.919)		-0.111 (0.111)		-0.0002 (0.570)	
Livestock infrastructure and support						
Hog density	10.537 (0.214)	14.627 (0.001)	0.00026 (0.018)		0.00016 (0.005)	0.00013 (0.000)
Hog growth rate 91-96	0.003 (0.945)		0.010 (0.684)	0.041 (0.031)	0.112 (0.340)	
Share of population in rural areas	-0.066 (0.391)		0.010 (0.869)		-0.082 (0.012)	-0.057 (0.006)
Share of land in agriculture	-0.010 (0.927)		-0.225 (0.059)		-0.061 (0.297)	
Share of labour force in agriculture	0.201 (0.523)		-0.195 (0.530)		0.112 (0.473)	
Adjusted R-Square	0.266	0.384	0.495	0.380	0.653	0.678
F-Value	1.93	9.54	3.460	9.194	5.713	17.848

[1]*Note: p-values are in parentheses under the estimated regression coefficient*

The threshold livestock density of the proposed farm operation at which a nutrient management plan is required does provide some limited support to the pollution haven hypothesis. Farms are more likely to build in municipalities where the stocking density restriction requiring a NMP is not as strict. However, the major driver on the relative number of barn building permits issued by a municipality appears to be livestock infrastructure and support.

The positive effect of hog density in the prior period provides support for the findings of Roe *et al.* (2002) on the importance of livestock infrastructure. Many of the permits issued during this period were by municipalities within Perth and Huron counties, which are the most livestock intensive counties in the province. These regions are more likely to have direct and indirect support for livestock farmers in the form of services such as feed and equipment dealers. The intensity of livestock production in these regions relative to other municipalities under consideration may also explain the significance of the other findings. Population density is greater in these livestock intensive regions which are located closer to major urban centres than most of the other municipalities.

The importance of agglomeration economies is supported by the results of the other two dependent variables. The prior density of hog production and the growth rate of hog production in the previous period are positively related to the density of permits issued for new facilities and swine barns. The correlation coefficient between the prior density of hog production and the subsequent growth rate in production is 0.7 suggesting that inventory levels have increased primarily in the hog intensive regions. These areas also tend to have higher farmland values reflecting the higher productivity of the land for agricultural purposes and also closer proximity to major population centres. The latter effect may explain the unexpected positive effect of population density.

There is limited empirical support for the pollution haven hypothesis with the two other specifications for the dependent variable. The significant variables related to environmental stringency all have signs inconsistent with the view that firms located where environmental regulations are laxest. The density of building permits for new barns and swine facilities increases with the size of manure storage requirements. Most new barns and/or swine barns normally involve storages large enough to hold the annual waste production so the restriction may not be a constraint in the location decisions of new hog operations. The minimum lot size was hypothesized to have an ambiguous effect on the density of building permits. While it can increase costs immediately, it may reduce future costs associated with potential conflicts with neighbours. The negative and statistically significant negative coefficient on minimum lot size implies the latter effect dominates the former in the case of swine barns.

National analysis

The regression results explaining the temporal changes in livestock inventories across US states are listed in Table 14.2. A fixed effects rather than a random effects model was estimated for each of the three livestock sectors on the basis of a Hausman specification test. The resulting coefficients explain the variability in production numbers well with the majority of the variables statistically significant.

Table 14.2 Regression results of explaining annual inventory changes across states in the US hog, dairy and fed-cattle sectors

	Hogs	Dairy	Fed-cattle
Environmental stringency			
Relative regulatory stringency	-0.0062 (0.020)	-0.0003 (0.023)	0.00267 (0.119)
Relative prices			
Output-corn price ratio	0.000096 (0.145)	-0.00022 (0.012)	0.00011 (0.000)
Energy price	0.00015 (0.327)	0.0001 (0.001	-0.00014 (0.141)
Farm labour wage	-0.00105 (0.02)	-0.0008 (0.005)	-0.0007 (0.258)
Farmland price	0.0000026 (0.02)	0.0000016 (0.000)	0.0000013 (0.045)
Business climate			
Unemployment rate	0.00005 (0.797)	0.0000013 (0.984)	0.00013 (0.242)
Farmland availability	-0.00000018 (0.235)	-0.00000018 (0.000)	-0.0000004 (0.000)
Livestock infrastructure			
Processing capacity	0.000004 (0.000)	0.000000005 (0.000)	0.000013 (0.000)
Economic importance of agriculture	0.09318 (0.001)	0.05359 (0.000)	0.00852 (0.645)
Rural population share	-0.000837 (0.002)	-0.00027 (0.000)	0.0007 (0.000)
Natural endowment			
Temperature	0.00035 (0.079)	-0.00012 (0.078)	-0.00001 (0.911)
Precipitation	0.000026 (0.573)	-0.00033 (0.389)	0.000044 (0.127)
Wald $\chi 2$ (12 D.F.)	7714 (0.0000)	208.2 (0.0000)	20278 (0.000)

p-value in parentheses

The coefficient for the environmental regulatory stringency is negative and statistically significant for the hog and dairy sectors with the relative effect largest in the hog sector. The results suggest that a state's regulatory stringency curtails the share of national hog and dairy numbers, thereby supporting the pollution haven hypothesis. The relationship runs counter to previous studies in agriculture that found livestock increases with environmental stringency (Metcalfe, 2001; Osei and Lakshminarayan, 1996; Mo and Abdalla, 1998). The results of those studies suggest that inventory levels increase first and regulations follow rather than the regulations being set *ex ante* with production decisions constrained by those laws. The support for the pollution haven hypothesis in this study could be due to the use of a time varying regulatory stringency measure and the accounting for possible endogeneity between the measure and livestock production shares.

In contrast to the hog and dairy sectors, there is no empirical support for the pollution haven hypothesis in the beef sector. The difference in effects across livestock types associated with the regulatory stringency index may be due to the nature of the production changes by sector. The increase in hog and dairy inventory has been in non-traditional production regions where environmental laws related to livestock farming appear to have had an effect on the location of production. In contrast, beef production increased in only the three states that had the largest numbers a generation ago. These remain relatively non-populated regions so that expansion may have been influenced by factors other than environmental regulations.

Relative prices generally have mixed effects on inventory levels depending on the sector. The output price-feed price ratio was statistically significant for both cattle and dairy but was the expected positive sign for only the former sector. Wage rates have the anticipated inverse relationship with growth rates in inventory levels across all sectors but are insignificant for fed-cattle. The price variable that has a statistically significant effect on production levels across states is the price of farmland. It was expected that increases in the value of farm real estate would curtail relative livestock production. The opposite result suggests farmers bid up the price of land as part of their expansion and potential concerns regarding land availability relative to the volume of manure generated. The result may also be due to locating in relatively productive agricultural regions after other factors including availability and demographics are considered.

Livestock infrastructure has significant effects on changes in production levels for all three livestock sectors. Hog slaughtering capacity is positively related to production increases, which is consistent with the findings of Roe *et al.* (2002) for 15 states. Similarly, beef slaughtering capacity had a statistically significant positive effect on relative cattle inventory levels which is consistent with the view that regional shifts in production have coincided with shifts in beef packing location. An increase in the processing capacity of the dairy sector was also positive implying that the access to markets is an important factor for dairy expansion as it has been hypothesized for the red meat sector. The results support the 'animal clusters' argument that states with a larger inventory density tend to have a greater processing capacity (Pagano and Abdalla, 1994).

States with a larger proportion of agricultural output in its gross state product tend to have a larger share of national livestock production but the size of the impact from this livestock infrastructure variable is much less than for processing capacity. This variable proxies both the level of services available to farmers and the level of support for agriculture within a state (and consequently the likelihood of acceptance to production expansion). Availability of common agricultural infrastructure (veterinary services, feed availability, other input supplies) is an important factor for livestock industry expansion. The results for the dairy sector are consistent with Weersink *et al.* (1995) who found that dairy farmers place a significant level of importance on the availability and quality of farm support services.

It was expected that states with a larger share of total population that is rural would be more likely to experience increases in relative livestock inventory levels across states. This assertion was accepted for the fed-cattle sector but the opposite effect was statistically significant for the other two sectors. A possible explanation for this result is that the likelihood for conflict between farmers and neighbours is enhanced by population levels in rural areas when all other factors are constant including land availability. Potential nuisance complaints from non-farm rural residents could deter the expansion of livestock production capabilities. In contrast, cattle production only increased in three states which have a relatively large rural population so the potential conflicts are not an issue.

Business climate and natural endowment variables have little explanatory power. The unemployment rate has a negative effect on inventory shares across all three sectors but the result is statistically insignificant. Farmland area is statistically significant but has an unexpected negative effect in both the dairy and fed-cattle sectors. The inverse relationship is consistent with the shift of dairy farming to the western states which have less agricultural land as a share of total area due to the mountains as compared to traditional dairy regions in the central and north-eastern regions of the country. The unexpected negative effect of farmland area on the fed-cattle sector suggests inventory numbers are increasing in states with smaller areas available for expansion. However, the increases in production shares have occurred in three traditional producing states with large amounts of agricultural land. The increase in production shares of fed-cattle by these states has been driven by relative prices and infrastructure support.

Conclusions

This chapter has reviewed two studies analysing the factors affecting farm location with particular emphasis on the influence of environmental pressure. Spatial differences in regulatory stringency can create the possibility of pollution havens in which livestock farms locate in the regions with the lowest regulatory compliance costs. The significant variation in the stringency of environmental regulations across municipalities within the Canadian province of Ontario and across states within the US created a natural experiment to test for the pollution haven hypothesis at the local and national level.

The local study by Weersink and Eveland (2006), which used actual building permit data rather than aggregate regional production numbers, found little support for the pollution haven hypothesis. Instead of locating to reduce environmental compliance costs, barns are being built largely where the livestock sector is concentrated within south-western Ontario suggesting the existence of agglomeration economies. The lack of a pollution haven effect could be due to the limited geographical area of analysis. There are limitations to spatial movement due to adjustment and moving costs from a farm's current location.

If the farm is considering a significant expansion or the environmental constraints are binding, these adjustment costs may be outweighed by the advantages of a move to a region with less restrictive environmental regulations. Examples include some movement by large hog operations to locate in the Prairie provinces from Ontario and particularly from Quebec where limits were placed on the expansion of pig breeding operations in June 2002. The move may also be prompted by more open area in the Prairies, which could also explain the differences in environmental stringency. Similarly, the move by Dutch dairy farmers to Ontario was prompted by increasingly restrictive environmental controls within the Netherlands (Wolleswinkel and Weersink, 2001).

The results at the national scale for the US livestock sector do provide support for the conjecture that the effect of environmental pressure on farm location depends on the scale. Previous studies at a larger spatial scale have tended to reject the pollution haven hypothesis in agriculture and instead have found that inventory levels appear to increase first and environmental regulations follow rather than the regulations being set and production decisions then being constrained by those laws. By developing state-specific measures of regulatory stringency over time and accounting for the potential endogeneity between environmental regulations and production decisions, the study by Herath *et al.* (2004) was able to conduct a more robust test of the pollution haven hypothesis. The results of the analysis suggest that regional production shares for hogs, and to a lesser extent dairy, have increased in those regions with relatively more lenient regulatory regimes. Combining environmental stringency at a detailed spatial level and assessing differences as conducted in the local analysis but with the temporal changes at a national scale would represent a contribution to the literature on the location effects of regulations.

In both studies, the location decisions of livestock facilities are largely determined by economic factors associated with livestock infrastructure support. Livestock clusters are occurring in North America and these are locating in agriculturally-intensive regions with available farmland and nearby processing capacity. These results suggest that government regulators can directly and indirectly have a major influence on the size of the animal industry in their jurisdiction. Tightening compliance requirements and enforcement can increase relative abatement costs to the point that livestock farmers may shift production to another location. Aside from this direct impact, regulators can alter the incentives faced by the processing sector to locate in their state. The location decisions of the fewer, but significantly larger, processors in turn have a major influence on the spatial production of the associated livestock sector.

References

Abdalla, C.W., L.E. Lanyon and M.C. Hallberg (1995), 'What We Know About Historical Trends in Firm Location Decisions and Regional Shifts: Policy Issues for an Industrializing Animal Sector', *American Journal of Agricultural Economics,* 77: 1229-36.

Apland, J. and H. Anderson (1996), 'Optimal Location of Processing Plants: Sector Modeling Considerations and an Example', *Review of Agricultural Economics,* 18: 491-504.

Bartik, T.J. (1988), 'The Effects of Environmental Regulation on Business Location in the US', *Growth and Change,* 19: 22-44.

Eberts, R.W. and D.P. McMillen (1999), 'Agglomeration Economies and Urban Public Infrastructure', in P. Cheshire and E.S. Mills (eds), *Handbook of Regional and Urban Economics*, New York: North-Holland, Volume 3 Applied Economics, Chap 38, 1455-95.

Eveland, C., A. Weersink, W. Caldwell and W. Yang (2005), 'Spatial Trends in Barn Building Permits', *Great Lakes Geographer,* 12: 19-27.

FitzGibbon, J., S. Hammel and A. Metrunec (2002), *Report on Nutrient Management by-laws in the Province of Ontario. A component of Profiles of Livestock Agriculture and Impacts of Regulation.* Supported by OMAFRA Special Projects Fund, School of Environmental Design and Rural Development, University of Guelph.

Gollehon, N., M. Caswell, M. Ribaudo, R. Kellogg, C. Lander and D. Letson (2001), *Confined Animal Production and Manure Nutrients*. Washington, DC: US Department of Agriculture, Economic Research Service, Agr. Econ. Rep. 717, June.

Hall, B. and M.L. Kerr (1991), *1991-1992 Green Index: A State-By-State Guide to the Nation's Environmental Health,* Island Press, Tucson, AZ.

Herath, D., A. Weersink and C. Line-Carpentier (2005), 'Spatial Dynamics of the Livestock Sector in the US: Do Environmental Regulations Matter?', *Journal of Agricultural and Resource Economics,* 30: 45-68.

Herath, D., A. Weersink and C. Line-Carpentier (2004), 'Spatial and Temporal Trends in US Livestock Production', *Review of Agricultural Economics,* 27: 49-69.

Isik, M. (2004), 'Environmental Regulation and the Spatial Structure of the US Dairy Sector', *American Journal of Agricultural Economics,* 86: 949-62.

Jafee, A.B., S.R. Peterson, P.R. Portney and R.N. Stavins (1995), 'Environmental Regulation and International Competitiveness: What does the Evidence Tell Us?', *Journal of Economic Literature,* 33: 132-63.
Kraft, M.E. and N.J. Vig (1994), 'Environmental Policy From the 1970s to the 1990s: Continuity and Change', in M.E. Kraft and N.J. Vig (eds), *Environmental Policy in the 1990s*. Second Ed. 3-29. Washington D.C.: Congressional Quarterly Inc.
Kunce, M. and J.F. Shogren (2002), 'On Environmental Federalism and Direct Emission Control', *Journal of Urban Economics,* 51: 238-45.
Levinson, A. (2000), 'The Missing Pollution Haven Effect Examining Some Common Explanations', *Environmental and Resource Economics,* 15: 343-64.
Martin, L.L. and K.D. Zering (1997), 'Relationships Between Industrial Agriculture and Environmental Consequences: The Case of Vertical Coordination in Broilers and Hogs', *Journal of Agricultural and Applied Economics,* 29: 45-56.
McConnell, V.D. and R.M. Schwab (1990), 'The Impact of Environmental Regulation on Industry Location Decisions: The Motor Vehicle Industry', *Land Economics,* 66: 67-81.
Metcalfe, M. (2000), 'State Legislation Regulating Animal Manure Management', *Review of Agricultural Economics,* 22: 519-32.
Metcalfe, M. (2001), 'US Hog Production and the Influence of State Water Quality Regulation', *Canadian Journal of Agricultural Economics,* 49: 37-52.
Mo, Y. and C.W. Abdalla (1998), 'Analysis of Swine Industry Expansion in the US: The Effect of Environmental Regulation', Staff Paper 316, Department of Agricultural Economics and Rural Sociology, Pennsylvania State University, University Park, PA.
Osei, E. and P.G. Lakshminarayan (1996), 'The Determinants of Dairy Farm Location', Livestock Series Report 7, Working Paper 96-WP 174 Center for Agricultural and Rural Development, Iowa State University, Ames, Iowa, December.
Pagano, A.P. and C.W. Abdalla (1994), 'Clustering in animal agriculture: economic trends and policy', *Balancing Animal Production and the Environment*. Proceedings of a conference of the Great Plains Animal Agriculture Task Force, October 19–21, Denver, CO.
Park, D., A. Seidl and S. Davies (2002), 'Environmental Policy and Industry Location: The Case of US Livestock Industry', *The Review of Regional Studies,* 32: 293-307.
Roe, B., E.G. Irwin and J.S. Sharp (2002), 'Pigs in Space: Modeling the Spatial Structure of Hog Production in Traditional and Nontraditional Production Regions', *American Journal of Agricultural Economics,* 84: 259-78.
Weersink, A. and C. Eveland (2006), 'The Siting of Livestock Facilities and Environmental Regulations', *Canadian Journal of Agricultural Economics,* 54: 159-174.
Weersink, A., P.H. Coughler, C. Nicholson and A. Novakavic (1995), 'Satisfaction with Farm Input Services by Dairy Farm Households in New York and Ontario', *Journal of International Food and Agribusiness Marketing,* 7: 65-88.
Wolleswinkel, A. and A. Weersink (2001), 'Farmer Emigration: The Case of Dutch Dairy Farmers Moving to Ontario', *Livestock Production Science*, 69: 197-205.

Chapter 15

Cost effectiveness of sediment abatement policies in potato production: assessing policies in Canada, the Netherlands and the US

John Baptist D. Jatoe, Alfons Weersink, B. James Deaton and Emmanuel K. Yiridoe

Introduction

The economic and environmental consequences of agriculture-induced pollution are of growing concern in regions with intensive agricultural production systems. Cropping systems based on potatoes may generate a wide range of residual problems – e.g., sediment and nitrogen pollution – because high levels of fertilizer, pesticides, and tillage are applied to, what are often, light-textured soils. Indeed, some potato-growing regions in the Netherlands have nitrate concentrations in excess of the European Commission's recommended limit of 50 mg NO_3 l^{-1}. Moreover, erosion is a problem in the north and south-east regions of the country (Spaan *et al.*, 2006; Fraters *et al.*, 1998).

These same residual issues confront policy makers in the Canadian province of Prince Edward Island (PEI) where potatoes are the dominant crop in most farming systems. Recently, a provincial task force examined means to reduce nitrate concentrations in groundwater with soil erosion generally acknowledged as the other major environmental problem (Savard *et al.*, 2007; Carter and Sanderson, 2001; Edwards, 1988; and Milburn *et al.*, 1997). These problems are also an issue in the US. For example, in the US state of Maine, potato cropping has been associated with severe soil erosion (Griffin *et al.*, 2007; Griffin and Porter, 2004; Grandy *et al.*, 2002) and with the excessive nitrate levels found in over 25% of wells sampled in the state (Williams *et al.*, 1987). The issue of residual runoff is important in all three settings because these countries are major potato producers. The US and Canada rank fourth and 13th, respectively, while the Netherlands ranks ninth among the world's top potato producers (FAO, 2008).

The three regions with similar potato production systems and environmental issues are using different agri-environmental policies to deal with the residuals, thereby resulting in a natural experiment on the effectiveness of those mechanisms. This study applies these different policy approaches to the same watershed to simulate differences in profits, costs, and sediment abatement levels. Our results compare the cost-effectiveness of three different policies. (We develop the conceptual approach and empirical methods in subsequent sections.) Our study is a departure from previous studies some of which examine the

 The Economics of Regulation in Agriculture: Compliance with Public and Private Standards (eds F.M. Brouwer, G. Fox and R.A. Jongeneel)

reasons for differences in environmental stringency and the determinants of policy success (Watkiss *et al.*, 2004; Dasgupta *et al.*, 2001; Caruso, 2000; Janicke, 1992).[1] Other studies compare compliance costs to agri-environmental policies imposed on the agricultural sector to assess the impact of these regulations on the relative competitiveness across countries or sectors. For example, Jongeneel *et al.* (2007) estimate the compliance cost associated with the Birds and Habitats Directive of the EU, for Spain (€33 ha^{-1}), France (€190 ha^{-1}) and the Netherlands (€160 ha^{-1}) and attribute differences to farm structure, management practices and biophysical conditions.

In contrast to the aforementioned literature, this study examines the cost effectiveness of actual policies employed across regions on the same watershed comprised of a number of farms. The various policy approaches – discussed more fully in section 3 – are imposed on a programming model that allows us to examine economic and environmental outcomes from agricultural production within a watershed. We derive the model using information about agricultural production, topography, and environmental characteristics using data from an actual watershed in PEI.

Conceptual framework

Our approach is to model economic and environmental outcomes under actual policy regimes used within three different jurisdictions. The environmental outcome of concern in the analysis is soil erosion and the amount of soil sediment deposited into a waterway differs by farm with a given watershed area. This difference might be due to the conservation practices employed – e.g. row crops versus cereal crops. Alternatively, the same practices could result in different erosion levels depending on site characteristics such as topography, soil type, and proximity to streams. In the remainder of this section, we characterize our method by assuming two farms, A and B. This is a simplification of our empirical approach, which examines many farms within a given watershed.

In an unconstrained scenario – where no policies are in place to reduce the residuals associated with agricultural production – both farm A and B are assumed to maximize profit by employing inputs until the marginal value product gained from an incremental increase in the input is just equal to its marginal cost. In addition, we assume that the farms are homogeneous with respect to production technology, input costs, and output price. The implication is that returns are the same across farms for a given practice – this assumption is relaxed in the empirical analysis. However, the level of residual associated with a given technology does differ across farms. Allow, for example, Farm A to produce a greater amount of residual, per unit of output, than Farm B. This difference may be due to different locations relative to a river.

In the unconstrained scenario, profits in the watershed are maximized. Indeed, by conventional assumption any policy that alters this input choice should reduce aggregate profits. Moreover, by assumption, each farm produces the same level of output with the same technology. However, Farm A deposits a higher amount of sediment into the water course per unit of output because it is located next to the river. While the unconstrained scenario generates the highest level of return from farming within the watershed, it is not

---

[1] These studies identify income per capita, degree of urbanization and industrialization, security of property rights and general administrative efficiency as the major factors influencing the likelihood and effectiveness of environmental policy.

socially optimal – i.e., maximize the present value of net-benefits for society – if the resulting residuals from agricultural production cause damages.

One approach to this problem is to impose a set of production practices on both farms in the watershed. In actual application, such a policy might require all farms to use the same cropping rotation. We refer to such a policy as the 'uniform policy approach' because it requires uniform production practices across all farms. Compared to the unconstrained scenario, net revenue in the watershed would decline and so would residuals under the assumption that the imposed crop rotations were effective. While total residuals decline, the level of abatement will differ between farms. One cost-effective measure – the one we employ in our empirical approach – is to estimate the 'policy cost' as the difference between profits between the unconstrained and constrained scenarios divided by the reduction in level of residuals across the two scenarios.

An alternative policy approach is the 'targeted approach'. Targeted policies identify farms associated with relatively high runoff (i.e., Farm A in our scenario) and apply specific regulations to those farms. Following a similar logic as the above scenario, when compared to the unconstrained scenario, there will be a decline in the profits and residuals associated with the targeted approach. The residuals associated with production will again differ across the two farms. However, unlike the uniform policy, the distribution of costs associated with the policy will differ. The cost-effectiveness of the targeted approach can be estimated following a similar logic as above: compare profits in the constrained and unconstrained scenario and divide by the reduction in residuals.

In the following sections, our conceptual approach is developed and applied to a single watershed. The specific policies, detailed in the next section, are currently in force throughout Europe and North America. Our empirical results provide one measure to assess policy differences.

Sediment abatement policies

The general nature of the sediment abatement policies for the three regions studied are presented in an order based on the degree of targeting. The discussion begins with Maine, which has the most uniform policy, followed by PEI and concludes with the Netherlands, which has the most targeted set of abatement policies. The specific practices imposed by the policies are outlined following the presentation of the three regions.

Maine, USA

The US Federal Agricultural and Improvement Reform (FAIR) Act of 1996 expanded conservation and environmental coverage of previous farm bills, by establishing the Environmental Quality Incentives Program (EQIP). EQIP is a voluntary, agri-environmental programme that provides technical, financial and educational assistance to farmers through the Natural Resources Conservation Service (NRCS). The 2002 Farm Bill reauthorized EQIP through 2007 with greater funding and flexibility (NRCS, 2004). National priorities under EQIP include reduction in soil erosion and sedimentation from agricultural land, and reduction of non-point source pollutants such as sediment, nutrients, pesticides or excess salinity (NRCS, 2004).

EQIP is used in conjunction with state legislation to address agri-environmental issues at the local level. In 1997, Maine enacted the Erosion and Sediment Control Law to establish baseline measures of erosion control for use in watersheds throughout the state (State of Maine Department of Environmental Protection, 1998). The appropriate conservation priorities and abatement measures by region in the state are implemented with the aid of EQIP in Maine through 16 Local Work Groups (LWGs) (NRCS, 2007). The LWGs, which consist of staff from NRCS and other government agencies, determine the resource concern, eligible practices to address the concern and the cost-share rates and limits for those practices. The funded measures are uniformly recommended for the entire watershed without discrimination by spatial location.

Prince Edward Island, Canada

Canada's Agricultural Policy Framework (APF), launched in June 2002, is based on five pillars, one of which is the environment. Several agri-environmental programmes are funded through this pillar including Environmental Farm Planning (EFP), the National Farm Stewardship Program and the Greencover Canada Program (AAFC, 2005). The Environmental Farm Plan (EFP) programme is voluntary and allows farmers to set priorities for actions that address on-farm environmental concerns while the other two programmes provide financial incentives to farmers who have or are implementing EFPs and are eligible for assistance to implement the practices identified in their action plans (AAFC, 2005). PEI provides provincial funding opportunities for agri-environmental programmes under the PEI Enhanced Environmental Farm Plan (Government of PEI, 2006). Both the Canada-PEI Agriculture Stewardship Program and the Sustainable Resource Conservation Program provide technical and financial incentives to farmers to implement measures or Action Plans identified under the EFP program (AAFC, 2005).

In addition to the cost-sharing arrangements stemming from voluntary adoption of EFPs, PEI has imposed design-based standards on potato farmers in an attempt to maintain and improve the quality of soil and water (both surface and ground), as well as preserve soil productivity. The PEI Agricultural Crop Rotation (ACR) Act, which was passed in 2002, regulates the crops grown with potatoes, and the frequency and sequence of specific crops in such rotation systems (Government of PEI, 2002). The ACR Act also prohibits row cropping on farmland with a slope greater than 9% and stipulates that regulated crops may be grown in a field no more than once in three years (except with the completion of an EFP).

The Netherlands

The influence the CAP once had on patterns of production through production related payments and market interventions has significantly decreased, and will decrease further over the coming years. Support within the CAP has started to place a greater emphasis on sustainability, the environment and rural development, encouraging the provision of public or non-marketed goods. Dutch efforts to control water erosion and sedimentation have focused on a multi-pronged approach using environmental legislation and incentive programmes, as well as cooperation among stakeholders. One such program dubbed 'desired grassland' was introduced in 2004 and pays farmers for maintaining permanent grassland or for converting arable land into permanent grassland in environmentally sensitive areas or critical zones. The major programmes, however, are the Erosion Ordinance and the Farm Erosion Management Plan (FEMP), which are designed to control soil erosion through changes in land use and land management practices particularly for crops susceptible to high erosion such as potatoes, maize and sugar beet (Spaan *et al.*, 2006). The Erosion Ordinance measures restrict practices related to tillage, cover crop, row length, buffer strips and crop rotation based on the degree of slope in the field. Under the FEMP, farmers are scored on the basis of their adoption of beneficial management practices categorized by crop/tillage, landscape infrastructure and drainage system. Those who meet threshold scores qualify for exemptions from certain measures under the Erosion Ordinance with the scores varying spatially. While the Dutch measures are a mixture of regulations and voluntary cost-share that are design-based similar to PEI, the Dutch Erosion Ordinance is a more targeted policy than the PEI ACR Act.

Empirical model

Study area

The cost-effectiveness of the soil erosion policies across the three regions were examined for one location – the Upper Wilmot River watershed, which is a relatively small agricultural watershed of 8,339 ha located in the potato belt of PEI (see Figure 15.1). Over 70% of the watershed is allocated to agriculture, and row-crop potato farming is the dominant agricultural land use in the watershed (Savard *et al.*, 2007). The watershed was selected for study because of its current environmental problems stemming from agriculture and the availability of reliable data on observed stream flow for SWAT model calibration and validation.

The watershed was subdivided, based on surface hydrology, into 45 sub-basins using a digital elevation model on a GIS platform (Figure 15.1). The sub-basins ranged in size from 1.02 ha to 438.68 ha, with an average of 99.44 ha and contained 353 individual hydrologic response units. Each sub-basin represents a contiguous land area that delivers its pollution to a common outlet.

The division of the watershed into spatial units was done using the Soil and Water Assessment Tool (SWAT), which was developed by the Agricultural Research Service of the US Department of Agriculture (USDA-ARS) to assist resource managers assess the long-term effects of farming practices on sediment and chemical yields in large watersheds and river basins. SWAT uses input data such as climate, land cover, digital elevation and stream network location to simulate average impacts of changes in land use and management practices on water, sediment and agricultural chemical yields in large, complex watersheds with varying soil, land use and management conditions over long periods of time (DiLuzio *et al.*, 2002; Neitsch *et al.*, 2002). SWAT model outputs include stream flow or water yield, sediment and nutrient yields, which can be traced across the watershed. SWAT estimated the sediment loads at the hydrological response unit for the cropping systems imposed under each policy regime as described in the next section.

Cropping systems

A potato-corn-potato-corn-clover rotation system under conventional tillage, without surface residue and winter cover crop management was used as the base cropping system across the three policy regimes (see Table 15.1). This cropping system generated the highest agricultural returns for the whole watershed across 21 possible systems, when no residual loading constraint was imposed on agricultural activities (Jatoe, 2008). Each of the 21 cropping systems was characterized by a crop rotation, tillage system and use or otherwise of soil cover and surface residue management. A continuous potato crop rotation was not considered among the 21 systems for agronomic reasons despite the high potential returns.

Figure 15.1 Upper Wilmot River watershed showing delineated sub-basins

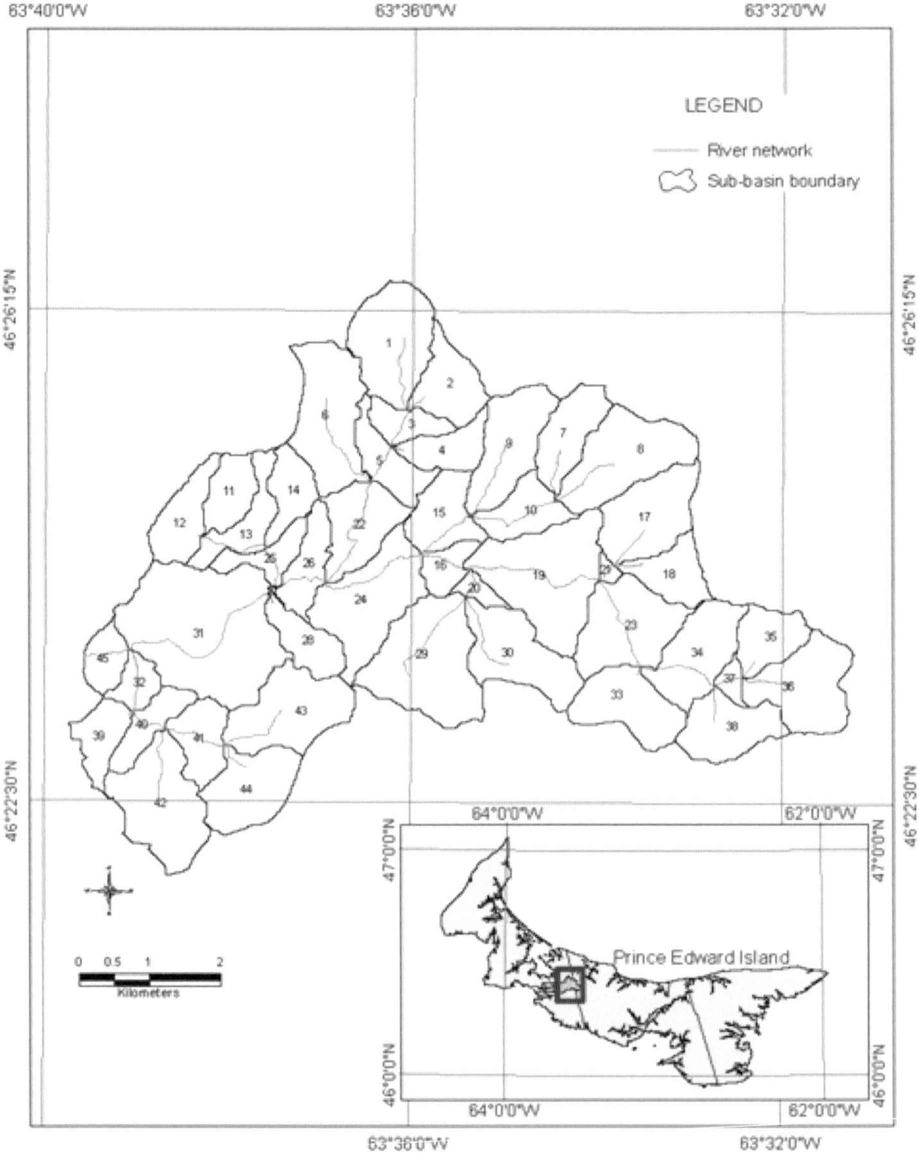

Table 15.1 Base profit-maximizing system and additional practices imposed by region

Scenario	Practices imposed by agri-environmental policy
Profit-maximizing base	-potato-corn-potato-corn-clover crop rotation -conventional tillage -no surface residue and winter cover crop
Maine	-surface residue (mulch) on potato and corn at planting -winter cover crop planted after harvest
PEI	-winter cover crop planted after potato -conservation tillage after cereal harvest in fall -only first cut hay removed -no row cropping on farmland with a slope greater than 9%
Netherlands	-surface residue (mulch) on potato and corn at planting -winter cover crop planted after harvest -exemption for sub-basins with a slope less than 2% -limit on crop row length on area with slope greater than 9% -grasslands on farmland with a slope greater than 18%

Gross margins across the potential cropping systems varied across sub-basins primarily due to the spatial variability in crop yields. Gross margins were derived as gross revenue minus the estimated variable cost of production for each crop and tillage/cover crop practice by soil class. Gross revenue for a given cropping system was the product of crop yield and respective market prices. Yield data for each crop was obtained from the PEI Department of Agriculture, Fisheries and Forestry and adjusted for differences in soil productivity, thereby allowing yields to vary across sub-basins within the watershed. Crop output prices were the average real prices in PEI for 1996 to 2005. Input costs were based on enterprise budgets for the crops compiled from the PEI Department of Agriculture, Fisheries and Forestry (1995; 1994) and the Ontario Ministry of Agriculture, Food and Rural Affairs (2005). Annual gross margins for each cropping system were calculated by adding the annual gross margin for each crop in the rotation and dividing by the length of the rotation (5 years).

Within a sub-basin, there is substantial variability in profitability and sediment loads across cropping systems. There are relatively minor differences in residual levels resulting from alterations in the crop rotation but net returns are greatest when corn is included in rotation with potatoes followed by wheat and then barley. The intensity of tillage is inversely related to soil erosion levels and positively related to profitability. The greatest changes between cropping systems arise from the inclusion of cover crops after harvest of the potato crop. While cover crop management decreases gross margins by approximately \$50 ha^{-1}, it significantly cuts sediment level.

The base cropping system incorporates some of the BMPs selected from the Maine BMP manual to qualify for government payments under EQIP, including crop rotation. The only modification to the base system for the Maine scenario is the inclusion of a winter cover crop for all sub-basins in the watershed. The crop rotation is also acceptable under the PEI ACR Act as no sub-basins have slopes greater than 9%, which would have imposed stricter PEI regulations. The additional general practice added for the PEI scenario was conservation tillage to provide the 30% surface residue after planting conventionally tilled potato, required by the policy. Key characteristics of the Dutch Erosion Ordinance simulated included spring tillage (which replaced tillage in autumn after harvest), a winter cover crop and row crops seeded directly in mulch on all sub-basins with an average slope greater than 2%. The practices imposed spatially on the model for each jurisdiction's policies are summarized in Table 15.1.

The analysis considered two main types of costs associated with the individual policies. One was the direct cost of implementing the management practices required or prescribed under each policy. For example, the annual cost of chopping straw and spreading as mulch was \$125 ha^{-1} while the inclusion of a cover crop increased costs by \$100 ha^{-1}. The other type of cost considered was imputed cost or the value of lost output due to policy-induced decreases in yield for each scenario. The total abatement cost of each policy or policy-induced cost was estimated as the difference between the gross margin for the benchmark potato-grain corn-potato-grain corn-clover rotation system using conventional tillage with no surface residue and winter cover crop and the gross margin for the same cropping system but incorporating adjustments to management practices as required by the particular policy.

Results

Environmental effectiveness

The soil erosion levels for each of the 45 sub-basins in the watershed under the base cropping system and the three policy regimes are listed in Table 15.2. Under the benchmark cropping system, total sediment load for the entire watershed was 11,170 metric tons $year^{-1}$ but there was substantial spatial variability across the sub-basins in the watershed. The difference in sediment pollution across the watershed was due largely to differences in land and soil characteristics such as soil texture, drainage and slope and elevation, as well as location of sub-basin relative to the watershed outlet. For example, while both sub-basins 44 and 45 have well drained sandy soils, sub-basin 45 has a sediment load that is approximately five times higher than sub-basin 44 because it is located at the main outlet of the watershed while sub-basin 44 does not drain directly into surface water. In contrast, sub-basins 1 and 12 are located close within the watershed but sub-basin 1 consists primarily of well-drained, sandy loam soils that are more prone to erosion than the heavier soils in sub-basin 12.

Total sediment loads under each policy regime were lower than the unconstrained base system. The uniform use of cover crops across all sub-basins (i.e., the Maine policy) resulted in the lowest sediment load for the entire watershed at 5229 metric tons $year^{-1}$. The more targeted but restrictive policies in PEI and the Netherlands generated higher levels of sediment than the Maine scenario at 10,015 metric tons $year^{-1}$and 7752 metric tons $year^{-1}$ respectively. Annual average sediment load was 1.17 metric tons ha^{-1} under the policy for Maine, 1.73 metric tons ha^{-1} for the Netherlands and 2.24 metric tons ha^{-1} for PEI. The differences in sediment load under the policies were due to differences in frequency of use of surface residue and winter cover crop, area covered by the surface residue and winter cover crop when used (uniformity of coverage) and use of fall (conservation) tillage.

The policy for Maine, with the lowest sediment load, used surface residue and winter cover crops on the entire cropland in four out of five years in each rotation cycle. There were no requirements for the use of conservation tillage, which is a common recommended practice to reduce soil erosion. Instead, the focus was on the uniform use of surface residue (mulch) after planting and winter cover crop after harvesting both potato and grain corn for farmland in the entire watershed. In this very flat watershed, the use of these practices is more effective in reducing sediment than tillage and crop rotation options.

Table 15.2 Sediment loads (metric tons ha-1) by sub-basin under alternative policies

Sub-basin	Benchmark	Policy Scenario		
		Maine	**Netherlands**	**PEI**
1	3.15	1.42	1.96	2.75
2	2.75	0.95	2.38	2.62
3	4.54	2.05	3.32	4.15
4	1.60	0.69	1.15	1.45
5	1.66	0.79	0.96	1.43
6	2.29	1.06	1.65	2.10
7	2.60	0.94	2.11	2.44
8	4.20	1.99	2.93	3.81
9	3.77	1.49	2.97	3.48
10*	0.66	0.66	0.66	0.66
11	0.75	0.42	0.59	0.73
12	1.04	0.41	1.04	1.00
13	2.15	1.09	1.51	2.00
14	2.63	1.57	1.38	2.27
15	2.18	0.82	1.79	2.05
16	2.22	1.18	1.63	2.06
17	2.42	1.16	1.83	2.26
18	0.97	0.55	0.75	0.91
19	3.24	1.31	2.34	2.93
20	2.58	1.24	1.77	2.26
21*	0.04	0.04	0.04	0.04
22	1.10	0.46	0.61	0.93
23	2.22	1.03	1.85	2.11
24	2.06	0.92	1.53	1.90
25	1.06	0.59	0.82	1.01
26	1.96	1.03	1.29	1.79
27*	0.21	0.21	0.21	0.21
28	1.83	1.02	1.10	1.61
29	1.38	0.56	1.08	1.30
30	2.34	1.06	1.40	2.03
31	1.76	0.66	1.51	1.69
32	5.71	3.15	3.67	5.10
33	3.13	1.39	2.37	2.92
34	6.60	2.78	4.40	5.87
35	4.59	2.08	2.89	4.02
36	5.41	2.32	3.57	4.74
37	4.43	2.64	2.26	3.68
38	2.39	1.00	1.63	2.14
39	0.84	0.47	0.66	0.82
40	3.20	1.77	2.04	2.88
41	1.91	1.03	1.08	1.68
42	0.97	0.49	0.97	0.88
43	1.58	0.69	1.28	1.52
44	1.55	0.80	0.98	1.38
45	5.67	2.63	4.00	5.09
Mean	**2.47**	**1.17**	**1.73**	**2.24**

* Sub-basins 10, 21 and 27 have no farmland

The abatement policy for PEI had the highest sediment load because it used surface residue and winter cover crop on cropland in only two out of five years in each rotation cycle. The same crop rotation was used as in the Maine scenario but the restriction on cover crops was only applied to certain crops rather than uniformly to all. In addition, the PEI policy required conservation tillage but in the fall, along with removal of straw in the second grain year before planting potatoes in the third year of the rotation. The watershed did not contain any farmlands with slope greater than 9% and the subsequent tighter restrictions on practices were not imposed. While the PEI policy required a number of specific practices depending on the crop and land quality, the characteristics of the watershed resulted in less sediment abated than the use of cover crops regardless of crop and land characteristics as specified in the policy for Maine.

The Dutch Erosion Ordinance had the potential to be the most restrictive, targeted policy. Crop row length is limited for lands with slopes greater than 5% and grassland obligatory for slopes greater than 18%. None of the sub-basins in the Wilmot watershed is hilly enough to meet these criteria. However, two sub-basins are flatter than 2% and thus do not require a cover crop in the winter or row crops to be sown directly into mulch. Consequently, the erosion levels for those two regions are higher. In addition, conservation tillage was required on the sub-basins with farmland slopes greater than 2%.

Cost-effectiveness

The gross margin for the base scenario in which a potato-corn-potato-corn-clover rotation is grown under conventional tillage on all 1550 ha of farmland in the watershed is \$1.36 million or \$877 ha^{-1}. The abatement cost, or reduction in returns from the gross margin for the base scenario, is listed for each policy scenario by sub-basin in Table 15.3. The abatement costs include the direct costs of the practice prescribed by the policy as well as the indirect costs stemming from a potential reduction in yields.

The uniform policy of Maine imposes the same cost across all sub-basins but without negative impact on crop yields since it allowed conventional tillage. Therefore, the total abatement (policy-induced) cost was limited to the annual cost, estimated at \$100 ha^{-1}. The direct costs of a cover crop, along with chopping and spreading straw as mulch, is \$100 ha^{-1} or \$155,000 for the whole watershed (Table 15.3). This abatement cost is the lowest of the three policy scenarios. Given the Maine policy also results in the least erosion, its conservation mechanism is the most cost effective with an annual cost per unit of sediment abated at \$77 metric ton^{-1} (=\$100 ha^{-1} /(2.47 metric ton ha^{-1} – 1.17 metric ton ha^{-1})) as listed in Table 15.4.

The targeted approaches of PEI and the Netherlands resulted in higher abatement costs (Table 15.3). Total abatement cost of the policy for PEI was \$185,675 due largely to the requirement of conservation tillage in three out of five years. It is a significant change to the farming system with higher operating costs and lower yields. Given the relatively low level of sediment reduction, the PEI program is the least cost-effective with annual cost per unit of sediment abated at \$507 metric ton^{-1} (=\$119.79 ha^{-1} / (2.47 metric ton ha^{-1} – 2.24 metric ton ha^{-1}).

The average abatement cost per unit of sediment for the Netherlands is \$193 metric ton^{-1}, which falls between the results for the other two regions (Table 15.4). The absolute level of abatement costs is the highest among the three approaches at \$221,437 for the entire watershed. While these total abatement costs are approximately 20% higher than under the PEI scenario, soil erosion was 30% less than the base scenario compared to 10% less for the PEI policy. The policy for the Netherlands targeted mulch and cover crops on row crops but also prescribed conservation tillage on cropland having a slope above 2%, which resulted in yield loss. The more frequent use of surface residue and winter cover crop under the policy for the Netherlands increased its effectiveness in sediment reduction compared to the PEI option, while the yield loss due to conservation tillage reduced its cost-effectiveness relative to the policy for Maine.

Table 15.3 Policy-induced farm costs under alternative policies (\$ ha^{-1})

Sub-basin #	Gross Margin Base Model (\$ ha^{-1})	Abatement Cost (\$ ha^{-1}) by Policy Scenario		
		Maine	Netherlands	PEI
1	1017.70	100.00	151.17	126.17
2	1063.69	100.00	153.70	128.70
3	995.91	100.00	149.84	124.84
4	1102.19	100.00	155.32	130.32
5	1014.04	100.00	150.51	125.51
6	804.55	100.00	140.87	115.87
7	787.85	100.00	139.36	114.36
8	743.83	100.00	137.97	112.97
9	782.19	100.00	139.64	114.64
11	748.85	100.00	139.72	114.72
12	571.20	100.00	99.21	105.48
13	962.76	100.00	148.13	123.13
14	1001.40	100.00	150.13	125.13
15	818.17	100.00	141.00	116.00
16	764.70	100.00	138.63	113.63
17	745.27	100.00	138.20	113.20
18	760.63	100.00	140.43	115.43
19	827.15	100.00	142.11	117.11
20	991.86	100.00	147.94	122.94
22	865.48	100.00	143.07	118.07
23	921.73	100.00	145.85	120.85
24	977.74	100.00	149.87	124.87
25	783.00	100.00	139.45	114.45
26	1073.68	100.00	153.42	128.42
28	1040.78	100.00	152.66	127.66
29	986.95	100.00	149.41	124.41
30	892.66	100.00	145.40	120.40
31	869.04	100.00	144.53	119.53
32	943.29	100.00	147.41	122.41
33	871.87	100.00	144.63	119.63
34	608.64	100.00	131.77	106.77
35	652.76	100.00	134.02	109.02
36	703.96	100.00	136.33	111.33
37	765.67	100.00	137.75	112.75
38	966.10	100.00	148.40	123.40
39	999.25	100.00	148.81	123.81
40	984.03	100.00	149.19	124.19
41	1057.72	100.00	153.42	128.42
42	962.90	100.00	99.21	123.91
43	981.31	100.00	149.38	124.38
44	1039.96	100.00	152.23	127.23
45	778.36	100.00	140.11	115.11
Mean	**886.45**	**100.00**	**142.86**	**119.79**
Total	**1.36 million**	**155,000.00**	**221,437.25**	**185,675.08**

Table 15.4 Cost-effectiveness* of alternative sediment control policies ($ metric ton^{-1})

Sub-basin	Maine	Netherlands	PEI
1	28.90	211.07	252.94
2	27.78	685.69	797.72
3	20.08	204.79	256.00
4	54.95	567.38	702.13
5	57.47	357.87	437.01
6	40.65	376.37	478.28
7	30.12	488.49	558.51
8	22.62	187.38	225.56
9	21.93	299.54	309.09
11	151.52	1498.28	4486.25
12	79.37	-	2012.02
13	47.17	387.70	654.19
14	47.17	200.11	278.15
15	36.76	617.95	699.99
16	48.08	404.46	553.95
17	39.68	403.72	551.23
18	119.05	1092.88	1507.23
19	25.91	269.01	297.11
20	37.31	306.09	306.05
22	78.13	496.06	547.48
23	42.02	664.47	871.39
24	43.86	471.45	624.17
25	106.38	997.73	1789.08
26	53.76	378.24	608.38
28	61.73	346.11	466.65
29	60.98	831.37	1242.63
30	39.06	261.06	307.74
31	45.45	978.12	1350.42
32	19.53	121.28	159.69
33	28.74	321.88	450.62
34	13.09	105.35	112.01
35	19.92	137.66	147.40
36	16.18	128.44	128.85
37	27.93	109.56	117.01
38	35.97	326.85	393.61
39	135.14	1382.26	4940.38
40	34.97	214.82	309.96
41	56.82	305.33	449.67
42	104.17	-	1098.97
43	56.18	831.26	1656.29
44	66.67	442.51	601.36
45	16.45	143.78	155.37
Mean	**77**	**193**	**507**

* Cost effectiveness (policy cost/ sediment reduction); - denotes division by a zero.

Conclusions

The study evaluated the cost-effectiveness of sediment reduction policies for potato-based farming systems used by the US state of Maine, the Canadian province of PEI and the Netherlands. The practices imposed by the actual policies in each jurisdiction were simulated under a biophysical model of a representative agricultural watershed to estimate sediment loads and under an economic model to estimate abatement cost. The Maine policy uniformly requires the use of cover crops and mulch across all farm fields in the watershed. The policies in PEI and the Netherlands are targeted with restrictions on cropping choice and tillage practice depending on spatial location and land characteristics, particularly slope of the farmland. The degree of potential targeting is greatest with the Dutch system.

The policies used in all three jurisdictions reduced sediment loads at relatively low cost from the benchmark system but, surprisingly, the most cost-effective choice was the uniform policy used in Maine. The legislative approach in Maine requires all farmers to use mulch after planting, and establish winter cover after harvesting row crops (potatoes and grain corn). The cover crop practice is low cost without negative effects on crop yield, and reduces erosion better than more costly alternatives such as tillage management. The targeted policies of the Netherlands and PEI impose tighter, and more costly, erosion control practices with spatial location, particularly farmland slope. The costs of the tighter and targeted controls even without the monitoring and enforcement costs are greater, and generate a higher sediment load than the uniform policy focusing on the most cost-effective practice for the watershed.

The paradoxical finding regarding the cost-effectiveness of the uniform over the targeted policy is due to two primary reasons. One is the nature of the watershed analysed. Cover crops were the best means to reduce sediment loading but other practices such as tillage and crop rotation could be a more appropriate means to cut erosion in another location, particularly one with rolling topography. Second is the pollutant target of the policy. This study has assumed sediment is the residual of focus and, consequently, cover crops are the low-cost means for cutting back on that residual. Potato farming systems, however, generate pollutants other than soil erosion, including nitrates. The choice of practices may reflect an attempt to balance reducing multiple pollutants rather than an incorrect choice of a single pollutant. While cover crops effectively reduce sediment, nitrate levels are often higher so that a policy attempting to reduce both pollutants may not require cover crops on all locations at all times in a rotation.

References

Agriculture and Agri-Food Canada. (2005), *Agricultural Policy Framework – Environment Programs,* Agriculture and Agri-Food Canada, Accessed May 2008, available at http://www4.agr.gc.ca/AAFC-AAC/display-afficher.do?id=1182359506411&lang=e.

Carter, M.R. and J.B. Sanderson (2001), 'Influence of Conservation Tillage and Rotation Length on Potato Productivity, Tuber Disease and Soil Quality Parameters on a Fine Sandy Loam in Eastern Canada', *Soil & Tillage Research,* 63: 1 – 13.

Caruso, B.S. (2000), 'Comparative Analysis of New Zealand and US Approaches for Agricultural Nonpoint Source Pollution Management', *Environmental Management,* 25(1): 9 – 22.

Dasgupta, S., A. Mody, S. Roy and D. Wheeler (2001), 'Environmental Regulation and Development: A Cross-country Empirical Analysis', *Oxford Development Studies,* 29 (2): 173 – 87.

Di Luzio, M., R. Srinivasan, J.G. Arnold and S.L. Neitsch (2002), *ArcView Interface for SWAT 2000, USER's Guide.* TWRI Report TR-193, Texas Water Resources Institute, College Station, Texas.

Edwards, L. (1988), 'The effects of slope position and cropping sequence on soil physical properties in Prince Edward Island', *Canadian Journal of Soil Science*, 68: 763 - 774.

Food and Agriculture Organization (FAO). (2008), *International Year of the Potato 2008,* Food and Agriculture Organization. Accessed August 2009, Available online at: http://www.potato2008.org/en/world/index.html

Fraters, D., L. J. M. Boumans, G. van Drecht, T. de Haan and W. D. de Hoop (1998), 'Nitrogen Monitoring in Groundwater in the Sandy Regions of the Netherlands', *Environmental Pollution,* 102(S1): 479 – 485.

Government of Prince Edward Island (2006), *PEI Enhanced Environmental Farm Plan, Fact Sheet,* Government of Prince Edward Island. Accessed March 2008, available at http://www.peiefp.ca/pdf/provincial_funding_opportunities.pdf.

Government of Prince Edward Island (2002), *Agricultural Crop Rotation Act,* Government of Prince Edward Island. Accessed February 2008, available at http://www.gov.pe.ca/royalgazette.

Grandy, A.S., G.A. Porter and M.S. Erich (2002), 'Soil amendment and rotation crop effects on the recovery of soil organic matter and aggregation in potato cropping systems', *Soil Science Society of America Journal,* 66:1311–1319.

Griffin, T. S. and G. A. Porter (2004), 'Altering Soil Carbon and Nitrogen Stocks in Intensively Tilled Two-Year Rotations', *Biology and Fertility of Soils,* 39: 366-374.

Griffin, T.S., G.C. Starr, C.W. Honeycutt and M. Williams (2007), 'Options for Reducing Erosion and Phosphorus Movement in Potato Systems', Maine Potato Conference Abstracts, CD-ROM.

Janicke, M. (1992), 'Conditions for Environmental Policy Success: An International Comparison', *The Environmentalist,* 12(1): 47 – 58.

Jatoe, J.B.D. (2008), 'Policy Options for Multiple Environmental Benefits in Agricultural Watersheds', PhD Thesis, University of Guelph. August 2008.

Jongeneel, R., F. Brouwer, M. Farmer, R. Mussner, K. de Roest, X. Poux, G. Fox, A. Meister, Z. Karaczun, J. Winsten and C. Ortega (2007), *Compliance with Mandatory Standards in Agriculture; A Comparative Approach of the EU vis-à-vis the US, Canada and New Zealand,* The Hague, LEI, 2007. Report 6.07.21; 113pp.

Milburn, P. H., J. MacLeod and B. Sanderson (1997), 'Control of Fall Nitrate Leaching from Early Harvested Potatoes on Prince Edward Island', *Canadian Journal of Agricultural Engineering,* 39: 263 - 71.

Natural Resources Conservation Service (2007), *How the Environmental Quality Incentives Program Works in Maine,* Natural Resources Conservation Service. Accessed May 2008, available at http://www.me.nrcs.usda.gov/programs/HowEQIPWorks07.html.

Natural Resources Conservation Service (2004), *Farm Bill 2002 – Environmental Quality Incentives Program. Key Points,* Natural Resources Conservation Service. Accessed May 2008, available at http://www.nrcs.usda.gov/programs/farmbill/2002/pdf/EQIPKyPt.pdf.

Neitsch, S.L., J.G. Arnold, J.R. Kiniry, R. Srinivasan and J.R. Williams (2002), *Soil and Water Assessment Tool User's Manual Version 2000.* GSWRL Report 02-02; BRC Report 02-06: TR-192, Texas Water Resources Institute, College Station, Texas.

Ontario Ministry of Agriculture, Food and Rural Affairs (2005) *Publication 60- Ontario Field Crop Budgets.* Accessed July 2012. http://www.omafra.gov.on.ca/english/busdev/facts/pub60.htm

Prince Edward Island (PEI) Department of Agriculture, Fisheries and Forestry (1995; 1994) *Crop Enterprise Budgets.* Goverment of PEI.

Savard, M.M., D. Paradis, G. Sommers, S. Liao and E. van Bochove (2007), 'Winter Nitrification Contributes to Excess Nitrate in Groundwater of an Agricultural Region: a Dual-Isotope Study', *Water Resources Research,* 43, w06422, doi: 10.1029/2006WR005469.

Spaan, W. P., H. J. Winteraeken and M.J.P.M. Riksen (2006), 'Dutch policy and practices on erosion control: Then and now', *Archives of Agronomy and Soil Science,* 52(2): 233 – 41.

State of Maine Department of Environmental Protection (1998), *DEP Information Sheet. Erosion and Sedimentation Control Law,* State of Maine Department of Environmental Protection, May 1998. Accessed May 2008, available at http://www.maine.gov/dep/blwq/docstand/is-esclaw.htm.

Watkiss, P., D. Forster, A. Hunt, A. Smith and T. Taylor (2004), *Study No. 1. A Comparison of EU Air Pollution Policies and Legislation with Other Countries: Review of the Implications for the*

Competitiveness of European Industry, European Commission, January 2004. Accessed May 2008, available at http://ec.europa.eu/enterprise/environment/reports_studies/reports/study1.pdf.

Williams, J.S., C.D. Neil and T.K. Weddle (1987), 'Influence of Agricultural Practices on Groundwater Quality in Maine', Proceedings of the Fourth Annual Eastern Regional Groundwater Conference, Burlington Vermont, July 14-16, 1987, National Water Well Association, Dublin, OH. pp. 329-341.

Chapter 16

Environmental compliance costs in developed country agricultures: implications for trade and competitiveness

Glenn Fox and Floor Brouwer

General themes

The research reported in this book reflects a recognition that agriculture, in the OECD countries in particular, is increasingly subject to environmental regulation. Increased income per caput, stronger public interest in environmental stewardship and more awareness of the origins and the attributes of the food supply have all contributed to a policy transformation that transcends national boundaries. At the same time, modes of agricultural production have become more specialized and input intensive and fewer and fewer people have direct contact with primary agricultural production. Primary agriculture still occupies a large share of the overall land base in most OECD countries, but generates only a small percentage of Gross Domestic Product (GDP). Farm consolidation in response to economies of size, scale and scope, often encouraged by commodity support policies, has changed the social matrix within which agriculture functions.

After a long period of being treated as an exemption, agriculture is now irrevocably engaged in the long term secular trend toward trade liberalization under the GATT/WTO institutional structure. Patterns of regional specialization in agricultural production are emerging as trade barriers and transportation costs fall and as processing, distribution and retailing functions have exploited their own size, scale and scope economies and as logistics have improved. This increasingly economically and socially integrated developed country agricultural economy is also being weaned from a not-yet forgotten dependence on government price supports and transfer payments. Within this evolving context, farmers are becoming more and more aware of the competitiveness and trade implications of domestic policies and regulations. One especially contentious topic is the cost of compliance with more and more demanding environmental regulations. The purpose of the research presented in this book is to provide an overview of the nature of these compliance costs for some of the important agricultural commodities produced in the agricultural economies of the OECD countries and to assess the magnitude of those costs and their impacts on competiveness and trade. In the pursuit of this purpose, however, the international team of researchers involved in this project discovered other important findings that are related to this primary purpose, but that were not anticipated at the outset of the project. But this is often the case in policy research.

Initially, this research focussed on the cost and competitiveness effects of cross compliance policy in the EU, where adherence to Statutory Management Requirements (SMRs) has been imposed as a condition on eligibility for the Single Farm Payments (SFPs). However, apart from previous applications of a cross compliance in the US during the 1980s, other OECD jurisdictions have generally not used this model. The economic explanation for this situation is simple. For export oriented commodities, such as dairy in New Zealand and wheat in Canada, there is no programme payment to which compliance with environmental regulations can be easily attached. So, as our research progressed, the focus of the international competitiveness analysis turned to the costs of compliance with environmental regulations directed toward agriculture generally. The research reported in this volume documented the importance of enforcement mechanisms and implementation, subjects which have been undervalued in previous research on agricultural environmental policies.

The particular focus of agricultural environmental policies studied in this research has been on measures to reduce emissions of nitrates, pesticide policies, food safety measures, animal welfare standards and wildlife habitat protection policies. This emphasis reflects largely an EU perspective. In North America, in many state and provincial jurisdictions, phosphorus emissions have emerged as a more important policy priority than nitrates. Most of the available empirical evidence on agricultural environmental compliance costs is related to nutrient management, pesticide policies, animal welfare standards and food safety measures. Less empirical research has been done on the costs of compliance with wildlife and habitat protection policies, although that is starting to change.

Reference standards for environmental performance in agriculture

A common feature of environmental regulation of agriculture across the countries considered in this book is the specification of a reference level of environmental performance for the farming sector. Most national policies directed toward improving the environmental performance of agriculture generally, implicitly or explicitly, invoke a type of threshold level of stewardship that separates environmental performance that farmers are expected to achieve at their own expense and performance that exceeds the threshold and for which farmers are partially or fully compensated. The levels at which these thresholds have been set seem to be the product of unique historical contexts and there is no general consistency across commodities or jurisdictions, creating considerable tension among farm groups, especially for traded agricultural commodities, as well as environmental constituencies.

Each reference level or threshold represents a particular structure of agricultural property rights. If a farmer is required to achieve a level of environmental performance sufficient to satisfy a threshold, that implies that someone possesses a right to enjoy that level of environmental performance. If a farmer is compensated for environmental performance in excess of the reference level, then this implies that no one has a right to enjoy that enhanced level of environmental performance. But agricultural environmental problems are not generally viewed from a consistent property rights perspective[1]. There are a number of possible explanations for the limitations of the agricultural environmental

[1] Brubaker (1995, 2007) and Bate (2001) are noteworthy exceptions.

policy literature on the question of property rights. One contributing factor is the generally low level of awareness of competing property rights theories in the discipline (see Fox, forthcoming). Agricultural and environmental economists writing on property rights issues appeal to at least five competing theories of property rights, theories which are incompatible on several dimensions. Clarification of property rights issues in the interaction between agricultural production practices and environmental outcomes is needed before significant progress can be made on normative aspects of agricultural environmental policy. And lack of resolution on these property rights controversies can derail policy action politically, as rural landowners organize against what they perceive to be regulatory takings of property. Even the justification of enforcement of environmental policy measures rests on a foundation of property rights theory. Air and water borne emissions from agricultural production systems constitute physical invasions of the property or the persons of neighbouring property owners. Odour, pesticide drift, nutrients and pathogens, even stray genetic material, if treated consistently as trespasses, nuisances or public nuisances, would create liability for damages or even trigger injunctive relief against the agricultural practice responsible for these torts. On the other hand, farmers who manage their land in a fashion that reduces its productivity as wildlife habitat do not commit torts, since there is no physical invasion of the property or persons of their neighbours.

Another aspect of reference standards that became apparent during the course of the research on this project has to do with general subjective perceptions of what constitutes natural. Under a farm subsidy system where payments are decoupled from agricultural output, there is a clear incentive for farmers to accept a payment and produce no agricultural product. Anticipating this outcome, the EU introduced requirements that made receipt of single farm payments conditional on maintaining existing agricultural land in Good Agricultural and Environmental Conditions (GAECs) so that lands for which payments are received are maintained for agricultural purposes. Within the European context, for various social and cultural reasons related to income from tourism and to aesthetic preferences for certain types of landscapes, the prospect of existing agricultural land being abandoned to forest was considered unacceptable. Under this view, payments to farmers could not be justified if land is idled. The GAECs, therefore, represent reference standards or thresholds. In practice, these standards vary across Member States, indicating differences in local preferences. However, the perceptions and preferences that are the motivation for this policy in the EU seem to be quite different from those in the North American context, where abandonment of farmland to forest has occurred, at least in selected areas, and is generally regarded as a favourable environmental outcome. Of course, this land abandonment to forest in the North American context has occurred largely as a result of market forces, and not because of policy. Whether a single farm payment system would be acceptable in North America without an equivalent GAEC standard is therefore an open question. Nevertheless, differences in consumers' and citizens' preferences and perceptions, at least as those preferences and perceptions are translated into trade actions, appear to differ across the jurisdictions considered in this book and also appear to lie at the heart of the long standing differences between the EU and North America on the use of hormones in beef production and on the use of genetically modified organisms in agriculture.

Voluntary private standards

Traditionally, economic research on environmental compliance costs has assumed that government was the institution doing the regulating. Increasingly, however, businesses in the supply chain that transform agricultural commodities for consumption are setting environmental performance standards. These are sometimes referred to as voluntary standards, which they are, but, for many farmers, these standards are a condition of access to retail markets and food processing companies, without which, their farms would no longer be economically viable. These voluntary standards are based on the perceptions, generally of larger retailers, of consumer preferences for non-price attributes of food. These may be process or product attributes, or a combination of both. Voluntary standards may or may not coincide with mandates set by government regulations. Typically, where voluntary standards are set on process or product attributes that are already regulated by governments, they set higher standards than those required by governments. There appear to be distinct differences between the monitoring and auditing standards supporting private voluntary standards and those maintained by government regulators. Inspection and auditing at the farm level typically takes place at least annually for private standards. Inspection rates for compliance with government regulations are much lower, typically a small sampling fraction being inspected in any given year. Voluntary or private standards tend to be commodity specific, and not necessarily whole farm based. Since these standards are technically voluntary, some farms might elect to not meet them; however, this is becoming less and less viable as private voluntary standards become more established.

A second additional dimension of testing of private standards has not been generally recognized in the literature. Standards related to the environment, food safety and animal welfare are largely about requirements regarding production methods and farm management. Such process standards may vary largely across countries, depending on farm structures, population density, links between urban and rural population, biophysical conditions and public preferences. Product standards such as rules regarding product quality, appearance and other physical attributes of food, like residues of pesticides, are largely governed by trade agreements. However, this is far more difficult to implement in process standards. Process standards are hard to compare among countries.

In any case, these emerging private standards must meet a profitability test. Being based on retailers' perceptions of consumer preferences, when products possessing the attributes required by the standard come to the market, actual consumer choices either confirm or contract the retailers' perceptions of consumer preferences. If consumers fail to see value in these novel product attributes, the retailers will have committed an entrepreneurial error, and will suffer accordingly in the competitive food retailing environment. On the other hand, if consumers' choices confirm retailers' perceptions, those retailers will be rewarded, affording them the opportunity to recoup whatever costs they incurred in establishing the private standards. From a farmer's point of view, however, private standards can represent a compliance cost just as much as government regulation can.

Functional equivalence of environmental regulation

International assessments of differences in compliance costs of environmental regulations applied to agriculture need to recognize the implications of differences in constitutional and legal systems across jurisdictions, if they are to avoid misleading comparisons. Environmental policymaking and responsibility for agricultural policy can take place at different levels of government, depending on the constitutional structure of a given country. In federal systems, divisions of authority exist that differentiate between powers of the national government and the state or provincial governments. In non-federal policies, divisions of powers can exist that delegate certain regulatory activities to local or municipal governments. In the EU, the European Commission has initiated a range of agricultural environmental policies, including the Nitrates Directive. In North and South America, however, there is no regulatory agency comparable to the Commission. But it would be incorrect to conclude that environmental regulation of agricultural practices is absent. National and state or provincial governments, and even local or municipal governments, are active participants in setting environmental policy that applies to agriculture. Comparisons of environmental compliance costs for agricultural production need to recognize the multiple levels of government that can initiate such regulations, as well as the roles of civil or common law liability, the role of private environmental standards in supply chains as well as the possibility of voluntary industry standards initiated by farm organizations. It is only when all levels of government policy as well as private initiatives are taken into account that we get a complete picture of the incentives that farmers face to achieve particular environmental performance levels and the costs that they incur to achieve that performance. Taking this systemic and comprehensive approach, as we have attempted to do in this book, reveals that every one of the countries included in this analysis has responded to concerns about excess nutrients, pesticide use, animal welfare, food safety and wildlife habitat, as well as numerous other issues, within its jurisdiction. The perceived and perhaps actual severity of national problems in each of these areas, however, does differ across the countries included in this research.

In addition to regulation by supra-national, national, regional and local governments, civil and common law liability for trespass and nuisance also creates incentives for farmers to meet environmental performance standards. In the UK, the Anglers Cooperative Association (Bate, 2001) has a long history of using litigation, or the threat of litigation, to mitigate water pollution problems in rivers.

Input intensity and environmental regulatory response

As illustrated by Goklany's (1999) insightful review of US air pollution policy, environmental policy is generally reactive, not pro-active. Eveland *et al.* (2005) have confirmed this generalization for the case of local regulations of large scale livestock operations in Ontario, Canada. Environmental regulation responds to emerging conflicts, generally over water and air quality issues, as competing demands for the use of air and water come into conflict. In the case of agriculture, Rajsic *et al.* (Chapter 4 of this volume) and Rajsic (2011) suggest that agricultural input intensity, coupled with increased human population density, and perhaps with increases in income per caput, are correlated with the emergence of such conflicts. As livestock density and nutrient application rates increase, especially in areas of high human population density, conflicts over nutrient use and odours from livestock manure application tend to rise. The hypothesis that follows from this view is that environmental regulation is likely to emerge in higher input intensity locations. The simple version of the pollution haven hypothesis is that polluting firms seek out locations where environmental regulation is lax and hence environmental compliance costs are low. A more dynamic theory is needed to explain the data that indicate that firms are often located in areas with higher standards of environmental regulation. Goklany's theory offers this sort of dynamic explanation, in that the regulation is a response to the operations of the firms, not that the firms locate based on differential regulatory standards.

This perspective has important implications for agricultural environmental regulation. As the data presented by Rajsic *et al.* (Chapter 4 of this volume) show, there are substantial differences in aggregate measures of input intensity among developed country agricultural economies. If we accept those measures of input intensity as indicators of emerging or existing environmental conflicts, then we would expect higher levels of environmental regulation in those jurisdictions with those higher levels of input intensity. Data on agricultural input intensity indicate that higher measures of that intensity often are observed in regions where agricultural price supports and subsidies have, historically, been high. Of course, there are and have been variations in support levels across countries and across commodities, historically, and, within the EU, where there have often been similar support levels across countries, there is considerable variability in input intensity, so a simple generalization that higher support levels leads to higher input intensities does not do justice to the more complex microeconomic incentives and effects. It is not difficult to understand, however, that farmers respond to higher output prices, sometimes linked with input price subsidies, by farming land more intensively. Over time, this higher level of input intensity can contribute to environmental pressures and conflicts. And it is in precisely those contexts where we could expect, following Goklany, environmental regulation of agriculture to be most vigorously applied. This dynamic perspective is needed to understand many of the emerging tensions involving agricultural environmental compliance costs.

Part II - Trade, competitiveness and public environmental concerns – a commodity approach

Chapters in part II surveyed environmental regulation compliance issues for a range of agricultural commodities with significant trade volumes. Specific regulatory compliance issues were studied. The specific issues varied over commodities, naturally, as different environmental conflicts have emerged in different agricultural production systems. Nitrate management, pesticide residue control, animal welfare, food safety, groundwater quantity management and use of hormones in livestock production and protection of wildlife habitat were the most frequently identified environmental policy priorities with implications for agricultural production in the countries and regions considered in this research.

The main trade related policy concern addressed was the impact of agricultural environmental compliance costs on competitiveness and trade. Empirical research on environmental compliance costs in agriculture has lagged behind research in manufacturing and resource extraction industries. This volume offers some preliminary findings on agricultural environmental compliance costs to redress this gap in knowledge. Cross compliance in the EU, per se, was not found to have a significant effect on trade, largely because the measures involved were already largely embodied in and enforced through various national regulations, so cross compliance did not represent appreciable additional costs to farmers. But even when overall environmental compliance costs were compared, the effects on trade were generally minor. This was especially true when the idea of functional equivalence discussed earlier in this chapter was applied and when environmental compliance costs were modelled as multilateral, not unilateral policy measures. Clearly, if one country imposes an environmental regulation on its own farmers and farmers in a competitive country are not subject to that same regulation, this could put the regulated farmers at a competitive disadvantage. But our research has demonstrated that parallel agricultural environmental regulatory processes are in place in all developed country agricultural economies, so multilateral, not unilateral, regulation is the norm. This means that only the differential environmental compliance costs would be expected to influence trade volumes.

Our research found that agriculture is much like other economic sectors in that environmental compliance costs, overall, are small relative to production costs in total. Environmental compliance costs are generally less than 5% of total unit costs, which means that the incentive for pollution haven type relocation of production to jurisdictions with more lenient environmental standards is weak. Other cost categories, such as land and labour costs, logistics for input supplies or output processing, distribution and retailing, proximity to markets and infrastructure, along with soil and climate factors, would appear to play a more significant role in determining the location of primary agricultural production than environmental compliance costs.

Production costs and quality of produce are critical to competitiveness in international markets and agricultural entrepreneurs will need to continue to search for competitive advantages, either by reducing costs and/or by increasing the quality of produce. The reward for this innovation takes the form of higher revenues or access to markets. A focus on sustainability could provide competitive advantages relative to other producers, by targeting at those consumers who are prepared to pay a bonus for food with more sustainable production methods and/or higher quality of the final product. Such a strategy seems to be most viable for niche markets and targeted on sales within (continental) markets. Another opportunity might be knowledge-export of sustainable practices. It might be less applicable for commodities traded globally.

Efforts to promote sustainability in agriculture could be risky for bulk products that are largely traded on the basis of price without much differentiation of product or process attributes on the world market. Across the globe, there are many different views on sustainable agriculture, and more specifically on the dimensions ‘people’ and ‘planet’. Selling food commodities on the world market that are promoted as sustainable agriculture is a complex undertaking. In some markets, price is more decisive than in other markets. This can be explained by differences related to both ‘people’ and ‘planet’ aspects. Different countries could have a wide range of views on labour conditions and on poverty. Another example is the use of GMOs. The EU has only recently lifted a ban on the use on GMOs in food and feed, whereas genetically modified corn and soybeans have become widely accepted in Argentina and the USA. Also, restrictions with respect to the environment differ from region to region (Brouwer *et al.*, 2000).

Part III - Emerging policy perspectives

Beyond public goods and externalities

Most economic analysis of environmental problems associated with agricultural production has been conducted using either public goods theory or externality theory. More recently, work using club goods theory has offered an alternative view on the provision of environmental goods and services provided by agricultural and other rural lands. This is becoming a more important option for promoting the provision of wildlife habitat on privately owned rural land. Several non-governmental organizations in North America, sometimes organizations linked to hunting, fishing and outdoor recreation, but also conservation organizations, are using a club goods approach to create incentives for farmers to participate in production practices that are valued for their environmental benefits. Ducks Unlimited, Delta Waterfowl and Trout Unlimited have developed programmes which use membership contributions and subscriptions to compensate farmers for the costs of installing conservation practices and/or the lost income that they incur when they provide wildlife habitat on their farms.

Legalized nuisance and regulation – Market versus non-market failure

The concept of legalized nuisance was introduced to the environmental policy and economics literature by Coase (1960), although little notice was taken of this concept at the time. Coase criticized Pigou's treatment of environmental externalities as a category of market failure based on the familiar narrative of the steam locomotive belching out smoke and sparks as it travels across the countryside. Pigou, and, for that matter, many economics textbook writers who followed him, used this apparent historical example from the economic history of the United Kingdom as an example of an externality. The classic Pigouvian remedy: emission taxes or subsidies to internalize the apparent externality. Coase pointed out, however, that an important legal and institutional dimension was ignored in Pigou's treatment of this example. Railroads operating in the United Kingdom, at the time of the putative historical events in his illustration, operated under royal charters. Among other things, this absolved railroads from liability for what, in the absence of these charters, would have been actionable nuisances by farmers and other property owners along the railroad lines. Consequently, what economists have often used as the quintessential illustration of an externality market failure is actually an example of non-market or policy failure. Coase's exposition of this systemic misdiagnosis are worth quoting directly (Coase, 1960, pp. 26, 27).

> Most economists seem to be unaware of all this. When they are prevented from sleeping at night by the roar of jet planes overhead (publicly authorized and perhaps publicly operated), are unable to think (or rest) in the day because of the noise and vibration from passing trains (publicly authorized and perhaps publicly operated), find it difficult to breathe because of the odour from a local sewage farm (publicly authorized and perhaps publicly operated) and are unable to escape because their driveways are blocked by a road obstruction (without any doubt, publicly devised), their nerves frayed and mental balance disturbed, they proceed to declaim about the disadvantages of private enterprise and the need for Government regulation. While

> most economists seem to be under a misapprehension concerning the character of the situation with which they are dealing, it is also the case that the activities which they would like to see stopped or curtailed may well be socially justified. It is all a question of weighing up the gains that would accrue from eliminating these harmful effects against the gains that accrue from allowing them to continue. Of course, it is likely that an extension of Government economic activity will often lead to this protection against action for nuisance being pushed further than is desirable. For one thing, the Government is likely to look with a benevolent eye on enterprises which it is itself promoting. For another, it is possible to describe the committing of a nuisance by public enterprise in a much more pleasant way than when the same thing is done by private enterprise.

Charles Wolf (1979) went on to expand Coase's concept of legalized nuisance to a more general condition that he called derived externalities. According to Wolf's definition, a derived externality resembles a traditional externality, at least physically, in that it involves the imposition of an uncompensated cost on a third party through the introduction of some harmful physical substance or force onto the property of that third party. But a derived externality has a different institutional or legal origin to an externality. Derived externalities arise because of some prior policy action on the part of government. So what does this have to do with environmental compliance costs in agriculture? Elizabeth Brubaker's (2007) *Greener Pastures* documents how the concept of legalized nuisance has been a contributing factor to pressure for policy change regarding the environmental effects of farming. One method that several jurisdictions have used to respond to rising pressures, especially between farm and non-farm neighbours, has been agricultural zoning or what is sometimes called 'Right to Farm' statutes. Under this approach, situations that would normally be treated as farm-based nuisances are treated in much the same way that a royal charter treated the nuisance effects of UK railroads, by absolving the perpetrator of the nuisance of liability. Of course, such statutory enactments resolve emerging conflicts over neighbouring land uses, at least temporarily, by changing property rights through legislative law, but they do not address the underlying cause of the emerging tensions. And, in many cases, protection of farmers under 'Right to Farm' policies is now often conditional on compliance of those farmers with nutrient and habitat management regulations.

Technical barriers to trade in agricultural products were not the primary focus of this research; however, restrictions on imports of beef produced with hormones and on GMO crops to the EU have been a long standing source of conflict between particularly the EU and the US and Canada. These tensions appear to be being resolved, but the analysis in Chapter 12 indicates that technical barriers to trade may be a more enduring obstacle for less developed countries. Scientific, technical and legal expertise is required to participate effectively in international fora, especially in WTO disputes addressing technical barriers, and less developed countries are often undersupplied with these forms of human capital. And regulation of perceived or suspected environmental and human health risks is difficult in the face of the scientific uncertainty involved. Demonstration of clear causality is often beyond the scope of current scientific knowledge. But policy is not driven exclusively or maybe even primarily, by science. Policy is often driven by preferences, perceptions and expectations. Harmonization of standards and procedures is difficult and monitoring compliance with those standards and procedures across international borders requires higher levels of transparency and trust than are often available. And the outcome of all of the influence of all of these factors, often, is limited access of less developed countries to developed country markets.

Consumer demands for health and safety attributes, as well as for environmental stewardship attributes, are strengthening. Animal welfare concerns, as well, are increasingly being expressed by consumers, through opinion surveys as well as through purchasing decisions. Retailers and food processers are demanding higher and audited performance of farming systems in response to changed consumer demands. In doing so, they promote sustainable practices in agriculture. Therefore, agriculture must respond to and work with others in the agrifood chain. Public-private partnerships may be the way forward for meeting societal demands from the agricultural sector. The incorporation of environmental concerns in marketing strategies from retailers could change farming practices and also contribute to reduce efforts needed for meeting public policy objectives.

It remains to be seen if these social trends toward stronger demands for non-traditional attributes of food will be deflected by fluctuations in the economic fortunes of consumers. Since the financial crisis of 2008, concerns about unemployment, government debt and stagnant economic growth have supplanted many expressions of concern about the environmental footprint of the food system. It remains to be seen if the return to price and value considerations is a temporary or a longer term development in food consumption and retailing. Fears of rising food prices, driven by general inflationary pressures and enhanced by demand for agricultural materials for biofuel production have changed the perceptions of the agriculture and food system from even a few years ago, when prosperity was more generally taken for granted and environmental and animal welfare concerns were being expressed more fully.

Consumers and civil society are vital to changes in agriculture and the social dimension will continue to have a major impact on the economic conditions and environmental constraints of sustainable development of the sector. A 'licence to produce' can be defined as the specifically stated conditions from outside and within agriculture, which determine the boundaries for agricultural activity. So, the concept of licence to produce indicates the existence right of the agricultural sector, from an internal and external point of view (Mureau, 2000; after definitions adopted from Shell). Currently, grazing dairy cows in Dutch meadows are part of the licence to produce in dairy farming. In the EU context, they are often seen as an icon of rural (farming) life and the accompanying landscape. Citizens associate them with romantic impressions of agriculture. They consider it important for animal health and also landscape matters (Van den Pol-van Dasselaar *et al.*, 2002). But it remains unknown how viable such systems are over time. Increasingly, dairy cows in the Netherlands and elsewhere are kept housed inside and the products of the pasture are delivered to them. This trend is mainly to further rationalize production and costs per unit of output. Rationalization of production impacts the structure of agriculture (larger holdings with more cows per farm and a reduction in the number of holdings), technological progress (milking robot) and environmental regulations. We see responses in society in some regions where a local premium is offered to the supply of milk from animals grazing outside (CONO kaasmakers, Beemster). Wakker Dier launched a campaign in which they asked municipalities to require local farmers to graze their cows. Some consumers indicate that they are willing to pay a bonus for additional process attributes of food. However, the cooperatives find that usually the cheapest milk sells best in the supermarket. The question remains whether this discussion about grazing the cows, and other similar questions related to structural aspects of modes of agricultural production, are temporary or permanent. Will consumers accept the loss of such icons as grazing dairy cows? From a longer historical perspective, changes in the agricultural landscape have occurred during the last century. The adaptive capacity of citizens seems to be rather large. There is no guarantee that the next generation will have the same concerns about grazing cows. Preferences on sustainable production are subject to continuous change and grazing cows might not remain part of the licence to produce in the long run.

Can the agricultural sector benefit from this discussion on an international level? It remains uncertain whether such a licence to produce could strengthen the agriculture sector by promoting recreational activities. Can they sell grazing cows in Holland, like wooden shoes or tulips (Keukenhof and Koeienhof) (Brouwer *et al.*, 2007)? This remains uncertain. The social dimensions of sustainable development create tensions between producers and consumers and involve value judgements regarding landscape values, intensity of farm production and the economic conditions of agriculture.

The research reported in this volume, among other things, documents the range of environmental issues associated with agricultural production in developed country agricultures as well as the range of responses that have emerged in different jurisdictions. But these various policy responses also illustrate what in other contexts has been called the environmental policy paradox. The environmental policy paradox is sometimes stated as follows. Environmental protection is generally viewed favourably by survey respondents and voters. Environmental policy measures in the forms of environmental legislation and regulation, since the 1970s, have embodied some of the more powerful policy instruments, at least on paper, available to governments. For example, the US federal Endangered Species Act trumps even national security, as habitat protection for the red-cockaded woodpecker has curtailed military operations on important bases in the US.

Nevertheless, and here is where the paradox becomes clear, objective assessments of progress achieved by these theoretically powerful and broadly politically supported policy initiatives have concluded that we have little to show for our efforts. This has led some analysts to propose a reassessment of the overall approach to environmental policy, which has for two generations generally adopted a centralized top-down approach. Hasnas (2009), Pardy (2005), Morris (2008) and Ackerman and Stewart (1985), in different ways, have all suggested that decentralized local approaches may do a better job of mobilizing scarce environmental protection resources. These discussions have generally focussed on environmental policies in the general economy. But they may have important insights for future agricultural environmental policy.

References

Ackerman, B. and R. Stewart (1985) 'Reforming Environmental Law' *Stanford Law Review*, 37(5): 1333-1365.

Bate, R. (2001) *Saving Our Streams*, Institute for Economic Affairs, London, UK.

Brouwer, F., D. Baldock, C. Carpentier, J. Dwyer, D. Ervin, G. Fox, A. Meister and R. Stringer (2000) Comparison of environmental and health-related standards influencing the relative competitiveness of EU agriculture vis-à-vis main competitors in the world market. The Hague, Agricultural Economics Research Institute (LEI), Report 5.00.07.

Brouwer, F.M., H. Leneman and R.A. Groeneveld (2007) Exploring the international policy dimension of sustainability in Dutch agriculture. Wageningen, WOt Natuur en Milieu, Rapport 14.

Brubaker, E. (1995) *Property Rights in the Defence of the Environment*, Earthscan Publications, Toronto.

Brubaker, Elizabeth (2007) *Greener Pastures: Decentralizing the Regulation of Agricultural Pollution*, Monograph Series, University of Toronto Centre for Public Management.

Coase, R. (1960) 'The Problem of Social Cost'*Journal of Law and Economics* 3(2):1-44.

Eveland, C., A. Weersink, W. Caldwell and W. Yang (2005), 'Spatial Trends in Barn Building Permits', *Great Lakes Geographer,* 12: 19-27.

Fox, Glenn (forthcoming) 'The Origins, Nature and Content of the Right to Property: Five Economic Solitudes' *Canadian Journal of Agricultural Economics*.

Goklany, I. (1999) *Clearing the Air: The Real Story of the War on Air Pollution*, Cato Institute.

Hasnas, J. (2009) 'Two Theories of Environmental Regulation' *Social Philosophy and Policy*, 26 (2, July): 95-129.

Morris, Julian (2008) 'Private vs Public Regulation of the Environment' in *The Legal Foundations of Free Markets*, edited by Stephen Copp, Institute for Economic Affairs, London, UK.

Mureau, N. (2000) The concept of 'Licence to Produce': Definition and application to dairy farming in the Netherlands. Paper presented at the International Conference: European Rural Policy at the Crossroads, 29 June – 1 July 2000.

Pardy, B. (2005) 'Environmental Law and the Paradox of Environmental Citizenship' *Environments* 33(3): 25-36.

Rajsic, P. (2011) *Linking Comparative Advantage, Supply Management and Environmental Externalities: Lessons from an Integrative Economic Approach*, PhD Dissertation, University of Guelph.

Van den Pol-van Dasselaar, A., W.J. Corré, H. Hopster, G.C.P.M. van Laarhoven and C.W. Rougoor (2002) Importance of grazing. Praktijkonderzoek Veehouderij, Lelystad.PraktijkRapport Rundvee 14.

Wolf, Charles (1979) 'A Theory of Non-Market Failure: Framework for Implementation Analysis' *Journal of Law and Economics,* 21(1):107-139.

Index